Flavors for Nutraceutical and Functional Foods

Nutraceuticals: Basic Research/Clinical Applications
Series Editor: Yashwant Pathak, PhD

Antioxidant Nutraceuticals: Preventive and Healthcare Applications
Chuanhai Cao, Sarvadaman Pathak, and Kiran Patil

Nutrigenomics and Nutraceuticals: Clinical Relevance and Disease Prevention
Yashwant Pathak and Ali M. Ardekani

Marine Nutraceuticals: Prospects and Perspectives
Se-Kwon Kim

Nutraceuticals and Health: Review of Human Evidence
Somdat Mahabir and Yashwant V. Pathak

Handbook of Metallonutraceuticals
Yashwant V. Pathak and Jayant N. Lokhande

Nanotechnology in Nutraceuticals: Production to Consumption
Shampa Sen and Yashwant Pathak

Herbal Bioactives and Food Fortification: Extraction and Formulation
D. Suresh Kumar

Flavors for Nutraceuticals and Functional Foods
M. Selvamuthukumaran and Yashwant Pathak

Flavors for Nutraceutical and Functional Foods

Edited by
M. Selvamuthukumaran
Yashwant V. Pathak

CRC Press
Taylor & Francis Group
Boca Raton London New York

CRC Press is an imprint of the
Taylor & Francis Group, an **informa** business

CRC Press
Taylor & Francis Group
6000 Broken Sound Parkway NW, Suite 300
Boca Raton, FL 33487-2742

© 2019 by Taylor & Francis Group, LLC
CRC Press is an imprint of Taylor & Francis Group, an Informa business

No claim to original U.S. Government works

Printed on acid-free paper

International Standard Book Number-13: 978-1-138-06417-1 (Hardback)

This book contains information obtained from authentic and highly regarded sources. Reasonable efforts have been made to publish reliable data and information, but the author and publisher cannot assume responsibility for the validity of all materials or the consequences of their use. The authors and publishers have attempted to trace the copyright holders of all material reproduced in this publication and apologize to copyright holders if permission to publish in this form has not been obtained. If any copyright material has not been acknowledged please write and let us know so we may rectify in any future reprint.

Except as permitted under U.S. Copyright Law, no part of this book may be reprinted, reproduced, transmitted, or utilized in any form by any electronic, mechanical, or other means, now known or hereafter invented, including photocopying, microfilming, and recording, or in any information storage or retrieval system, without written permission from the publishers.

For permission to photocopy or use material electronically from this work, please access www.copyright.com (http://www.copyright.com/) or contact the Copyright Clearance Center, Inc. (CCC), 222 Rosewood Drive, Danvers, MA 01923, 978-750-8400. CCC is a not-for-profit organization that provides licenses and registration for a variety of users. For organizations that have been granted a photocopy license by the CCC, a separate system of payment has been arranged.

Trademark Notice: Product or corporate names may be trademarks or registered trademarks, and are used only for identification and explanation without intent to infringe.

Library of Congress Cataloging-in-Publication Data

Names: Selvamuthukumaran, M., editor.
Title: Flavors for nutraceutical and functional foods / editors, M. Selvamuthukumaran, Yashwant Pathak.
Description: Boca Raton : Taylor & Francis, 2018. | Series: Nutraceuticals : basic research/clinical applications | Includes bibliographical references.
Identifiers: LCCN 2018004874 | ISBN 9781138064171 (hardback : alk. paper)
Subjects: LCSH: Functional foods--Sensory evaluation. | Flavor.
Classification: LCC QP144.F85 F53 2018 | DDC 613.2--dc23
LC record available at https://lccn.loc.gov/2018004874

Visit the Taylor & Francis Web site at
http://www.taylorandfrancis.com

and the CRC Press Web site at
http://www.crcpress.com

I wholeheartedly thank God, my family, my friends, and everyone who has encouraged and supported me in my work on this book.

Selvamuthukumaran M

I dedicate this book to all the rushies, sages, shamans, medicine men and women, and people of ancient traditions and cultures who have contributed to the development of drugs and nutraceuticals worldwide and who have kept the science of health alive for the past several millennia.

Yashwant V. Pathak

Contents

Series Preface ix
Preface xi
Editors xv
Contributors xvii

1 **Introduction to Functional Foods and Nutraceuticals** 1
   M. Selvamuthukumaran and Yashwant V. Pathak

2 **Flavoring of Pediatric Nutritional Supplements and Pediatric Compliance: A Perspective.** 33
   Amit M. Pant, Rupesh V. Chikhale, and Pramod B. Khedekar

3 **Flavors in Probiotics and Prebiotics** 51
   Deepak Yadav, Kummaripalli Srikanth, and Kiran Yadav

4 **Natural Ingredients/Botanical Extracts for the Nutraceutical Industry** 75
   Rahul Maheshwari, Kaushik N. Kuche, Ashika Advankar, Namrata Soni, Nidhi Raval, Piyoosh A. Sharma, Muktika Tekade, and Rakesh Kumar Tekade

5 **Taste-Masking Techniques in Nutraceutical and Functional Food Industry** 123
   Shankar D. Katekhaye, Bhagyashree Kamble, Ashika Advankar, Neelam Athwale, Abhishek Kulkarni, and Ashwini Ghagare

6 **The Effect of Bitter Components on Sensory Perception of Food and Technology Improvement for Consumer Acceptance** 145
   Geeta M. Patel and Yashwant Pathak

7 **Sensory Qualities and Nutraceutical Applications of Flavors of Terpenoid Origin** ... 167
   Ana Clara Aprotosoaie, Irina-Iuliana Costache, and Anca Miron

8 **The Biopsychology of Flavor Perception and Its Application to Nutraceuticals** 201
   Richard J. Stevenson

9 **Flavor Nanotechnology: Recent Trends and Applications** 217
   Komal Parmar, Jayvadan Patel, and Navin Sheth

10 **Nanoencapsulation of Flavors: Advantages and Challenges** 235
   Farhath Khanum, Syeda Juveriya Fathima, N. Ilaiyaraja, T. Anand, Mahanteash M. Patil, Dongzagin Singsit, and Gopal Kumar Sharma

11 **Nanoencapsulated Nutraceuticals: Pros and Cons** 273
   T. Anand, N. Ilaiyaraja, Mahantesh M. Patil, Farhath Khanum, and Rakesh Kumar Sharma

Index 311

Series Preface

Consumers have likings for specific flavors and that is reflected in their buying preferences. This is applicable to nutraceuticals, food products, pharmaceuticals, cosmetics, and other personal products. This perception is based not only on the flavor but also the taste of the products. After the award of the 2004 Nobel Prize to Dr. Richard Axel and Dr. Linda Buck for their discovery of olfactory receptors, we have developed a better understanding of the molecular mechanisms involved in taste and smell. We now understand that there are more than 36 taste receptors and hundreds of smell receptors. The olfactory system consists of approximately 350 functional receptors, and humans can detect thousands of different odorants. The range of flavors perceived by humans is strongly influenced by the odor of foods.

The nutraceutical market is growing rapidly and is expected to exceed US$207 billion by 2020. There is huge growth potential in the functional foods, beverages, and supplements market. Consumers are looking for tasty, nutritious food and drink products. The functional foods market alone is expected to surpass US$70 billion by 2020. Functional foods and beverages allow manufacturers to differentiate their products in a highly competitive marketplace.

Flavor and taste make nutraceuticals and functional foods attractive and increase consumer acceptance. Flavoring substances should be generally recognized as safe (GRAS). Regulations for their use vary, depending on country, regulatory authority, and application. Masking of taste and smell of the natural products is challenging. Flavors are often added to products to reduce the perception of off-flavors. Compounds are also added that interfere with the perception of bitterness at the receptor level. Flavoring of a functional food or nutraceuticals is very customized.

It is mostly performed in collaboration with a flavor company, so an open communication between the product developers and the flavor company will greatly improve the likelihood of a successful product.

Approximately 7000 flavor volatiles have been identified and catalogued from various foods and beverages. The majority of these are plant volatiles produced during fruit or vegetable ripening or maturation. Only a small number of the volatile compounds produced by a fruit or vegetable contribute to what is sensed as food flavors. For example, only a few of the more than 400 volatiles produced by tomatoes impact the flavor profile detected by humans. Studies on the biochemical synthesis of important fruit and vegetable flavors have revealed that the majority are derived from essential nutrients, vitamins, or other health-promoting compounds, collectively referred to as micronutrients.

This book addresses various aspects of the flavoring of nutraceuticals and functional foods. Topics include natural and botanical flavoring ingredients, taste masking in nutraceuticals, sensory qualities and flavors of terpenoid origin, and the biophysiology of flavor perception.

This book is a volume in the series Nutraceuticals: Basic Research and Clinical Applications. This new addition covers important aspects of nutraceuticals and functional foods. We welcome reader feedback and any suggestions that will improve the next edition.

Yashwant V. Pathak

Preface

Functional foods and nutraceuticals play a major role in mitigating lifestyle-related disorders. These foods are formulated to complement the treatment of specific diseases. They have huge market potential. This volume explores the importance of flavor in functional foods and nutraceuticals.

Chapter 1, Introduction to Functional Foods and Nutraceuticals, provides an overview of the importance of flavor in foods. Flavor technology helps take food products to new levels of performance, palatability, and consumer appeal. Flavor is one of the most important attributes that contributes to better acceptance of food by the consumer.

Chapter 2, Flavoring of Pediatric Nutritional Supplements and Compliance: A Perspective, discusses the formulation of pediatric nutritional supplements, which has long been a challenging task for the food and drug industries as well as for regulatory authorities worldwide. Most nutritional supplements intended for pediatric use include vitamins A, D, E, and B, all of which are very bitter in taste. Common mineral supplements, such as iron and zinc, are very metallic in the mouth, and calcium is chalky in taste. Various flavoring agents are used to mask and conceal the odor, texture, and taste of these supplements to make them more acceptable to the pediatric population.

The market for probiotic and prebiotic food products is expanding worldwide. Probiotic products have altered sensory characteristics that can be improved by adding different flavors such as chocolate, vanilla, or fruit-based flavors.

Chapter 3, Flavors in Probiotics and Prebiotics, discusses dairy-based probiotic products in detail, including their formulation and processing characteristics that influence their flavors. Measures to prevent the development of off-flavors are also examined. The manufacturing characteristics, components, and flavors of fermented nondairy-based probiotic products are also described.

Chapter 4, Natural Ingredients/Botanical Extracts for the Nutraceutical Industry, addresses the global market for these ingredients and extracts. More than 35 of the most commonly used herbs in the nutraceutical industry and their roles in improving our health and well-being are explored.

Taste is paramount for consumers. Various methods and approaches such as taste masking, inhibition and suppression, and encapsulation are used to improve taste to increase consumer acceptance.

Chapter 5, Taste-Masking Techniques in the Nutraceutical and Functional Foods Industry, examines conventional, recent, and patented techniques used for masking taste. Bitter taste perception is natural and induces insensitive reactions. Some toxic compounds, including synthetic chemicals, secondary plant metabolites, rancid fats, and inorganic ions have a bitter taste. Perception of bitterness in food may be a defense mechanism against potential poisons ingested in the form of foods. Chapter 6, Effect of Bitter Components on Sensory Perception of Food and Technology Improvement for Consumer Acceptance, focuses on aspects of bitterness in food and the industry-wide problem of debittering of foods.

Volatile terpenes, mainly terpenoids, such as citral, geraniol, linalool, menthol, carvone, and thymol, are valuable natural flavors that play a prominent role in nutrition by enhancing the sensory appeal of foods and beverages. Chapter 7, Sensory Qualities and Nutraceutical Applications of the Flavors of Terpenoid Origin, provides an overview of the potential applications of the main terpene flavors as nutraceuticals. The sensory qualities of these flavors are described and the biological properties of representative compounds are discussed with reference to recent studies.

Seeing, smelling and hearing food allow us to build expectancies of what it will be like to eat. Once the food is placed in the mouth, the sense of taste, smell, and touch combine to generate flavor, the sensory experience that accompanies eating and drinking. Chapter 8, The Biopsychology of Flavor Perception and Its Application to Neutraceuticals, considers the processes involved in generating flavor and its acceptability and then examines the practical implications of this knowledge for nutraceuticals.

Retention of taste, flavor, and aroma in foods remains a challenge to the food industry. Fragrances and aromas tend to fade by evaporation, interaction with other compounds, oxidation, and chemical degradation. Nanotechnology provides techniques of encapsulation, protection, and controlled release of these essential constituents. Chapter 9, Recent Trends in Flavor Nanotechnology and Its Applications, reviews recent trends in flavor nanotechnology and applications in the food industry. Chapter 10, Nanoencapsulation of Flavors: Advantages and Challenges, explores processes such as microencapsulation, which is used primarily to stabilize flavors/aroma chemicals, and taste-transforming liquid oils and food flavors that are transformed into emulsions and dry powders that improve encapsulation efficiency, mask unpleasant odors and tastes, and protect and stabilize flavors during storage. The aim is to prevent volatile losses and enable high-impact, long-lasting taste and extend the shelf life of products.

Nanoencapsulation of bioactive compounds has versatile advantages for targeted site-specific delivery and efficient absorption through cells. Chapter 11, Nanoencapsulated Nutraceuticals: Pros and Cons, provides details about encapsulating materials, bioactive compounds, detergents, encapsulating equipment, and techniques used for encapsulating various different bioactive constituents.

We would like to express our sincere thanks to the chapter authors. Without their support this book would not have come to fruition. We also thank Steve Zollo and the other members of the CRC Press team for their work on this book.

M. Selvamuthukumaran
Yashwant V. Pathak

Editors

M. Selvamuthukumaran, Ph.D., is Associate Professor in the Department of Food Science & Post Harvest Technology, Institute of Technology, Haramaya University, Dire Dawa, Ethiopia. He received his Ph.D. in food science from Defence Food Research Laboratory, affiliated with the University of Mysore, India.

Dr. Selvamuthukumaran's core area of research is processing of underutilized fruits for development of antioxidant-rich functional food products. He has transferred to Indian firms several technologies that were the outcome of his research work. He has received several awards and citations for his research work.

Dr. Selvamuthukumaran has published several international papers and book chapters in the areas of antioxidants and functional foods. He has supervised several national and international postgraduate students in the area of food science and technology.

Yashwant V. Pathak, Ph.D., is Professor and Associate Dean for Faculty Affairs at the College of Pharmacy, University of South Florida, Tampa, Florida. He received his EMBA and master's in conflict management from Sullivan University, Louisville, Kentucky, and his Ph.D. in pharmaceutical technology from Nagpur University, India.

Dr. Pathak has extensive experience in both industry and academia. He has more than 100 publications, including 16 edited volumes, on nanotechnology, nutraceuticals, and drug delivery systems. He holds two patents and has three additional patent applications pending. He has also written several books on cultural studies and conflict management.

Dr. Pathak was a Fulbright Senior Fellow in Indonesia (2017) for 6 months. He visited seven countries and lectured at more than 20 universities. He was a keynote speaker at the World Hindu Wisdom meeting in Bali, Indonesia, in June 2017. He also co-chaired the meeting with the Governor of Bali.

Contributors

Ashika Advankar
Department of Natural Products
National Institute of Pharmaceutical
  Education and Research (NIPER)
Ahmedabad, India

T. Anand
Department of Nutrition,
  Biochemistry & Toxicology
Defence Food Research Laboratory
Mysore, India

Ana Clara Aprotosoaie
Department of Pharmacognosy
Faculty of Pharmacy
University of Medicine and
  Pharmacy Grigore T. Popa
Iasi, Romania

Neelam Athwale
Department of Natural Products
National Institute of Pharmaceutical
  Education and Research (NIPER)
Ahmedabad, India

Rupesh V. Chikhale
Division of Pharmacy
  and Optometry
School of Health Sciences
University of Manchester
Manchester, U.K.

Irina-Iuliana Costache
Department of Internal
  Medicine
Faculty of Medicine
Sf. Spiridon University Hospital
and
University of Medicine and
  Pharmacy Grigore T. Popa
Iasi, Romania

Syeda Juveriya Fathima
Nutrition, Biochemistry
  and Toxicology
Defence Food Research
  Laboratory
Mysore, India

Ashwini Ghagare
Department of Natural Products
National Institute of
 Pharmaceutical Education
 and Research (NIPER)
Ahmedabad, India

N. Ilaiyaraja
Department of Nutrition,
 Biochemistry & Toxicology
Defence Food Research
 Laboratory
Mysore, India

Bhagyashree Kamble
Department of Natural Products
National Institute of
 Pharmaceutical Education
 and Research (NIPER)
Ahmedabad, India

Shankar D. Katekhaye
Department of Molecular Biology
 and Biotechnology
Institute of Advanced Research
Gandhinagar, India

Farhath Khanum
Nutrition, Biochemistry
 and Toxicology
Defence Food Research
 Laboratory
Mysore, India

Pramod B. Khedekar
Department of Pharmaceutical
 Sciences
Rashtrasant Tukadoji Maharaj
 Nagpur University
Nagpur, India

Kaushik N. Kuche
National Institute of
 Pharmaceutical Education
 and Research (NIPER)
Gujarat, India

Abhishek Kulkarni
Department of Natural Products
National Institute of Pharmaceutical
 Education and Research (NIPER)
Gujaret, India

Rahul Maheshwari
National Institute of
 Pharmaceutical Education
 and Research (NIPER)
Gujarat, India

Anca Miron
Department of Pharmacognosy
Faculty of Pharmacy
University of Medicine and
 Pharmacy Grigore T. Popa
Iasi, Romania

Amit M. Pant
Vitane Pharma Ltd. (Vitane
 Pharma GmbH, Germany)
Mumbai, India

Komal Parmar
ROFEL
Shri G.M. Bilakhia College
 of Pharmacy
Gujarat, India

Geeta M. Patel
Department of Pharmaceutics and
 Pharmaceutical Technology
Shree S. K. Patel College of
 Pharmaceutical Education
 and Research
Ganpat University
Ganpat Vidyanagar, India

Jayvadan Patel
Faculty of Pharmacy
Nootan Pharmacy College
Gujarat, India

Yashwant V. Pathak
College of Pharmacy
University of South Florida
Tampa, Florida

Mahantesh M. Patil
Department of Nutrition,
 Biochemistry & Toxicology
Defence Food Research
 Laboratory
Mysore, India

Nidhi Raval
National Institute of
 Pharmaceutical Education
 and Research (NIPER)
Gujarat, India

Gopal Kumar Sharma
Grain Science Technology
Defence Food Research
 Laboratory
Mysore, India

Piyoosh A. Sharma
TIT College of Pharmacy
Technocrats Institute of
 Technology Campus
Bhopal, India

Rakesh Kumar Sharma
Department of Nutrition,
 Biochemistry & Toxicology
Defence Food Research
 Laboratory
Mysore, India

Navin Sheth
Gujarat Technological University
Gujarat, India

Dongzagin Singsit
Nutrition, Biochemistry
 and Toxicology
Defence Food Research Laboratory
Mysore, India

Namrata Soni
BM College of Pharmaceutical
 Education & Research
Indore, India

Kummaripalli Srikanth
Department of Pharmaceutics
National Institute of
 Pharmaceutical Education and
 Research (NIPER)
Raebareli, Uttar Pradesh, India

Richard J. Stevenson
Department of Psychology
Macquarie University
Sydney, Australia

Muktika Tekade
Sri Aurobindo Institute
 of Pharmacy
Indore, India

Rakesh Kumar Tekade
National Institute of
 Pharmaceutical Education
 and Research (NIPER)
Gujarat, India

Deepak Yadav
School of Pharmacy
Hadassah Medical Centre
The Hebrew University
 of Jerusalem
Jerusalem, Israel

and

Department of Pharmaceutics
National Institute of
 Pharmaceutical Education and
 Research (NIPER)
Raebareli, Uttar Pradesh, India

Kiran Yadav
Institute of Pharmaceutical
 Sciences
Kurukshetra University
Kurukshetra, India

1

Introduction to Functional Foods and Nutraceuticals

M. Selvamuthukumaran and Yashwant V. Pathak

Contents

1.1 Introduction 2
1.2 Functional foods 2
    1.2.1 The concept of functional foods 2
    1.2.2 World market for functional foods 3
1.3 Nutraceuticals 4
    1.3.1 The concept of nutraceuticals 4
    1.3.2 Market scenario for nutraceuticals 5
    1.3.3 Safety aspects of nutraceuticals 6
1.4 Sources of functional foods 6
    1.4.1 Grains and cereals 6
        1.4.1.1 Wheat 6
        1.4.1.2 Oats 7
        1.4.1.3 Barley 7
        1.4.1.4 Flaxseed 7
        1.4.1.5 Psyllium 9
        1.4.1.6 Brown rice 9
        1.4.1.7 Buckwheat 9
        1.4.1.8 Sorghum 10
        1.4.1.9 Maize 10
        1.4.1.10 Finger millet 10
        1.4.1.11 Breakfast cereals 10
1.5 Dairy-based functional foods 11
1.6 Margarine 12
1.7 Prebiotics 12
1.8 Confectionery-based functional foods 12
    1.8.1 Cocoa 12
    1.8.2 Food bars 12
    1.8.3 Chocolate 13
1.9 Beverage-based functional foods 13
    1.9.1 Tea 13

1.10 Legume-based functional foods 13
    1.10.1 Soybeans 13
1.11 Fruit- and vegetable-based functional foods 14
    1.11.1 Garlic 14
    1.11.2 Tomatoes 15
    1.11.3 Broccoli and other cruciferous vegetables 15
    1.11.4 Citrus fruits 17
    1.11.5 Cranberries 17
    1.11.6 Red wine and grapes 17
    1.11.7 Strawberries 18
    1.11.8 Figs 18
    1.11.9 Cordyceps mushrooms 18
    1.11.10 Pumpkin 19
1.12 Animal-Based functional foods 19
    1.12.1 Fish 19
    1.12.2 Beef 21
    1.12.3 Eggs enriched with PUFAs 21
1.13 Conclusion 22
References 22

1.1 Introduction

Functional foods and nutraceuticals play a major role in combating and mitigating various lifestyle-related illnesses and disorders. These foods contain dietary ingredients that help maintain a healthy lifestyle and may even cure certain diseases. A food can be described as functional if it has a significant health effect that extends beyond basic traditional nutrition. Nutraceutical products are derived from foods that contain essential components which, like functional foods, have therapeutic effects. Their beneficial components may be isolated and purified from plant, animal, or marine sources. This chapter focuses on the basic concepts of functional foods and nutraceuticals and examines some of their sources, including fruits, vegetables, cereals, and legumes. Functional foods and nutraceuticals have attracted worldwide attention and they constitute an expanding global market.

1.2 Functional foods

1.2.1 The concept of functional foods

A functional food is defined as any food or food ingredient that can provide a health benefit beyond basic nutrition (International Food Information Council, 1999). Such foods reduce the risk of lifestyle-related disorders by achieving physiological functions beyond nutritional effects. These foods are designed either to prevent or to cure disease (Roberfroid, 2000). The term *functional food* originated in Japan, which was also the first country to enact legislation

Table 1.1 Sources of Foods with Their Functional Effects

Source of Food	Phytoconstituents/Beneficial Microorganisms Present	Functional Effects Achieved
Oat	β-Glucan	Reduction in the level of glucose; Improved gastrointestinal function
Barley	Fibre	Type-2 diabetes
Brown rice bran	Tricin	Anticancer effect
Buckwheat leaves	Rutin	Increased vascular permeability and fragility
Sorghum	Deoxyanthocyanidins	Skin, liver, colon and breast cancers
Yogurt	*Lactobacillus* and *Bifidobacteria*	Lactose intolerance reduction; Colon cancer reduction; Cholesterol lowering properties
Honey	Oligosaccharides	Suppression of harmful bacteria residing in colon
Cocoa	Polyphenols	Diabetes, Anaemia and Coronary Heart Diseases
Soyabean	Isoflavones	Cancer eradication, osteoporosis
Garlic	Allicin	Stomach cancer lowering effects, Blood pressure lowering effects
Tomato	Lycopene	Cancer lowering effects
Cruciferous vegetables	Glucosinolates	Cancer lowering effects
Fish oil	Omega-3 Fatty acids	Cardiovascular effect

that brought functional food products to the market. These foods carry the label FOSHU, i.e., Food for Specified Health Uses in Japan, which underscores the therapeutic focus of these products.

A food is made functional in several ways. The functional component is added, removed, or modified during processing or via genetic engineering, resulting in new products that are then introduced into the market. Day et al. (2009) reported that the major challenge for functional foods is to ensure that the bioactive constituents remain stable during processing and storage. Vitamins, fiber, omega-3 fatty acids, minerals, bacterial cultures, and flavonoids are components that can add functionality to any kind of food that is produced (Kesarvani et al., 2010).

The regular consumption of such foods will help efficiently manage diseases such as cardiovascular disease (CVD), cancer, diabetes, hypertension, and osteoporosis. Table 1.1 summarizes sources of functional foods and their potential health effects.

1.2.2 World market for functional foods

The market potential for functional foods is growing rapidly. The total global market value of functional foods was approximately US$34 billion in 2004, $73 billion in 2003, $81 billion in 2005, $168 billion in 2010, and $252 billion in 2013 (Kotilainen et al., 2006; Just-food, 2007; Euromonitor, 2010, 2013). The global revenue generated from the functional food market accounts for approximately US$299.32 billion by the end of 2017 and it was forecasted to reach $441.56 billion by 2022. The primary market for functional foods is the Asia Pacific region. Asia and the Pacific Islands together constitute approximately 34% of total worldwide functional food market revenue. The Japanese market potential is huge. Market potential is also great in China, India, Southeast Asia, and Brazil. The United States and Canada constitute approximately 25% of the total market. For economic reasons the potential in areas such as the Middle East and Africa is poor.

There is great investor interest in the functional food sectors. Brazil is one of the leading countries both in production and consumption of functional foods. Socio-economic conditions in Brazil have attracted companies eager to invest in this arena. Brazil has shown a 10% growth in market potential for functional foods compared to conventional foods.

The medicinal effects of functional foods have been explored for thousands of years (Wildman, 2001). The ancient Egyptian, Chinese, and Sumerian civilizations used functional foods to cure and prevent various diseases. In Asian countries, natural herbs and spices have long been widely used as folk medicines. Food scientists no longer limit food analysis to flavor and nutritional value. Active ingredient analysis plays a major role in making the food functional in treating various health style–related disorders.

Functional foods are similar to conventional foods in appearance. In 1980, the Japanese began developing and marketing functional foods commercially. Japan markets more than 200 functional foods as FOSHU (Farnworthe, 1997). The U.S. Food and Drug Administration (FDA) has approved a total of 15 (Food and Drug Administration, 1999).

In 1991 Japan became the first country to introduce functional foods to the market under the FOSHU label. Initially 69 foods were approved. By 2016 more than 400 products were being marketed under the FOSHU label. FOSHU foods contain ingredients that have met criteria to be officially approved to claim physiological effects for human health. Japanese companies continue to develop functional foods that are marketed under the FOSHU label.

1.3 Nutraceuticals

1.3.1 The concept of nutraceuticals

Stephen L. DeFelice, the founder and chairman of the Foundation of Innovation Medicine, coined the term *nutraceutical* in 1989. The term, a hybrid of *nutrient* and *pharmaceutical*, underscores the intersection of the food and pharmaceutical industries. Nutraceuticals contain nontoxic food components components that can cure or prevent disease or an unsafe condition (Ross, 2000).

Table 1.2 Commonly Available Nutraceuticals and Their Health Effects

Trade Name	Component	Functional Effect
Soylife	Phytoestrogen	Essential for development of healthy bones
Xangold	Lutein esters	Helps maintain healthy eyes
Betatene	Carotenoids	Improves immunity
Cholestaid	Saponin	Reduces cholesterol level
Teamax	Green tea extract	Scavenge formation of free radicals
Genivida	Isoflavone	Bone health effects and relief from menopause symptoms
Oatwell	β-Glucan	Reduces blood cholesterol and glucose level
Peptopro	Peptides	Muscle protein synthesis
Floraglo	Lutein	Reduces age-related macular degeneration
Elavida	Polyphenols	Scavenge formation of free radicals

The concept is by no means new. Almost 2,500 years ago the Greek physician Hippocrates, the father of medicine, wrote "Let food be thy medicine and medicine be thy food."

Nutraceuticals play a role in biological processes such as cell proliferation, gene expression, and antioxidant defense. Nutraceuticals can delay the aging process and reduce the risk of conditions such as cancer, heart disease, hypertension, excessive weight, high cholesterol, diabetes, osteoporosis, arthritis, insomnia, cataracts, constipation, indigestion, and many other lifestyle-related disorders. Nutraceuticals may be isolated and purified from plant, animal, or marine sources. Advantages of nutraceuticals include a longer half-life period, immediate activity upon intake, ready availablity, and few side effects. Table 1.2 shows commonly available nutraceuticals and their potential health effects. The product soylife helps in formation of healthy bones, xangold in maintaining healthy eyes, betatene in immunity improvement, cholestaid and oatwell in blood cholesterol level reduction, and peptopro in synthesis of muscle protein.

1.3.2 Market scenario for nutraceuticals

The nutraceutical market has expanded rapidly in the past few years (World Nutraceuticals, 2006). The United States has the largest share of the global market. The U.S. FDA Act of 1997 paved the way for approving several therapeutic products, which were subsequently marketed in the United States. The U.S. nutraceutical market sector has risen to 5.8% annually to a total of US$15.5 billion in 2010. After the United States, China and India have emerged as the most rapidly expanding markets. Herbal and nonherbal extracts have risen to 6.5% annually, while nutrients, vitamins, and minerals rose to 6.5% in 2005. In the year 2010 the demand for nutraceuticals rose to 46% (http://www.freedoniagroup.com/brochure/20xx/2083smwe).

There are two types of nutraceuticals. The first is traditional natural foods with significant health effects. The second is nontraditional nutraceuticals, which can be obtained through breeding, genetic engineering, or addition of nutrients or ingredients (http://www.aboutbioscience.org/pdfs/Nutraceuticals.pdf 2011). Herbal-based nutraceutical product formulation has several health benefits. Herbal-based bioactive ingredients can prevent cancer, CVD, gastrointestinal disorders, and other diseases.

1.3.3 Safety aspects of nutraceuticals

Safety and efficacy of nutraceutical products are very important considerations. A significant drawback of nutraceutical products is a lack of clinical evidence of their efficacy. Some nutraceutical products used as nutritional and medical substitutes have been found to contain toxic substances that may cause problems. The food and drug regulating authorities in countries such as the United States, Canada, China, India, and the European Union have strict regulations for manufacturing, marketing, and servicing such products. Good manufacturing practices, GRAS status (Generally Recognized As Safe), analytical methods, and validation should be adopted to ensure the safety aspects of using neutraceuticals (Bagchi, 2006; Ohama et al., 2006).

The nutraceutical market is projected to grow to US$250 billion in 2018 (PRNewswire-iReach, 2012). Nutraceuticals are generally sold in in pill, powder, and liquid forms and have been shown to have specific health effects. In the United States, these products are regulated as drug and dietary supplements. Two thirds of the American population takes at least one nutraceutical health product daily.

1.4 Sources of functional foods
1.4.1 Grains and cereals
1.4.1.1 Wheat

Wheat is one of the most widely cultivated cereal crops. Wheat has been cultivated for 10,000 years (Shewry, 2009). It is used primarily in baked goods. There are two types of wheat: red wheat and white wheat. Red wheat contributes more antioxidant activity than white wheat. The baked goods prepared from whole wheat flour contain more antioxidant capacity than refined flour products (Yu et al., 2013).

Whole wheat is an important source of dietary antioxidants. The bran portion contains several rich sources of antioxidants, in particular vitamin E. It also contains phenolic acids, which exhibit great antioxidant activity when wheat is subjected to acid treatment and enzymatic hydrolysis (Baublis et al., 2000). Several insoluble fibers found in wheat bran contribute significantly to maintaining the digestive system in a healthy way.

1.4.1.2 Oats

Oats (*Avena sativa*) is a minor cereal crop but an important functional food source that contains high concentrations of protein and fiber together with multiple minerals and vitamins (Peterson, 1992; Welch, 1995). It also contains phytic acid and several phenolic compounds. The oat bran layer contains several functional components. Consumption of oat bran–rich food products reduces the blood cholesterol level; the fiber present in oats reduces LDL cholesterol. Oat-rich foods are highly beneficial for diabetic patients (Truswell, 2002).

The cholesterol-lowering properties of oats may be due to the presence of soluble fiber, i.e., β-glucan. Several research studies using human models show a significant relationship between β-glucan consumption and glucose level reduction as a result of oat consumption (Casiraghi et al., 2006; Tosh et al., 2008).

Butt et al. (2008) reported that flour and bran obtained from oats are used as a substitute food for celiac disease patients. Malkki and Virtanen (2001) demonstrated several physiological effects, including delayed gastric emptying, diminished nutrient absorption, and prolonged satiety, which also affects small bowel motility.

Oats improve gastrointestinal function and glucose metabolism (Malkki and Virtanen, 2001). Consumption of 60 g of oatmeal/40 g of oat bran containing 3g of β-glucan significantly reduces serum cholesterol level up to 5% (Hasler, 1998). Oat hull fiber contains more total dietary fiber, 900–970 g/kg dry wt, than oat grout (60–90 g/kg). The gum obtained from oats contains more β-glucan, i.e., 600–800 g/kg dry wt, than oat grout (35–50) and oat bran (55–90) (Wani et al., 2014; Malkki and Virtanen, 2001).

1.4.1.3 Barley

Barley (*Hordeum vulgare*) is a grain used for food, malt, and animal feed (Das et al., 2012). Barley is a rich source of insoluble and soluble dietary fiber. Barley fiber has several functional effects. Regular consumption helps cure cardiovascular disease, type 2 diabetes, and various cancers (Brennan and Cleary, 2005; Wood, 2010).

Barley cultivars also contain β-glucan, a soluble fiber that helps reduce the risk of colon cancer. It also had hypocholesterolemic and hypoglycemic effects in several experimental animal studies. Barley bran has cholesterol lowering properties (Truswell, 2002).

1.4.1.4 Flaxseed

Flax (*Linum usitassimum*) is used primarily for medicinal purposes. It has been cultivated for more than 5,000 years (Figure 1.1c) (Tolkachev and

Figure 1.1 Cereal-based functional foods. (a) Brown rice; (b) psyllium; (c) flaxseed; (d) buckwheat; (e) food bar.

Zhuchenko, 2000; Singh et al., 2011). Flax contains several phyto-constituents as well as abundant linolenic acid and lignans. It is a good source of soluble fiber and protein. Adding flax seed to the diet prevents fatty acid deficiency (Oomah, 2001; Rapport and Lockwood, 2001). Ingestion of 10 g of flaxseed/day demonstrated hormonal changes associated with a reduced risk of breast cancer (Phipps et al., 1993). It also reduces total and LDL cholesterol (Bierenbaum et al., 1993; Cunnane et al., 1993) and platelet aggregation (Allman et al., 1995).

Rodents fed flaxseed showed a reduction in tumors of the colon, mammary glands, and lungs (Thompson, 1995; Yan et al., 1998). Oil obtained from flaxseed has been found to have an antihepatotoxicity property (Hendawi et al., 2016); Muscular dystrophy has been treated with dietary supplements of flax in several in vitro and in vivo models (Carotenuto et al., 2016). Several human and animal model studies showed that consumption of flaxseed phytoestrogens such as omega-3s and lignin can achieve a chemoprotective effect (Westcott and Muir, 1996; Adlercreutz, 1995).

1.4.1.5 Psyllium

Psyllium is a dietary fiber used to produce the laxative Metamucil™ (Figure 1.1b). Health benefits include antidiarrheal effect, prevention of cardiovascular disease, and weight reduction. The seed contains a considerable amount of fiber, which increases water content and stool weight. Psyllium has cholesterol-lowering properties. It is used to prevent chronic constipation and to maintain intestinal system health and balance. It also helps cure irritable bowel syndrome. Winston (2000) reported that dietary sources of psyllium help improve the blood glucose and lipid levels in certain individuals.

1.4.1.6 Brown rice

Brown rice is a food staple for most of the world's population (Figure 1.1a). It contains good quality of protein compared to wheat and corn (Das et al., 2012). Wang et al. (1999) reported that the amino acid composition of rice protein was better than that of casein and soy protein, which is why it is used in infant foods (Gurpreet and Sogi, 2007). Fermented foods derived from rice sources such as idli and dosa provide high energy and carbohydrates and are palatable and easily digestible (Steinkraus, 1994). Chung et al. (2016) reported that white rice can be used as a functional food when it is fortified with 8% pigmented giant embryonic rice, which has high anti-obesity and hypolipidemic effects. The extract obtained from germinated brown rice has excellent anti-diabetic properties (Choi et al., 2009).

Bran obtained from brown rice has been found to have excellent health benefits (Jariwalla, 2001). Brown rice is recommended over white rice because contains a greater amount of tricin, which has an anticancer effect, than white rice (Hudson et al., 2000). Bran can be fortified with several products to make it a functional food for the consumer (Jariwalla, 2001).

1.4.1.7 Buckwheat

Buckwheat is an important raw material (Figure 1.1d) used in the formulation of various functional foods because it contains flavonoids, flavones, phytosterols, proteins, thiamine-binding proteins, and other compounds. It exhibits strong biological activity that contributes to antihypertension, anticholesterolic effect, anti-obesity, and constipation improvement.

The British Herbal Pharmacopoeia (1990) has approved buckwheat as a hypotensive and antihemorrhagic drug. It is used to cure circulatory disorders. It is an efficient vasculo-protector. When used with lime, it can be the best remedy for curing hemorrhagic retinopathy (Iserin et al., 1997; Gimenez-Bastida and Zielinski, 2015). Buckwheat leaves contain 3–8% rutin (Bruneton, 1999), which has been proved to be an potent antioxidant with several functional effects such as edema protection, increased vascular permeability, fragility and hyaluronidase inhibition, anti-inflammatory effects, and leg edema

(Ihme et al., 1996). In vitro and animal studies demonstrated that bioactive compounds found in buckwheat viz. d-chiro-inositol (DCI), proteins, and some flavonoids (i.e., rutin and quercetin) have shown potent prebiotic and antioxidant properties (Gimenez-Bastida and Zielinski, 2015).

1.4.1.8 Sorghum

Sorghum (*Sorghum bicolor*) is consumed by humans worldwide (Lazaro and Favier, 2000). It is a major source of protein and carbohydrates. The pigmented sorghum is a good source of multiple antioxidants such as flavones and deoxyanthocyanidins. Several researchers have reported that the 3-deoxyanthocyanidins in sorghum have a preventive effect against skin, liver, colon, breast, esophagus, liver, and bone marrow cancers (Shih et al., 2007; Devi et al., 2011). Sorghum is also used to treat diseases such as diabetes, hypertension, obesity, and inflammation (Cardoso et al., 2015).

1.4.1.9 Maize

Maize is called the queen of cereals because of its high genetic yield potential. Next to rice and wheat, it is the most important cultivated crop used for food purposes, contributing approximately 5% to the world's dietary energy supply (Saikumar et al., 2012). Maize is widely used in the pharmaceutical industry in the form of pregelatinized maize starch for tablet release formulation.

1.4.1.10 Finger millet

Finger millet, one of the oldest crops cultivated in India, is considered the staple food crop of both Africa and India. It is typically ground, malted, and fermented to yield several products, including beverages, idli, dosa, and roti (Malathi and Nirmalakumari, 2007). Upon germination, finger millet produces statins with antihypocholestrolemic effects (Venkateswaran and Vijayalakshmi, 2010). It also has blood glucose and cholesterol reducing effects (Shobana et al., 2009). It is highly recommended for anemia sufferers. It has good antioxidant effects (Chandrasekara and Shahidi, 2011) and antimicrobial properties (Chethan and Malleshi, 2007). Compared to other cereals and millets it is a rich source of potassium (408 mg/100g) and calcium (344 mg/100g).

1.4.1.11 Breakfast cereals

Breakfast cereal foods can be enriched with inulin or psyllium. Brighenti et al.'s (1999) research studies show that consumption of 50 g of such inulin-fortified breakfast cereals for a period of one month significantly reduced total cholesterol by 7–9% and total triglycerides by up to 21.2% in human test subjects. Another study conducted by Roberts et al. in 1994 showed that breakfast cereal containing 50 g of psyllium, oats, and barley significantly decreased blood LDL and total cholesterol content in hypercholesterolemic men.

1.5 Dairy-based functional foods

Probiotics contain live microorganisms with various health benefits when consumed as part of the daily diet. The advantages/benefits of probiotic ingestion result in nutrient bioavailability improvement in host immune system modulation and a reduction in lactose intolerance. Probiotic species with beneficial effects include *Lactobacillus* and *Bifidobacterium*.

Million et al. (2012) reported that probiotic bacteria play an important role in suppressing obesity in farm animals. HFD-induced obese mice fed yoghurt with *L. acidophilus* NCDC 13 strain did not exhibit significant changes in adiposity markers. Dietary supplement beverages enriched with *L. casei, L. bulgaricus,* and *Streptococcus thermophilus* have been shown to improve natural defenses.

Probiotic foods are a good source of calcium, a nutrient that helps prevent osteoporosis and colon cancer. A total of 400 bacterial species reside in the gastrointestinal tract. They are classified into two categories: one is bifidobacterium and lactobacillus and the other is harmful bacteria such as Clostridium and Enterobacteriaceae. Maasai tribesmen in Africa who consumed fermented milk products, which were found to have hypocholesterolemic effects, had a greater life span (Mann and Spoerry 1974).

Several human clinical studies showed that such probiotic-based fermented milk products have significant cholesterol reducing properties (Sanders, 1994). Probiotic foods reduce the risk of colon cancer (Mital and Garg, 1995).

Fecal enzymes such as nitroreductase and azoreductase play important roles in preventing cancer. *Lactobacillus* has a major role in altering the enzyme activity in feces, which indirectly helps reduce the incidence of colon cancer. Fermented dairy products help reduce hypertension. They can alter and modify the microflora composition in the intestinal gut (Fuller, 1994). Consumption of probiotic foods reduces gastrointestinal disorders such as diarrhea and bowel disorders. A wide array of microorganisms, including *L. rhamnosus, L. reuteri, L. casei, L. acidophilus, L. plantarum,* and *Saccharomyces boulardii*, are used in developing fermented dairy products.

Fermented dairy products have been shown to cure diseases such as cancer (Kumar et al., 2010), AIDS (Trois et al., 2008), respiratory and urinary tract infections (Kaur et al., 2009), allergies (Yao et al., 2009), obesity, and type 2 diabetes (Douglas and Sanders, 2008). The probiotic effect can be achieved if the count exceeds $>10^6$ cfu/ml in the small bowel and $>10^8$ cfu/g in the colon. Proper selection of strain and its dose incorporation in the product are essential to achieve probiotic effect.

1.6 Margarine

In the United States margarine is considered a functional food because it is enriched with plant sterols. Hendriks et al. (1999) reported that margarine reduced the blood cholesterol levels in mildly hypercholesterolemic test subjects. Hallikainen and Uusitupa (1999) reported similar results of LDL cholesterol level reduction to 10–15%, even when three servings/day are consumed. The risk of heart disease can be drastically reduced by 25% with consumption of sterol-enriched margarine products (Law, 2000).

1.7 Prebiotics

Gibson and Roberfroid (1995) defined prebiotics as nondigestible food ingredients such as starches, dietary fiber, sugar alcohol, and oligosaccharides, which can beneficially affect the host by promoting the activity of one or a limited number of bacterial species in the colon, thereby improving the health of the host. Oligosaccharides have proved to be a highly prebiotic ingredient with multiple health benefits (Tomomatsu, 1994). Fruits and vegetables, including bananas, garlic, onions, and artichokes, as well as milk and honey, have been found to contain high amounts of oligosaccharides.

1.8 Confectionery-based functional foods

1.8.1 Cocoa

Cocoa (*Theobroma cacao*) is a product of the cacao tree, which is native to Central America (Baharum et al., 2016). Cocoa is an important ingredient used in the chocolate industry. It has several functional effects, including reduction of blood pressure, insulin resistance, and vascular platelet function. It contains a large amount of polyphenol (Cooper et al., 2008); the bean contains 6–8% of total polyphenol by dry weight. It is efficient for managing diabetes, anemia, CVD, obesity, kidney stones, and neurodegenerative diseases (Dillinger et al., 2000; Pearson et al., 2002; Othman et al., 2007). In the study reported by Ried et al. (2012) cocoa-rich foods were found to reduce blood pressure significantly.

1.8.2 Food bars

The Corazonas® HEARTBAR™ was developed to help manage vascular disease (Figure 1.1e). The HEARTBAR has a soy protein base and is fortified with vitamins C, E, B_6, and B_{12}, niacin, folate, and L-arginine. It is effective in patients with coronary artery disease. Maxwell et al. (2000) reported that consumption of two bars/day continuously for a period of 14 days significantly improved pain-free walking distance. There was a 23% increase in total walking distance for patients with lower limb atherosclerosis. These bars play a significant role in reducing the excess weight and it is specially meant to be used for overweight or obese people with diabetes (Craig, 2013).

1.8.3 Chocolate

Chocolate contains procyanidins as flavonoids, which can reduce oxidative stress on LDL cholesterol. Wan et al. (2001) reported that subjects who consumed chocolate and cocoa powder with approximately 466 mg procyanidin intake/day enhanced the time for LDL cholesterol oxidation to 8% compared to subjects who only followed a normal diet.

1.9 Beverage-based functional foods

1.9.1 Tea

Tea is a beverage with potent polyphenols that act as antioxidants. It is widely consumed throughout the world. Polyphenols account for 30% of total dry weight of fresh tea leaves. Graham (1992) reported that tea leaves contain catechins, natural phenols and antioxidants with chemopreventive effects, the most potent of which is epigallocatechin-3-gallate (EGCG). Nakachi et al. (1998) reported that drinking five cups of green tea/day may reduce incidence of breast cancer in Japanese women. Some research studies show that tea consumption decreases the risk of CVD. This may be because of the presence of various flavonoids, such as quercetin, kaempferol, myricetin, apigenin, and luteolin. Regular consumption of green tea is a significant factor in reducing risk of CHD and CVD (Tijburg et al., 1997).

Consumption of green tea significantly reduces cancer risk. One brewed cup of green tea contains approximately 200 mg of EGCG. Polyphenols obtained from black tea consumption may reduce platelet activation in vitro. Its anti-inflammatory effects may prevent CVD (Steptoe et al., 2007). EGCG has been shown to reduce the arachidonic acid-induced inhibitory effect of peroxinitrite on platelet aggregation.

1.10 Legume-based functional foods

1.10.1 Soybeans

Consumption of soyafoods dates back centuries (Xiao, 2008). Soy is a fiber-rich food that helps alleviate symptoms of several diseases. Soybean (*Glycine max*) consists of heterocyclic phenols known as isoflavones, which have various health-promoting effects (Jackson et al., 2002). It helps treat cancer, osteoporosis, and cardiovascular disease. Anderson et al. (1995) reported that soy protein consumption led to a significant decrease in LDL (12.9%), triglycerides (10.5%), and total cholesterol (9.3%). This effect is achieved when 20–25 g/day is consumed. Potter (1998) reported that isoflavones is the sole compound responsible for minimizing cholesterol in the biological system.

Soybeans contain many anticarcinogens such as saponins, phenolic acids, phytic acids, isoflavones, protease inhibitors, and phytosterols (Messina and Barnes, 1991). Of the various isoflavones present, daidzein, and genistein play significant roles in promoting health. Erdman and Potter (1997) reported that intake of 40 g of isolated soy protein containing 90 mg total isoflavones significantly increased bone mineral content and density after 6 months among post-menopausal women. They concluded that soy intake is essential for maintaining healthy bones.

Davidson (2008) reported that soy protein decreases LDL synthesis and the insulin/glucagon ratio in the liver. Data from several clinical studies showed that soy prevents different types of cancer, including breast, stomach, intestine, and prostate cancer.

1.11 Fruit- and vegetable-based functional foods

1.11.1 Garlic

Garlic (*Allium sativum*) is a species of the onion family that contains several organosulfur components such as allylic sulfides and allicin. Clinical studies have shown that garlic helps lower blood pressure and incidence of stomach cancer (Silagy and Neil, 1994; Takezaki et al., 1999). Garlic has cardioprotective properties due to its cholesterol reducing effect. Warshafsky et al. (1993) reported that consumption of 900 mg of garlic daily significantly reduced total serum cholesterol levels to 9%. Stevinson et al. (2000) reported that consumption of garlic in the form of steam distilled oil (10 mg) or garlic powder (600–900 mg) reduced total cholesterol content up to 4–6% in 13 placebo-controlled double blind trials. Garlic is widely used as a flavoring agent in cooking. It is considered to be a functional food that may promote both mental and physical health (Slowing et al., 2001). Garlic bulb has a high amount of allicin, which has chemo-preventive properties (Hasler, 1998).

Garlic has antidiabetic, antihypertensive, anti-aging, anti-gastric, antioxidant, and antibiotic effects. It also promotes blood circulation (Dorant et al., 1993; Srivastava et al., 1995; Borlinghaus et al., 2014). Various preclinical trials have exhibited good anti-tumorigenesis activity (Reuter et al., 1996). Allium from garlic can alleviate stomach cancer (You et al., 1988). It plays a major role in reducing cancer in human models (Dorant et al., 1993). Garlic consumption has been shown to reduce the risk of colon cancer by 50% (Steinmetz et al., 1994). Ernst (1997) reported that garlic has the potential to reduce gastrointestinal tract cancers. Garlic is well known for its antimicrobial activity (Shuford et al., 2005; Low et al., 2008). An aqueous extract derived from garlic reportedly contains a antimicrobial agent that can prevent dental caries and periodontitis (Houshmand et al., 2013).

Garlic has free radical scavenging activity, immune stimulation, and anti-infectious properties (Srivastava et al., 1995; Borek, 2006; Singh et al., 2007).

Garlic boosts the immune system (Amagase et al., 2001). Garlic consumption increased white blood cell count and bone marrow cellularity (Kuttan, 2000).

Garlic components had a suppressing effect on tumor growth in various in vivo models (Reuter et al., 1996). Organosulfur compounds play a major role in such effects. The medicinal properties of garlic may be due to the presence of such oil and water-soluble sulfur-containing components. Crushed garlic has excellent medicial properties. It contains an active constituent known as allicin, which has chemopreventive activity. Srivastava et al. (1995) and Silagy and Neil (1994) also reported that garlic has antibiotic, anti-hypertensive, cancer chemopreventive, and cholesterol-reducing properties.

1.11.2 Tomatoes

The tomato is considered a functional food because it contains a large amount of carotenoids, particularly lycopene, which helps reduce the risk of cancer (Weisburger, 1998) and prostate cancer in the in vivo human models (Giovannucci et al., 1995). It can also scavenge the formation of free radicals because of increased antioxidant activity. Di Mascio et al. (1989) reported that lycopene exhibits the highest singlet oxygen quenching ability in biological systems for all dietary carotenoids.

Other bioactive constituents present in tomatoes include β-carotene, phenolic compounds, flavonoids, glycoalkaloids, and the provitamins A, C, and E (Canene-Adams et al., 2005; Juroszek et al., 2009). Studies have shown that EPG protein expression through lycopene-rich tomato consumption prevents prostate cancer (Graff et al., 2016). Breast, skin, cervix, bladder, and digestive tract cancer can also be prevented (Clinton, 1998; Weisburger, 1998). Lycopene has nitrogendioxide and hydrogen peroxide scavenging capability (Bohm et al., 1995). The bioavailability of lycopene is superior in processed products compared to raw/unprocessed ones (Hsu et al., 2008). Lee et al. (2015) reported the hypolipidemic effect of processed tomato juice from an in vivo model. Das et al. (2010) reported that consumption of tomato juice can reduce DNA damage in lymphocyte-caused oxidative stress by 42%.

1.11.3 Broccoli and other cruciferous vegetables

Several epidemiological studies have shown that routine consumption of cruciferous vegetables reduces the risk of cancer significantly. The cancer reducing effect was as high as 30–70% as a result of consuming vegetables such as cabbage, broccoli, cauliflower, and Brussels sprouts (Figure 1.2a). These results may be due to the presence of glucosinolates. Sulforaphane, an isothiocyanate isolated from broccoli, significantly induces the phase II enzyme quinine reductase, which helps reduce the risk of cancer. Tender broccoli sprouts were found to contain 10–100 times more glucoraphanin than mature plants, which helped reduce the risk of cancer to a greater extent (Fahey et al., 1997).

Figure 1.2 *Fruit- and vegetable-based functional foods. (a) Cruciferous vegetables; (b) red wine and grapes; (c) cranberry juice.*

Table 1.3 Citrus Fruits and Their Functional Effects

Citrus Fruit	Phytoconstituent	Functional Effects
Orange	Hesperidin	Antioxidant and anti-inflammatory effects
Lemon	Vitamin C	Protects against immune system deficiencies, cardiovascular disease, and skin wrinkling
Lime	D-limonene	Anticancer effects
Grapefruit	Lycopene	Antitumor effects

1.11.4 Citrus fruits

The risk of human cancer can be reduced by consuming citrus fruits such as oranges, lemons, limes, and grapefruit, which contain important nutrients such as folate, vitamin C, and fiber. Table 1.3 shows the functional effects achieved from citrus fruit consumption. These fruits also contain limonoids, phytochemicals that have cancer-preventive properties (Hasegawa and Miyake, 1996). Several tumors were suppressed by these components (Crowell, 1997).

1.11.5 Cranberries

The cranberry belongs to the Ericaceae family. The major producers of cranberries are Canada and North America (Vattem et al., 2005). Cranberry juice contains benzoic acid, which is used in treating urinary tract infections. Avorn et al. (1994) showed that consumption of 300 ml of cranberry juice daily significantly reduced the incidence of bacteriuria after 6 months of feeding trials.

Cranberry juice cocktail is widely available (Figure 1.2c) and it is also used as a functional food. It can prevent plaque-forming bacteria in the mouth (Weiss et al., 1998). It is widely used in dental products such as toothpaste and mouthwash. Cranberries contain the biologically active component condensed tannin in the form of proanthocyanidin, which can prevent *E. coli*, which may adhere to the epithelial cells lining the urinary tract (Howell et al., 1998). It also has anti-adhesion properties.

Cranberries are effective in curing cardiovascular disease, lipoprotein oxidation, and atherosclerosis (McKay and Blumberg, 2007; Chu and Liu, 2005). Several research studies show the efficacy of cranberries in inhibiting cancer (Bomser et al., 1996; Krueger et al., 2000; Seeram et al., 2004) and increasing plasma antioxidant levels (Chu and Liu, 2005).

1.11.6 Red wine and grapes

Red wine obtained from red grape cultivars has the potential to minimize the risk of cardiovascular disease (Figure 1.2b). Red wine contains a high concentration of phenols, approximately 20–50 times more than white wine. This may be due to inclusion of the grape skin—which contains a large amount of polyphenols—in the production process. Wine obtained from black seedless grape cultivars contains phenolic concentrations as high as 3200 mg/L. These phenolics prevent LDL oxidation (Frankel et al., 1993). Red wine consumption also increased total plasma antioxidant capacity (Serafini et al., 1998). Red wine also contains a larger amount of trans-resveratrol, which is abundant in red grape skin, and has gut estrogenic properties (Gehm et al., 1997) and carcinogenesis effect (Jang et al., 1997).

1.11.7 Strawberries

Strawberries have good antioxidant capacity (Meyer et al., 1998) and are well known for their antihypertensive, antihyperlipidemic, anti-inflammatory and antiproliferative effects (McDougall et al., 2008; Lee et al., 2008; Cheplick et al., 2010; Giampieri et al., 2015). They contain a significant amount of flavonoids. Cheplick et al. (2010), Giampieri et al. (2015) and Basu et al. (2010) reported that the strawberry protects against inflammation, hypertension, and cardiovascular mortality. The in vitro model of Pinto Mda et al. (2010) and McDougall et al. (2005) showed that strawberry consumption helps manage hypertension and hyperglycemia.

Promising results of antitumor effect were observed after feeding rodents freeze-dried strawberries (Stoner et al., 2010).

Strawberries also showed anticarcinogenic effects (Wang et al., 2005), likely due to the presence of ellagic acid in the fruit. It can also cure age-related neurodegenerative disorders (Joseph et al., 2009). Kasimsetty et al. (2010) and Stanojkovic et al. (2010) reported that strawberry tannins showed anticancer effects in human breast, cervix, and colon carcinoma cells.

1.11.8 Figs

The fig belongs to the Moraceae family, which has many medicinal properties. Several research studies show that figs have potent anti-inflammatory, anti-cancer, anti-viral, cardiovascular, and aphrodisiac effects (Rubnov et al., 2001; Jeong et al., 2005); antidiabetic and hypolipidemic activities (Perez et al., 1998; Canal et al., 2000); antibacterial activity (Al-Yousuf, 2012); antifungal activity (Aref et al., 2010); antipyretic activity (Patil et al., 2010); and free radical scavenging activity (Yang et al., 2009). Dried figs have a potent anti-infertility effect (Naghdi et al., 2016); help cure gastrointestinal and inflammatory disorders (Gilani et al., 2008); and have a hepatoprotective effect (Gond and Khadabadi, 2008).

1.11.9 Cordyceps mushrooms

The cordyceps mushroom is found widely in India, Nepal, Bhutan, and China. It has several bioactive components including mannitol, cordcepin, polysaccharides, adenosine, and ergosterol (Yue et al., 2013). Including these mushrooms in the diet will boost the immune system (Paterson, 2008). They are also used to treat various inflammatory disorders (Paterson, 2008). Han et al. (2011) and Zhu and Liu (1992) reported that cultivation of this mushroom on germinated soyabean extract, which was effective against allergy and Type I hypersensitive in various experimental animal model studies.

Nakamura et al. (2015) reported that cordycepin extracted from this mushroom is being used to treat cancer. It is a rich source of polyunsaturated fatty acids (PUFAs) (68–87%). It also contains various organic acids such as oxalic, citric, and fumaric acids. It is used as a tonic in traditional Chinese medicine (Li et al., 2006).

1.11.10 Pumpkin

Pumpkin (*Cucurbita pepo*) belongs to the Cucurbitaceae family. It is widely used as a functional food. It is a rich source of fatty acids such as palmitic, stearic, oleic, linoleic, and linolenic; sterols, proteins, polysaccharides, peptides, and para-aminobenzoic acid; essential fatty acids such as omega-6 and omega-9 phytosterols; and antioxidant compounds such as tocophenols, carotenoids, vitamin A, and vitamin E (Ardabili et al., 2011; Srbinoska et al., 2012).

Pumpkin has anti-viral, antimicrobial, antidiabetic, anti-ulcer and anticancer effects (Park et al., 2010; Gill and Bali, 2011; Rathinavelu et al., 2013).

Several experimental studies showed that pumpkin has hypoglycaemic activity that decreases triglycerides, LDL and CRP (C-reactive protein), and cholesterol (Sedigheh et al., 2011). Mice models have shown potent antidiabetic activity (Xia and Wang, 2007). Pumpkin can improve the function of pancreatic β cells, which increase insulin cells (Acosta-Patino et al., 2001). Pumpkin seeds have shown good functional effect against prostatic hyperplasia (Medjakov et al., 2016). They can cure hepatic damage induced by alcohol consumption (Fruhwirth and Hermetter, 2007).

1.12 Animal-based functional foods

Animals are sources of various physiologically active components, including the co-enzyme Q_{10}, α-lipoic acid, and vitamin-like substances. The co-enzyme Q_{10} plays an effective part in cellular energy generation in humans as well as a protective role against CVD (Overvad et al., 1999). In addition, α-lipoic acid, a good free radical scavenger, has potent antioxidant activity (Packer et al., 1995).

1.12.1 Fish

Fish oil contains several vitamins and minerals (Tacon and Metian, 2013) (Figure 1.3). Fish consumption can reduce the risk of thrombosis

Figure 1.3 *Fish liver oil.*

(Bessell et al., 2015) and lower triglyceride level (Harris, 1996) and CVD (Bang and Dyerberg, 1972). Wu et al. (2015) reported that supplementation of the diet with fish oil prolongs the lifespan without chronic heart failure problems. Type 2 diabetes is noticeably lower for higher fish intake patients compared to lower intake patients (Kutner et al., 2002). Pase et al. (2015) also reported the effects of n-3 fatty acids on cardiovascular and cognitive function.

Fish oil also contains omega-3 fatty acid, which plays an important role in the prevention of cardiovascular disease. Krumhout et al. (1985) correlated intake of fish consumption with cardioprotective effect. Daviglus et al. (1997) observed that intake of 35 g on a daily basis significantly reduced the risk of death in the United States from nonsudden myocardial infarction. Therefore, it is highly advisable to include fish as a part of one's daily diet to achieve the functional effects to a greater extent.

Fatty fish such as salmon, tuna, sardines, herring, and mackerel contain n-3 fatty acids. There are two different (n-3) fatty acids: eicosapentaenoic acid (EPA) and docosahexaenoic acid (DHA). The brain and eye retina contain DHA, which is needed for proper functioning of the brain and eye and is also important for the development of these organs in infants.

Several clinical studies reported that (n-3) fatty acids have a protective role against cancer, psoriasis, Crohn's disease, cardiovascular disease, cognitive dysfunction, and rheumatoid arthritis (Rice, 1999).

These (n-3) fatty acids showed a positive effect in reducing mortality due to myocardial infarction and sudden death in patients with coronary heart disease (Bucher et al., 2002). The FDA also suggests that omega-3 fatty acids will decrease the risk of coronary heart disease. The FDA has also approved the use of DHA in infant food formulations. To maintain a healthy heart, two servings of fatty fish/week are recommended (American Heart Association Dietary Guideline, 2000). The FDA approved the use of EPA and DHA as dietary supplements if daily intake does not exceed 2 g/day.

Research findings shows that the gestation period was lengthened as a result of fish consumption during pregnancy, resulting in higher birth weight (Daviglus et al., 2010). Pregnant women may require omega-3 PUFAs for developing the infant's central nervous system.

1.12.2 Beef

Meat products contain important constituents of protein, fat, and many functional food components (Pighin et al., 2016). Meat has good quality protein with significant biological value. Since 2008, in the United States conjugated linolic acid (CLA) obtained from meat has been deemed to be safe for further consumption (Kim et al., 2016). Maslak et al. (2015) reported that CLA isomers block lipogenic gene expression in rats. CLA is a mixture of isomers of linoleic acid. It consists of two isomers: cis-9, trans-11 and cis-12, trans-10.

Ha et al. (1987) first isolated CLA from grilled beef, which has a potent anticarcinogenic effect. It is found in the fat of ruminant animals such as cows and sheep. It occurs in all foods but there are greater quantities in dairy products and ruminant animals (Yurawecz et al., 1999). Uncooked beef contains 2.9–4.3 mg CLA/g fat; lamb, chicken, pork, and salmon contain approximately 5.6, 0.9, 0.6, and 0.3 mg CLA/g fat, respectively. The CLA content of dairy products is 3.1–6.1 mg/g fat (Chin et al., 1992). Watkins et al. (1999) reported bone density increase due to CLA intake. Beef fat was found to contain 57–85% of total CLA, which contributes approximately 3.1–8.5 mg CLA/g fat. Processed foods were found to contain more CLA than raw foods.

Ip and Scimeca (1997) reported that CLA helps prevent tumors and cancer in animal experimental models. It can be an effective anticarcinogen if it is incorporated into the diet in the amount of 0.1–1%. CLA modulates tumor growth. It manages weight, thereby helping change body composition. Park et al. (1997) reported that mice fed a diet supplemented with CLA of 0.5% showed body fat reduction of 60% and an increase in lean body mass of 14% compared to control rats. Their study shows that CLA further helps reduce the fat deposition and further increases lipolysis in adipocytes.

1.12.3 Eggs enriched with PUFAs

Eggs rich in PUFAs are marketed to cater for those whose diets have enriched PUFA content. Eggs with enriched PUFAs are obtained by feeding the hen a diet supplemented with fish oil/vegetable oil or an algal source of docosahexaenoic acid. Farrell (1998) reported that feeding of enriched egg with PUFAs once daily significantly increases blood omega-3 PUFAs and HDL cholesterol levels compared to controls. Body weight also increased simultaneously.

1.13 Conclusion

Consumption of functional foods and nutraceuticals helps maintain or promote better health or alleviate or control disease and thus enhance and extend the human lifespan. Functional foods and nutraceuticals are readily available in many common foods. These foods have rapidly growing market potential in much of the world.

References

Acosta-Patino, J.L., E. Jimenez-Balderas, M.A. Juarez-Oropeza, and J.C. Diaz-Zagoya. 2001. Hypoglycemic action of *Cucurbita ficifolia* on Type 2 diabetic patients with moderately high blood glucose levels. *J of Ethnopharmacol.* 77: 99–101.

Adlercreutz, H. 1995. Phytoestrogens: Epidemiology and a possible role in cancer protection. *Environ Health Perspect.* 103: 103–112.

AHA Scientific Statement: AHA Dietary Guidelines: Revision 2000: A Statement for Healthcare Professionals From the Nutrition Committee of the American Heart Association, Greenville Ave, Dallas, Texas, USA.

Al-Yousuf, H.H.H. 2012. Antibacterial activity of *Ficus carica* L. extract against six bacterial strains. *Int J. Drug Dev Res.* 4: 307–310.

Allman, M.A., M.M. Pena, and D. Pang. 1995. Supplementation with flaxseed oil versus sunflowerseed oil in healthy young men consuming a low fat diet: Effects on platelet composition and function. *Eur J Clin Nut* 49: 169–178.

Amagase, H., B. Petesch, H. Matsuura, S. Kasuga, and Y. Itakara. 2001. Impact of various sources of garlic and their constituents on 7, 12-dimethlybenz(a)anthracene binding to mammary cell DNA. *Carcinogenesis* 14: 1427–1631.

Anderson, J.W., B.M. Johnstone, and M.E. Cook-Newell. 1995. Meta-analysis of the effects of soy protein intake on serum lipids. *New Engl J Medi.* 333: 276–282.

Ardabili, G., R. Farhoosh, and M.H.H. Khodaparast. 2011. Chemical composition and physicochemical properties of pumpkin seeds (*Cucurbita pepo* subsp. *pepo* var. Styriaka) grown in Iran. *J Agri Sci and Techn.* 13: 1053–1063.

Aref, H.L., K.B. Salah, J.P. Chaumont, A. Fekih, M. Aouni, and K. Said. 2010. In vitro antimicrobial activity of four *Ficus carica* latex fractions against resistant human pathogens (antimicrobial activity of *Ficus carica* latex). *Park J Pharm Sci.* 23: 53–58.

Avorn, J., M. Monane, J.H. Gurwitz, R.J. Glynn, I. Choodnovskiy, and L.A. Lipsitz. 1994. Reduction of bacteriuria and pyuria after ingestion of cranberry juice—A reply. *J Am Med Assoc.* 272: 589–590.

Bagchi, D. 2006. Nutraceuticals and functional foods regulations in the United States and around the world. *Toxicology* 221:1–3.

Baharum, Z., A.M. Akim, T.Y. Hin, R.A. Hamid, and R. Kasran. 2016. *Theobroma Cacao*: Review of the extraction, isolation, and bioassay of its potential anti-cancer compounds. *Trop Life Sci Res.* 27: 21–42.

Bang, H.O., and J. Dyerberg. 1972. Plasma lipids and lipoproteins in Greenlandic west coast Eskimos. *Acta Medi Scand.* 192: 85–94.

Basu, A., M. Rhone, and T.J. Lyons. 2010. Berries: Emerging impact on cardiovascular health. *Nutr. Rev* 68: 168–177.

Baublis, A.J., C. Lu, F.M. Clydesdale, and E.A. Decker. 2000. Potential of wheat based breakfast cereals as a source of dietary antioxidants. *J Am Coll Nutr* 19: 308–311.

Bessell, E., M.D. Jose, and C. McKercher. 2015. Associations of fish oil and vitamin B and E supplementation with cardiovascular outcomes and mortality in people receiving haemodialysis: A review. *BMC Nephrol.* 16:143.

Bierenbaum, M.L., R. Reichstein, and T.R. Watkins. 1993. Reducing atherogenic risk in hyperlipemic humans with flax seed supplementation: A preliminary report. *J Am Coll Nutr* 12: 501–504.

Bohm, F., J.H. Tinkler, and T.G. Truscott. 1995. Carotenoids protect against cell membrane damage by the nitrogen dioxide radical. *Nat Med.* 1: 98–99.

Bomser, J., D.L. Madhavi, K. Singletary, and M.A. Smith. 1996. In vitro anticancer activity of fruit extracts from Vaccinium species. *Planta Medica* 62: 212–216.

Borek, C. 2006. Garlic reduces dementia and heart-disease risk. *J Nutr.* 136: 810S–812S.

Borlinghaus, J., F. Albrecht, M.C. Gruhlke, I.D. Nwachukwu, and A.J. Slusarenko. 2014. Allicin: Chemistry and biological properties. *Molecules* 19: 12591–12618.

Brennan, C.S., and L.J. Cleary. 2005. The potential use of cereal (103) (104)-b-D-glucans as functional food ingredients. *J Cereal Sci.* 42: 113.

Brighenti, F., M.C. Casiraghi, E. Canzi, and A. Ferrari. 1999. Effect of consumption of a ready-to-eat breakfast cereal containing inulin on the intestinal milieu and blood lipids in healthy male volunteers. *Eur J Clin Nutr* 53: 726–733.

British Herbal Pharmacopoeia. 1990. *British Herbal Pharmacopoeia.* Bournemouth, UK: British Herbal Medical Association.

Bruneton, J. 1999. *Pharmacognosie, Phytochimie, Plantes médicinales.* Paris: Techniques and Documentation.

Bucher, H.D., P. Hengstler, C. Schindler, and G. Meiter. 2002. N-3 PUFA in coronary heart disease: A meta-analysis of randomized controlled trials. *Am J Med.* 112: 298–304.

Butt, M.S., M.T. Nadeem, M.K.I. Khan, R. Shabir, and M.S. Butt. 2008. Oats: Unique among the cereals. *Eur J Nutr.* 47: 68–79.

Canal, J.R., M.D. Torres, A. Romero, and C. Perez. 2000. A chloroform extract obtained from a decoction of *Ficus carica* leaves improves the cholesterolaemic status of rats with streptozotocin-induced diabetes. *Acta Physiol Hung.* 87: 71–76.

Canene-Adams, K., J.K. Campbell, S. Zaripheh, E.H. Jeffery, and J.W. Erdman, Jr. 2005. The tomato as a functional food. *J Nutr.* 135: 1226–1230.

Cardoso, L.M., S.S. Pinheiro, S.H.D. Martino, and H.M. Pinheiro-Sant'Ana. 2015. Sorghum (*Sorghum bicolor* l): Nutrients, bioactive compounds, and potential impact on human health. *Crit Rev Food Sci Nutr.* 57: 372–390.

Carotenuto, F., A., Costa, M.C. Albertini, M.B. Luigi Rocchi, A. Rudov, D. Coletti, M. Minieri, P.D. Nardo, and L. Teodori. 2016. Dietary flaxseed mitigates impaired skeletal muscle regeneration: In vivo, in vitro and in silico studies. *Int J Med Sci.* 13: 206–219.

Casiraghi, M.C., M. Garsetti, G. Testolin, and F. Brighenti. 2006. Post-prandial responses to cereal products enriched with barley glucan. *J Am Coll Nutr.* 25: 313–320.

Chandrasekara, A., and F. Shahidi. 2011. Determination of antioxidant activity in free and hydrolyzed fractions of millet grains and characterization of their phenolic profiles by HPLC-DAD-ESI-MSn. *J Funct Foods.* 3: 144–158.

Cheplick, S., Y.I. Kwon, P. Bhowmik, and K. Shetty. 2010. Phenolic-linked variation in strawberry cultivars for potential dietary management of hyperglycemia and related complications of hypertension. *Bioresour Technol.* 101: 404–413.

Chethan, S., and N.G. Malleshi. 2007. Finger millet polyphenols: Characterization and their nutraceutical potential. *Am J Food Technol.* 2: 582–592.

Chin, S.F., W. Liu, J.M. Storkson, Y.L. Ha, and M.W. Pariza. 1992. Dietary sources of conjugated dienoic isomers of linoleic acid, a newly recognized class of anticarcinogens. *J Food Composit Analy.* 5: 185–197.

Choi, H.D., Y.S. Kim, I.U. Choi, Y.K. Park, and Y.D. Park. 2009. Hypotensive effect of germinated brown rice on spontaneously hypertensive rats. *Korean J Food Sci Technol.* 38: 448–451.

Chu, Y.F., and R.H. Liu. 2005. Cranberries inhibit LDL oxidation and induce LDL receptor expression in hepatocytes. *Life Sci* 77: 1892–1901.

Chung, S.I., C.W. Rico, S.C. Lee, and M.Y. Kang. 2016. Instant rice made from white and pigmented giant embryonic rice reduces lipid levels and body weight in high fat diet-fed mice. *Obes Res Clin Pract.* 10: 692–700.

Clinton, S.K. 1998. Lycopene: Chemistry, biology, and implications for human health and disease. *Nutr Rev.* 56: 35–51.

Cooper, K.A., J.L. Donovan, A.L. Waterhouse, and G. Williamson. 2008. Cocoa and health: A decade of research. *Br J Nutr.* 99: 1–11.

Craig, J. 2013. Meal replacement shakes and nutrition bars: Do they help individuals with diabetes lose weight? *Diabetes Spectr.* 26(3): 179–182.

Crowell, P.L. 1997. Monoterpenes in breast cancer chemoprevention. *Breast Cancer Res Treat.* 46: 191–197.

Cunnane, S.C., S. Ganguli, C. Menard, A.C. Liede, M.J. Hamadeh, Z.Y. Chen, T.M.S. Wolever, and D.J.A. Jenkins. 1993. High-linolenic acid flaxseed (*Linum usitatissimum*): Some nutritional properties in humans. *Br J Nutr.* 69: 443–453.

Das, A., U. Raychaudhuri, and R. Chakraborty. 2012. Cereal based functional food of Indian subcontinent: A review. *J Food Sci Technol.* 49: 665–672.

Das, D., R. Vimala, and N. Das. 2010. Functional foods of natural origin: An overview. *Indian J Nat Prod Resour.* 1: 136–142.

Davidson, M. 2008. A review of the current status of the management of mixed dyslipidemia associated with diabetes mellitus and metabolic syndrome. *Am J Cardiol.* 102: 19L–27L.

Daviglus, M., J. Stamler, A. Orencia, A.R. Dyer, K. Liu, P. Greenland, M.K. Walsh, D. Morris, and R.B. Shekelle. 1997. Fish consumption and the 30-year risk of fatal myocardial infarction. *N Eng J Med.* 336: 1046–1053.

Daviglus, M., J. Sheeshka, and E. Murkin. 2010. Health benefits from eating fish. *Comments Toxicol.* 8: 345–374.

Day, L., R.B. Seymour, K.F. Pitts, I. Konczak, and L. Lundin. 2009. Incorporation of functional ingredients into foods. *Trends Food Sci Technol.* 20: 388.

Devi, P.S., M.S. Kumar, and S.M. Das. 2011. Evaluation of antiproliferative activity of red sorghum bran anthocyanin on a human breast cancer cell line (mcf-7). *Int J Breast Cancer* 2011:1–6.

Di Mascio, P., S. Kaiser, and H. Sies. 1989. Lycopene as the most efficient biological carotenoid singlet oxygen quencher. *Arch Biochem Biophys.* 274: 532–538.

Dillinger, T.L., P. Barriga, S. Escarcega, M. Jimenez, D. Salazar-Lowe, and L.E. Grivetti. 2000. Food of the gods: Cure for humanity? A cultural history of the medicinal and ritual use of chocolate. *J Nutr.* 130: 2057S–72S.

Dorant, E., P.A. Van den Brandt, R.A. Goldbohm, R.J. Hermus, and F. Sturmans. 1993. Garlic and its significance for the prevention of cancer in humans: A critical view. *Br J Cancer* 67: 424–429.

Douglas, L.C., and M.E. Sanders. 2008. Probiotics and prebiotics in dietetics practice. *J Am Diet Assoc.* 108: 510–521.

Erdman, J.W., and S.M. Potter. 1997. Soy and bone health. *Soy Connect.* 5: 1–4.

Ernst, E. 1997. Can Allium vegetables prevent cancer? *Phytomedicine* 4: 79–83.

Euromonitor. 2010. Cardiovascular health: A key area of functional food and drinks development. Euromonitor International, London.

Euromonitor. 2013. http://www.portal.euromonitor.com/portal/default.aspx

Fahey, J.W., Y. Zhang, and P. Talalay. 1997. Broccoli sprouts: An exceptionally rich source of inducers of enzymes that protect against chemical carcinogens. *Proc Nat Acad Sci.* 94: 10366–10372.

Farnworthe, E.R. 1997. FOSHU foods in Japan. No. 11. http://www.medicinalfoodnews.com/vol01/issue3/foshu

Farrell, D.J. 1998. Enrichment of hen eggs with n-3 long-chain fatty acids and evaluation of enriched eggs in humans. *Am J Clin Nutr.* 68: 538–544.

Food and Drug Administration (FDA). 1999. Food labeling: Use on dietary supplements of health claims based on authoritative statements. Department of Health and Human Services. www.cfsan.fda.gov/~lrd/fr990121

Food and Drug Administration. 2005. Health claims: Soluble fibre from certain foods and coronary heart diseases (CHD). *Fed Regist.* 70: 76150–76162.

Frankel, E.N., J. Kanner, J.B. German, E. Parks, and J.E. Kinsella. 1993. Inhibition of oxidation of human low-density lipoprotein by phenolic substances in red wine. *Lancet* 341: 454–457.

Fruhwirth, G.O., and A. Hermetter. 2007. Seeds and oil of the Styrian oil pumpkin: Components and biological activities. *Eur J Lipid Sci Technol.* 109: 1128–1140.

Fuller, R. 1994. History and development of probiotics. In *Probiotics,* R. Fuller (ed.). Chapman & Hall, New York.

Gehm, B.D., J.M. McAndrews, P.Y. Chien, and J.L. Jameson. 1997. Resveratrol, a polyphenolic compound found in grapes and wine, is an agonist for the estrogen receptor. *Proc Natl Acad Sci.* 94: 14138–14143.

Giampieri, F., T.Y. Forbes-Hernandez, M. Gasparrini, J.M. Alvarez-Suarez, S. Afrin, S. Bompadre, J.L. Quiles, B. Mezzethi, and M. Battino. 2015. Strawberry as a health promoter: An evidence based review. *Food Funct.* 6: 1386–1398.

Gibson, G.R., and M.B. Roberfroid. 1995. Dietary modulation of the human colonic microbiota: Introducing the concept of prebiotics. *J Nutr.* 125: 1401–1412.

Gilani, A.H., M.H. Mehmood, K.H. Janbaz, A.U. Khan, and S.A. Saeed. 2008. Ethnopharmacological studies on antispasmodic and antiplatelet activities of *Ficus carica. J Ethnopharmacol.* 119: 1–5.

Gill, N.S., and M. Bali. 2011. Isolation of anti ulcer cucurbitane type triterpenoid from the seeds of *Cucurbita pepo. Res J Phytochem.* 5: 70–79.

Gimenez-Bastida, J.A., and H. Zielinski. 2015. Buckwheat as a functional food and its effects on health. *J Agric Food Chem.* 63(36): 7896–7913.

Giovannucci, E., A. Ascherio, E.B. Rimm, M.J. Stampfer, G.A. Colditz, and W.C. Willett. 1995. Intake of carotenoids and retinol in relation to risk of prostate cancer. *J National Cancer Inst.* 87: 1767–1776.

Gond, N.Y., and S.S. Khadabadi. 2008. Hepatoprotective activity of *Ficus carica* leaf extract on rifampicin-induced hepatic damage in rats. *Indian Journal of Pharmaceutical Sci.* 70: 364–366.

Graff, R.E., A. Pettersson, R.T. Lis, T.U. Ahearn, S.C. Markt, S. Finn, S.A. Kenfield, M. Loda, E.L. Giovannucci, B. Rosner, and L.A. Mucci. 2016. Dietary lycopene intake and risk of prostate cancer defined by ERG protein expression. *Am J Clin Nutr.* 103: 851–860.

Graham, H.N. 1992. Green tea composition, consumption and polyphenol chemistry. *Prev Med.* 21: 334–350.

Gurpreet, K.C., and D.S. Sogi. 2007. Functional properties of rice bran protein concentrates. *J Food Eng.g* 79: 592–597.

Ha, Y.L., N.K. Grimm, and M.W. Pariza. 1987. Anticarcinogens from fried ground beef: Health-altered derivatives of linoleic acid. *Carcinogenesis* 8: 1881–1887.

Hallikainen, M.A., and M.I.J. Uusitupa. 1999. Effect of two low-fat stanol ester-containing margarines on serum cholesterol concentration as part of a low-fat diet in hypercholesterolemic subjects. *Am J Clin Nutr.* 69: 403–410.

Han, E.S., J.Y. Oh, and H.J. Park. 2011. *Cordyceps militaris* extract suppresses dextran sodium sulfate-induced acute colitis in mice and production of inflammatory mediators from macrophages and mast cells. *J Ethnopharmacol.* 134: 703–710.

Harris, W.S. 1996. n-3 fatty acids and lipoproteins: Comparison of results from human and animal studies. *Lipids* 31: 243–252.

Hasegawa, S., and M. Miyake. 1996. Biochemistry and biological functions of citrus limonoids. *Food Rev Int.* 12: 413–435.

Hasler, C.M. 1998. Functional foods: Their role in disease prevention and health promotion. *Food Technol.* 52: 57–62.

Hendawi, M.Y., R.T. Alam and S.A. Abdellatief. 2016. Ameliorative effect of flaxseed oil against thiacloprid-induced toxicity in rats: Hematological, biochemical, and histopathological study. *Environ Sci Pollut Res Int.* 23: 11855–11863.

Hendriks, H.F., J.A. Weststrate, T. Van-Vliet, and G.W. Meijer. 1999. Spreads enriched with three different levels of vegetable oil sterols and the degree of cholesterol lowering in normocholesterolaemic and mildly hypercholesterolaemic subjects. *Eur J Clin Nutr.* 53: 319–327.

Howell, A.B., N. Vorsa, A. Der Marderosian, and L.Y. Foo. 1998. Inhibition of the adherence of P-fimbriated *Escherichia coli* to uroepithelial-cell surfaces by proanthocyanidin extracts from cranberries. *N Engl J Med.* 339(15):1085–1086.

Houshmand, B., F. Mahjour, and O. Dianat. 2013. Antibacterial effect of different concentrations of garlic (*Allium sativum*) extract on dental plaque bacteria. *Indian J Dental Research* 24: 71–75.

Hsu, Y.M., C.H. Lai, C.Y. Chang, C.T. Fan, C.T. Chen, and C.H. Wu. 2008. Characterizing the lipid-lowering effects and antioxidant mechanisms of tomato paste. *Bioscience, Biotechnol Biochem.* 72: 677–685.

http://www.freedoniagroup.com/brochure/20xx/2083smwe. pdf. Accessed 30 Nov 2011.

http://www.aboutbioscience.org/pdfs/Nutraceuticals.pdf. Accessed 5 Dec 2011.

Hudson, E.A., P.A. Dinh, T. Kokubun, M.S. Simmonds, and A. Gescher. 2000. Characterization of potentially chemopreventive phenols in extracts of brown rice that inhibit the growth of human breast and colon cancer cells. *Cancer Epidemiol, Biomarker Prev.* 9: 1163–1170.

Ihme, N., H. Kiesewetter, F. Jung, K.H. Hoffmann, A. Birk, A. Muller, and Grutzner, K.I. 1996. Leg oedema protection from buckwheat herb tea in patients with chronic venous insufficiency: A single-centre, randomised, doubleblind, placebo-controlled clinical trial. *Eur J Clin Pharmacol.* 50: 443–447.

International Food Information Council. 1999. Functional Foods Now. Washington, DC.

Ip, C., and J.A. Scimeca. 1997. Conjugated linoleic acid and linoleic acid are distinctive modulators of mammary carcinogenesis. *Nutr Cancer.* 27: 131–135.

Iserin, P., M. Masson, and J.P. Restellini. 1997. *Encyclopédie des Plantes médicinales: Identification, Préparations, Soins.* Larousse-Bordas, Paris.

Jackson, C.J.C., J.P. Dini, C. Lavandier, H.P.V. Rupasinghe, H. Faulkner, V. Poysa, D. Buzzell, and S. De-Grandis. 2002. Effects of processing on the content and composition of isoflavones during manufacturing of soy beverage and tofu. *Proc Biochem.* 37: 1117–1123.

Jang, M., J. Cai, G. Udeani, K.V. Slowing, C.F. Thomas, C.W.W. Beecher, H.H.S. Fong, N.R. Farnsworth, A.D. Kinghorn, R.G. Mehta, R.C. Moon, and J.M. Pezzuto. 1997. Cancer chemopreventive activity of resveratrol, a natural product derived from grapes. *Science* 275: 218–220.

Jariwalla, R.J. 2001. Rice-bran products: Phytonutrients with potential applications in preventive and clinical medicine. *Drugs Exp Clin Res.* 27: 17–26.

Jeong, M.R., J.D. Cha, and Y.E. Lee. 2005. Antibacterial activity of Korean Fig (*Ficus carica* L.) against food poisoning bacteria. *Korean J Food Cookery Sci.* 21: 84–93.

Joseph, J.A., B. Shukitt-Hale, and L.M. Willis. 2009. Grape juice, berries, and walnuts affect brain aging and behavior. *J Nutr.* 139: 1813S–1817S.

Juroszek, P., H.M. Lumpkin, R.Y. Yang, D.R. Ledesma, and C.H. Ma. 2009. Fruit quality and bioactive compounds with antioxidant activity of tomatoes grown on-farm: Comparison of organic and conventional management systems. *J Agric Food Chem.* 57: 1188–1194.

Just-food. 2007. Global market review of functional foods: Forecasts to 2012. Bromsgrove: Aroq Limited, UK. http://www.just-food.com/store/product.aspx?id=44028&lk=pop

Kasimsetty, S.G., D. Bialonska, M.K. Reddy, G. Ma, S.I. Khan, and D. Ferreira. 2010. Colon cancer chemopreventive activities of pomegranate ellagitannins and urolithins. *J Agric Food Chem.* 58: 2180–2187.

Kaur, I.P., A. Kuhad, A. Garg, and K. Chopra. 2009. Probiotics: Delineation of prophylactic and therapeutic benefits. *J Med Food.* 12: 219–235.

Keservani, R.K., N. Vyas, S. Jain, R. Raghuvanshi, and A.K. Sharma. 2010. Nutraceutical and functional food as future food: A review. *Pharm Lett.* 2: 106.

Kim, J.H., Y. Kim, Y.J. Kim, and Y. Park. 2016. Conjugated linoleic acid: Potential health benefits as a functional food ingredient. *Annu Rev Food Sci Technol.* 7: 221–244.

Kotilainen. L., R. Rajalahti, C. Ragasa, and E. Pehu. 2006. Health enhancing foods. Opportunities for strengthening the sector in developing countries. Agriculture and Rural Development. Discussion Paper 30. Washington, DC: The World Bank.

Krueger, C.G., M.L. Porter, D.A. Weibe, D.G. Cunningham, and J.D. Reed. 2000. Potential of cranberry flavonoids in the prevention of copper-induced LDL oxidation. *Polyphenols Commun.* 2: 447–448.

Krumhout, D., E.B. Bosschieter, and C. Coulander. 1985. The inverse relation between fish consumption and 20-year mortality from coronary heart disease. *N Eng J Med.* 312: 1205–1209.

Kumar, M., A. Kumar, R. Nagpal, D. Mohania, P. Behare, V. Verma, P. Kumar, D. Poddar, P.K. Aggarwal, C.J. Henry, S. Jain, and H. Yadav. 2010. Cancer-preventing attributes of probiotics: An update. *Int J Food Sci Nutr.* 61: 473–496.

Kutner, N.G., P.W. Clow, R. Zhang, and X. Aviles. 2002. Association of fish intake and survival in a cohort of incident dialysis patients. *Am J Kidney Disord.* 39: 1018–1024.

Kuttan, G. 2000. Immunomodulatory effect of some maturally occurring sulphur-containing compounds. *J Ethnopharmacol.* 72: 93–99.

Law, M. 2000. Plant sterol and stanol margarines and health. *Br Medical J.* 320: 861–864.

Lazaro, E.L., and J.F. Favier. 2000. Alkaline debranning of sorghum and millet. *Am Assoc Cereal Chem.* 77: 717–719.

Lee, I.T., Y.C. Chan, C.W. Lin, W.J. Lee, and W.H. Sheu. 2008. Effect of cranberry extracts on lipid profiles in subjects with Type 2 diabetes. *Diabetic Med.* 25: 1473–1477.

Lee, L.C., L. Wei, W.C. Huang, Y.J. Hsu, Y.M. Chen, and C.C. Huang. 2015. Hypolipidemic effect of tomato juice in hamsters in high cholesterol diet-induced hyperlipidemia. *Nutrients* 7: 10525–10537.

Li, S.P., G.H. Zhang, Q. Zeng, Z.G. Huang, Y.T. Wang, T.T.X. Dong, and K.W.K. Tsiml. 2006. Hypoglycemic activity of polysaccharide, with antioxidation, isolated from cultured *Cordyceps mycelia*. *Phytomedicine* 13: 428–433.

Low C.F., P.P. Chong, P.V. Yong, C.S. Lim, Z. Ahmad, and F. Othman. 2008. Inhibition of hyphae formation and SIR2 expression in Candida albicans treated with fresh Allium sativum (garlic) extract. *J Appl Microbiol.* 105(6): 2169–77.

Malathi, D., and A. Nirmalakumari. 2007. Cooking of small millets in Tamil Nadu. In K.T. Gowda, and A. Seetharam. Food Uses of Small Millets and Avenues for Further Processing and Value Addition. Indian Council of Agricultural Research. Project Coordination Cell, All India Co-ordinated Small Millets Improvement Project. Indian Council of Agricultural Research, UAS, GKVK, Bangalore, India.

Malkki, Y., and E. Virtanen. 2001. Gastrointestinal effects of oat bran and oat gum: A review. *LWT Food Sci. Technol.* 34: 337–347.

Mann GV, Spoerry A. 1974. Studies of a surfactant and cholesteremia in the Maasai. *Am J Clin Nutr.* 27(5): 464–469.

Maslak, E., E. Buczek, A. Szumn, W. Szczepnski, M. Franczyk-Zarow, A. Kopec, S. Chlopicki, T. Leszczynska, and R.B. Kostogrys. 2015. Individual CLA isomers, c9t11 and t10c12, prevent excess liver glycogen storage and inhibit lipogenic genes expression induced by high fructose diet in rats. *Biomed Res Int.* 2015: 1–10.

Maxwell, A.J., B.E. Anderson, and J.P. Cooke. 2000. Nutritional therapy for peripheral arterial disease: A double-blind, placebo-controlled, randomized trial of Heartbar™. *Vasc Med.* 5: 11–19.

McDougall, G.J., H.A. Ross, M. Ikeji, and D. Stewart. 2008. Berry extracts exert different antiproliferative effects against cervical and colon cancer cells grown in vitro. *J Agric Food Chem.* 56: 3016–3023.

McDougall, G.J., F. Shapiro, P. Dobson, P. Smith, A. Blake, and D. Stewart. 2005. Different polyphenolic components of soft fruits inhibit alpha-amylase and alpha-glucosidase. *J. Agric Food Chem.* 53: 2760–2766.

McKay, D.L., and J.B. Blumberg. 2007. Cranberries (*Vaccinium macrocarpon*) and cardiovascular disease risk factors. *Nut Rev.* 65: 490–502.

Medjakovic, S., S. Hobiger, K. Ardjomand-Woelkart, F. Bucar, and A. Jungbauer. 2016. Pumpkin seed extract: Cell growth inhibition of hyperplastic and cancer cells, independent of steroid hormone receptors. *Fitoterapia* 110: 150–156.

Messina, M., and S. Barnes. 1991. The role of soy products in reducing risk of cancer. *J Nat Cancer Inst.* 83: 541–546.

Meyer, A.S., M. Heinonen, and E.N. Frankel. 1998. Antioxidant interactions of catechin, cyanidin, caffeic acid, quercetin, and ellagic acid on human LDL oxidation. *Food Chem.* 61: 71–75.

Million, M., E. Angelakis, M. Paul, F. Armougom, L. Leibovici, and D. Raoult. 2012. Comparative meta-analysis of the effect of Lactobacillus species on weight gain in humans and animals. *Microb Pathog.* 53: 100–108.

Mital, B.K., and S.K. Garg. 1995. Anticarcinogenic, hypocholesterolemic, and antagonistic activities of *Lactobacillus acidophilus*. *Crit Rev Microbiol.* 21: 175–214.

Naghdi, M., M. Maghbool, M. Seifalah-Zade, M. Mahaldashtian, Z. Makoolati, S.A. Kouhpayeh, A. Ghasemi, and K.W.K. Tsim. 2016. Effects of common fig (*Ficus carica*) leaf extracts on sperm parameters and testis of mice intoxicated with formaldehyde. *Evidence-Based Complementary Altern Med.* 2016: 1–9.

Nakachi, K., K. Suemasu, K. Suga, T. Takeo, K. Imai, and Y. Higashi. 1998. Influence of drinking green tea on breast cancer malignancy among Japanese patients. *Jpn J Cancer Res.* 89: 254–261

Nakamura, K., K. Shinozuka, and N. Yoshikawa. 2015. Anticancer and antimetastatic effects of cordycepin, an active component of *Cordyceps sinensis*. *J Pharmacol Sci.* 127: 53–56.

Ohama, H., H. Ikeda, and H. Moriyama. 2006. Health foods and foods with health claims in Japan. *Toxicology* 221: 95–111.

Oomah, B.D. 2001. Flaxseed as a functional food source. *J Sci Food Agric.* 81: 889–894.

Othman, A., A. Ismail, N. Abdul-Ghani, and I. Adenan. 2007. Antioxidant capacity and phenolic content of cocoa beans. *Food Chem.* 100: 1523–1530.

Overvad, K., B. Diamant, L. Holm, G. Holmer, S.A. Mortensen, Stender S 1999. Coenzyme Q10 in health and disease. *Eur J Clin Nutr.* 53: 764–770.

Packer, L., E.H. Witt, and H.J. Tritschler. 1995. α-Lipoic acid as a biological antioxidant. *Free Radical Biol Med.* 19: 227–250.

Park, S.C., J.R. Lee, J.Y. Kim, I. Hwang, J.W. Nah, H. Cheong, Y. Park, and K.S. Hahm. 2010. Pr-1, a novel antifungal protein from pumpkin rinds. *Biotechnol Lett.* 32: 125–130.

Park, Y., K.J. Albright, W. Lu, J.M. Storkson, M.E. Cook, and M.W. Pariza. 1997. Effect of conjugated linoleic acid on body composition in mice. *Lipids* 32: 853–858.

Pase, M.P., N. Grima, R. Cockerell, C. Stough, A. Scholey, A. Sali, and A. Pipingas. 2015. The effects of long-chain omega-3 fish oils and multivitamins on cognitive and cardiovascular function: A randomized, controlled clinical trial. *J Am Coll Nutr.* 34: 21–31.

Paterson, R.R. 2008. Cordyceps: A traditional Chinese medicine and another fungal therapeutic biofactory? *Phytochemistry* 69: 1469–1495.

Patil, V.V., S.C. Bhangale, and V.R. Patil. 2010. Evaluation of antipyretic potential of *Ficus carica* leaves. *Pharm Sci Rev Res.* 2(2): 48–50.

Pearson, D.A., T.G. Paglieroni, D. Rein, T. Wun, D.D. Schramm, J.F. Wang, R.R. Hoosselin, H.H. Schmitz, and C.L. Keen. 2002. The effects of flavanol-rich cocoa and aspirin on ex vivo platelet function. *Thromb Res.* 106: 191–197.

Perez, C., E. Dominguez, J.M. Ramiro, J.E. Campillo, and M.D. Torres. 1998. A study on the glycemic balance in streptozotocin-diabetic rats treated with an aqueous extract of *Ficus carica* (fig tree) leaves. *Phytother Res.* 10: 82–83.

Peterson, D.M. 1992. Composition and nutritional characteristics of oat grain and products. In *Oat Science and Technology*. H.G. Marshall and M.E. Sorrells, eds., pp. 265–292. American Society of Agronomy, Madison, WI.

Phipps, W.R., M.C. Martini, J.W. Lampe, J.L. Slavin, and M.S. Kurzer. 1993. Effect of flax seed ingestion on the menstrual cycle. *J Clin Endocrinol Met.* 77: 1215–1219.

Pighin, D., A. Pazos, V. Chamorro, F. Paschetta, S. Cunzolo, F. Godoy, V. Messina, A. Pordomingo, and G.Grigioni. 2016. A contribution of beef to human health: A review of the role of the animal production systems. *Sci World J.* 2016: 1–10.

Pinto Mda, S., J.E. de Carvalho, F.M. Lajolo, M.I. Genovese, and K. Shetty. 2010. Evaluation of antiproliferative, anti-type 2 diabetes, and antihypertension potentials of ellagitannins from strawberries (Fragaria × ananassa Duch.) using in vitro models. *J Med Food.* 13(5): 1027–1035.

Potter, S.M. 1998. Soy protein and cardiovascular disease: The impact of bioactive components in soy. *Nutr Rev.* 56: 231–235.

PRNewswire-iReach. 2012. Global Health Movement Drives Market for Nutraceuticals to $250 bn by 2018; Probiotics to Touch $39.6 bn and Heart Health Ingredients Near $15.2 bn. Nov. 19, 2012 /PRNewswire-iReach/—Global Information Inc., Farmington, CT.

Rapport, L., and B. Lockwood. 2001. Flaxseed and flaxseed oil. *Pharma J.* 266: 287–289.

Reuter, H.D., H.P. Koch, and L.D. Lawson. 1996. Therapeutic effects and applications of garlic and its preparations. In *Garlic: The Science and Therapeutic Application of* Allium sativum *L*, H.P. Koch and L.D. Lawson (eds.). Williams & Wilkins, Baltimore, MD.

Rathinavelu, A., A. Levy, D. Sivanesan, and M. Gossell-Williams. 2013. Cytotoxic effect of pumpkin (*Cucurbita pepo*) seed extracts in LNCAP prostate cancer cells is mediated through apoptosis. *Curr Top Nutraceutical Res.* 11: 137.

Rice, R. 1999. Focus on omega-3 ingredients *Health Nutr.* 2: 11–15.

Ried, K., T.R. Sullivan, P. Fakler, O.R. Frank, and N.P. Stocks. 2012. Effect of cocoa on blood pressure. *Cochrane Database Syst Rev.* 8: CD008893.

Roberfroid, M.B. 2000. Prebiotics and probiotics: Are they functional foods? *Am J Clin Nutr* 71: S1682–S1687.

Roberts, D.C., A.S. Truswell, A. Bencke, H.M. Dewar, and E. Farmakalidis. 1994. The cholesterol-lowering effect of a breakfast cereal containing psyllium fibre. *Med J Aust.* 161:600–664.

Ross, S. 2000. Functional foods: The Food and Drug Administration perspective. *Am J Clinical Nutr.* 71: 1735–1738.

Rubnov, S., Y. Kashman, R. Rabinowitz, M. Schlesinger, and R. Mechoulam. 2001. Suppressors of cancer cell proliferation from fig (*Ficus carica*) resin: Isolation and structure elucidation. *J Nat Prod.* 64: 993–996.

Saikumar, R., B. Kumar, J. Kaul, and A. Kumar. 2012. Maize research in India historical prospective and future challenges. *Maize J.* 1: 1–6.

Sanders, M.E. 1994. Lactic acid bacteria as promoters of human health. In *Functional Foods—Designer Foods, Pharmafoods, Nutraceuticals*, I. Goldberg (ed.), pp. 294–322. Chapman & Hall, New York.

Sedigheh, A., M.S. Jamal, S. Mahbubeh, K. Somayeh, R.K. Mahmoud, A. Azadeh, and A. Fatemeh. 2011. Hypoglycaemic and hypolipidemic effects of pumpkin (*Cucurbita pepo* L.) on alloxan-induced diabetic rats. *Afr J Pharmacy Pharmacol.* 5: 2620–2626.

Seeram, N.P., L.S. Adams, M.L. Hardy, and D. Heber. 2004. Total cranberry extract versus its phytochemical constituents: Antiproliferative and synergistic effects against human tumor cell lines. *J Agric Food Chem.* 52: 2512–2517.

Serafini, M., G. Maiani, and A. Ferro-Luzzi. 1998. Alcohol-free red wine enhances plasma antioxidant capacity in humans. *J Nutr.* 128: 1003–1007.

Shewry, P.R. 2009. Wheat. *J Exp Bot.* 60: 1537–1553.

Shih, C.H., S.O. Siu, R. Ng, E. Wong, L.C. Chiu, I.K. Chu, and C. Lo. 2007. Quantitative analysis of anticancer 3-deoxyanthocyanidins in infected sorghum seedlings. *J Agric Food Chem.* 55: 254–259.

Shobana, S., Y.N. Sreerama, and N.G. Malleshi. 2009. Composition and enzyme inhibitory properties of finger millet (*Eleusine coracana* L.) seed coat phenolics: Mode of inhibition of α-glucosidase and α-amylase. *Food Chem.* 115: 1268–1273.

Shuford, J.A., J.M. Steckelberg, and R. Patel. 2005. Effects of fresh garlic extract on *Candida albicans* biofilms. *Antimicrob Agents Chemother.* 49: 473.

Silagy, C.A., and H.A. Neil. 1994. A meta-analysis of the effect of garlic on blood pressure. *J Hypertens.* 12: 463–468.

Singh, B. B., S.P. Vinjamury, C. Der-Martirosian, E. Kubik, L.C. Mishra, N.P. Shepard, V.J. Singh, M. Meier, and S.G. Madhu. 2007. Ayurvedic and collateral herbal treatments for hyperlipidemia: A systematic review of randomized controlled trials and quasi-experimental designs. *Altern Ther Health and Med.* 13:22–28.

Singh, K.K., D. Mridula, J. Rehal, and P. Barnwal. 2011. Flaxseed: A potential source of food, feed and fiber. *Crit Rev Food Sci Nutr.* 51: 210–222.

Slowing, K., P. Ganado, M. Sanz, E. Ruiz, and T. Tejerina. 2001. Study of garlic extracts and fractions on cholesterol plasma levels and vascular reactivity in cholesterol-fed rats. *J Nutr.* 131: 994S–999S.

Srbinoska, M., N, Hrabovski, V, Rafajlovska, and S.S. Fiser. 2012. Characterization of the seed and seed extracts of the pumpkins *Cucurbita maxima* and *Cucurbita pepo* L. from Macedonia. *Maced J Chem Chem Eng.* 31: 65–78.

Srivastava, K.C., A. Bordia, and S.K. Verma. 1995. Garlic (*Allium sativum*) for disease prevention. *S Afr J Sci.* 91: 68–77.

Stanojkovic, T.P., A. Konic-Ristic, Z.D. Juranic, K. Savikin, G. Zdunic, N. Menkovic, and M. Jadranin. 2010. Cytotoxic and cell cycle effects induced by two herbal extracts on human cervix carcinoma and human breast cancer cell lines. *J Med Foods.* 13: 291–297.

Steinkraus, K.H. 1994. Nutritional significance of fermented foods. *Food Res Int.* 27: 259.

Steinmetz, K.A., H. Kushi, R.M. Bostick, A.R. Folsom, and J.D. Potter. 1994. Vegetables, fruit, and colon cancer in the Iowa Women's Health Study. *Am J Epidemiol.* 139: 1–15.

Steptoe, A., E. Gibson, R. Vuononvirta, M. Hamer, J. Wardle, J. Rycroft, J. Martin, and J. Erusalimsky. 2007. The effects of chronic tea intake on platelet activation and inflammation: A double-blind placebo controlled trial. *Atherosclerosis.* 193: 277–282.

Stevinson, C., M.H. Pittler, and E. Ernst. 2000. Garlic for treating hypercholesterolemia. A meta-analysis of randomized clinical trials. *Ann Intern Med.* 133: 420–429.

Stoner, G.D., L.S. Wang, C. Seguin, C. Rocha, K. Stoner, S. Chiu, and A.D. Kinghorn. 2010. Multiple berry types prevent N-nitrosomethylbenzylamine-induced esophageal cancer in rats. *Pharm Res.* 27: 1138–1145.

Tacon, A.G.J., and M. Metian. 2013. Fish matters: Importance of aquatic foods in human nutrition and global food supply. *Rev Fish Sci.* 21: 22–38.

Takezaki, T., C.M. Gao, J.H. Ding, T.K. Liu, M.S. Li, and K. Tajima. 1999. Comparative study of lifestyles of residents in high and low risk areas for gastric cancer in Jiangsu Province, China with special reference to Allium vegetables. *J Epidemiol.* 9: 297–305.

Thompson, L.U. 1995. Flaxseed, lignans, and cancer. In *Flaxseed in Human Nutrition*, S. Cunnane and L.U. Thompson (eds.), pp. 219–236. AOCS Press, Champaign, IL.

Tijburg, L.B.M., T. Mattern, J.D. Folts, U.M. Weisgerber, and M.B. Katan. 1997. Tea flavonoids and cardiovascular diseases: A review. *Crit Rev Food Sci Nutr.* 37: 771–785.

Tolkachev, O.N., and A.A. Zhuchenko, Jr. 2000. Biologically active substances of flax: Medicinal and nutritional properties (a review). *Pharm Chem J.* 34: 360–367.

Tomomatsu, H. 1994. Health effects of oligosaccharides. *Food Technol.* 48: 61–65.

Tosh, S.M., Y. Brummer, T.M.S. Wolever, and P. Wood. 2008. Glycemic response to oat bran muffins treated to vary molecular weights of β-glucan. *Cereal Chem.* 85: 211–217.

Trois, L., E.M. Cardoso, and E. Miura. 2008. Use of probiotics in HIV-infected children: A randomized double-blind controlled study. *J Trop Pediatr.* 54: 19–24.

Truswell, A.S. 2002. Cereal grains and coronary heart disease. *Eur J Clinical Nutr.* 56: 1–14.

Vattem, D.A., R. Ghaedian, and K. Shetty. 2005. Enhancing health benefits of berries through phenolic antioxidant enrichment: Focus on cranberry. *Asia Pac J Clinical Nutr.* 14: 120–130.

Venkateswaran, V., and G. Vijayalakshmi. 2010. Finger millet (*Eleusine coracana*): An economically viable source for antihypercholesterolemic metabolites production by *Monascus purpureus*. *J Food Sci Technol*. 47: 426–431.

Wan, Y., J.A. Vinson, T.D. Etherton, J. Proch, S.A. Lazarus, and P.M. Kris-Etherton. 2001. Effects of cocoa powder and dark chocolate on LDL oxidative susceptibility and prostaglandin concentrations in humans. *Am J Clinical Nutr*. 74(5): 596–602.

Wang, S.Y., R. Feng, Y. Lu, L. Bowman, and M. Ding. 2005. Inhibitory effect on activator protein-1, nuclear factor-kappaβ, and cell transformation by extracts of strawberries (*Fragaria* × *ananassa* Duch.). *J Agric Food Chem*. 53: 4187–4193.

Wang, X.L., J. Liu, Z.H.B. Chen, F. Gao, J.X. Liu, and X.L. Wang. 2001. Preliminary study on pharmacologically effect of *Curcurbita pepo* cv Dayanggua. *J Tradit Chin Vet Med*. 20: 6–9.

Wani, S.A., T.R. Shah, B. Bazaria, G.A. Nayik, A, Gull, K, Muzaffar, and P. Kumar. 2014. Oats as a functional food: A review. *Univers J Pharm*. 3: 14–20.

Warshafsky, S., R.S. Kamer, and S.L. Sivak. 1993. Effect of garlic on total serum cholesterol. A meta-analysis. *Ann Int Med*. 119: 599–605.

Watkins, B.A, L. Yong, and M.F. Feifert. 1999. Bone metabolism and dietary conjugated linoleic acid. In *Advances in Conjugated Linoleic Acid Research*, Vol. 1, M.P. Yuraweca, M.M. Mossoba, J.K.G. Kramer, M.W. Pariza, and G.J. Nelson (eds.), pp. 253–275. AOCS Press, Champaign, IL.

Weisburger, J.H. 1998. International symposium on lycopene and tomato products in disease prevention. *Proc Soc Exp Biol Med*. 218: 93–143.

Weiss, E.I., R. Lev-Dor, Y. Kashamn, J. Goldhar, N. Sharon, and Ofek I. 1998. Inhibiting interspecies coaggregation of plaque bacteria with cranberry juice constituent. *J Am Dent Assoc*. 129: 1719–1723.

Welch, R.W. 1995. Oats in human nutrition and health. In *The Oat Crop. Production and Utilization*, R.W. Welch (ed.), pp. 433–479. Chapman & Hall, London.

Westcott, N.D., and A.D. Muir. 1996. Variation in the concentration of the flax seed lignan concentration with variety, location and year. In *Proceedings of the 56th Flax Institute of the United States Conference*, Flax Institute of the United States, Fargo, ND.

Wildman, R.E. 2001. *Handbook of Nutraceuticals and Functional Foods* (1st ed.). CRC Series in Modern Nutrition. CRC Press, Boca Raton, FL.

Winston, J.C. 2000. Psyllium: Soluble fiber to the rescue. *Vibrant Life*. 16: 40–41.

Wood, P. 2010. Oat and rye β-glucan: Properties and function. *Cereal Chem*. 87: 315–330.

World Nutraceuticals. 2006. Industry Study with Forecasts to 2010 & 2015. The Freedonia Group, Cleveland, OH.

Wu, C., T.S. Kato, R. Ji, C. Zizola, D.L. Brunjes, Y. Deng, H. Akashi, H.F. Armstrong, P.J. Kennel, T. Thomas, D.E. Formen, J. Hall, A. Chokshi, M.N. Bartels, D. Mancini, D. Seres, and P.C. Schulze. 2015. Supplementation of L-Alanyl-L-glutamine and fish oil improves body composition and quality of life in patients with chronic heart failure. *Circ Heart Failure*. 8: 1077–1087.

Wu, X., R.R. Santos, and J. Fink-Gremmels. 2015. Analyzing the antibacterial effects of food ingredients: Model experiments with allicin and garlic extracts on biofilm formation and viability of *Staphylococcus epidermidis*. *Food Sci Nutr*. 3: 158–168.

Xia, T., and Q. Wang. 2007. Hypoglycaemic role of *Cucurbita ficifolia* (Cucurbitaceae) fruit extract in streptozotocin induced diabetic rats. *J Sci Food Agric*. 87: 1753–1757.

Xiao, C.W. 2008. Health effects of soy protein and isoflavones in humans. *J Nutr*. 138: 1244S–1299S.

Yan, L., J.A. Yee, D. Li, M.H. McGuire, and L.U. Thompson. 1998. Dietary flaxseed supplementation and experimental metastasis of melanoma cells in mice. *Cancer Lett*. 124: 181–186.

Yang, X.M., W. Yu, Z.P. Ou, H.L. Ma, W.M. Liu, and X.L. Ji. 2009. Antioxidant and immunity activity of water extract and crude polysaccharide from *Ficus carica* L. fruit. *Plant Foods Hum Nutr*. 64: 167–173.

Yao, T.C., C.J. Chang, Y.H. Hsu, and J.L. Huang. 2009. Probiotics for allergic diseases: Realities and myths. *Pediatr Allergy Immunol.* 21: 900–919.

You, W.C., W.J. Blot, Y.S. Chang, A.G. Ershow, Z.T. Yang, Q. An, B. Henderson, G.W. Xu, J.F. Faumeni, and T.G. Wang. 1988. Diet and high risk of stomach cancer in Shandong, China. *Cancer Res.* 48: 3518–3523.

Yu, L., A.L. Nanguet, and T. Beta. 2013. Comparison of antioxidant properties of refined and whole wheat flour and bread. *Antioxidants (Basel)* 2: 370–383.

Yue, K., M. Ye, Z. Zhou, W. Sun, and X. Lin. 2013. The genus Cordyceps: A chemical and pharmacological review. *J Pharm Pharmacol.* 65: 474–493.

Yurawecz, M.P., M.M. Mossoba, J.K.G. Kramer, M.W. Pariza, and G.J. Nelson. 1999. *Advances in Conjugated Linoleic Acid Research*, Vol. 1. AOCS Press, Champaign, IL.

Zhu, J.L., and C. Liu. 1992. Modulating effects of extractum semen Persicae and cultivated *Cordyceps hyphae* on immuno-dysfunction of inpatients with posthepatitic cirrhosis. *Zhongguo Zhong Xiyi Jiehe Zazhi.* 12: 207–209.

2

Flavoring of Pediatric Nutritional Supplements and Pediatric Compliance: A Perspective

Amit M. Pant, Rupesh V. Chikhale, and Pramod B. Khedekar

Contents

2.1 Introduction 33
2.2 Child nourishment problems, related diseases, and nutritional supplementation solutions 35
2.3 Nutritional deficiencies that affect infants and children 36
    2.3.1 Vitamin D deficiency 36
    2.3.2 Iron deficiency 36
    2.3.3 Calcium deficiency 37
    2.3.4 Vitamin A deficiency 37
2.4 Effect of flavor on supplement intake and consumption in children 38
2.5 Flavoring agents: Types and applications 39
2.6 Regulatory consideration of pediatric flavoring agents 41
2.7 Success stories 44
    2.7.1 Advanced demasking of iron supplements 44
    2.7.2 Masking the fishy odor and taste of omega 3 fatty acids 45
    2.7.3 Masking the taste of calcium and magnesium supplements 45
2.8 Discussion and conclusion 46
References 48

2.1 Introduction

Nutrition has assumed worldwide importance in recent years. Nutritional deficiency is a serious public health concern in many countries. Good health is vital to humans and good nutrition is essential for good health. The foundation of good health begins with conception. The formative years of life are crucial in laying the foundation of good health, growth, and development. Well-balanced nutrition is essential for a growing child to become a healthy adult.

The relationship between nutrition and good health has been well known since ancient times. The importance of food was mentioned in detailed description in the ancient Sanskrit text Taittiriya Upanishad. Food is quoted as "Annam Brahma" which represents that the life initiates, exists, and amalgamates with "Anna"—the food. Similarly, the Ayurveda explained the importance of food in detail and offered advice about proper food for optimal health, growth, and development. The nations of the world have set the goal of a level of health for all citizens that would enable them to lead a socially and economically productive life. Good nutrition plays a pivotal role in this.

Good nutrition in the early phase of life forms the foundation of health and, in particular, for growth, development, survival, and lifelong maintenance of health. At this stage, certain specific biological and physiological needs must be met to ensure the survival and healthy development of the child and future adult. This can be achieved only by good and optimal nutrition of both mother and child. The stage after exclusive breastfeeding, i.e., 6 months to 2 years, is a high-risk age for malnutrition. This is due to poor understanding and lack of knowledge of the young child's food requirements and the common foods that can make up the intake deficit.

Nutritional status refers to the condition of the essential biomolecular environment in the body. Nutritional status depends on three factors: (1) the kind of food one eats; (2) the amount of food one eats; and (3) the body's ability to make use of these foods. In poor families and in rural and urban slums, this is compounded by unavailability of essential and nutritious foods. In children nutrition, growth, development, and infections are interdependent. Poor nutritional supply results in poor optimal growth and development. To overcome this, balanced nutrition either in food or supplementation plays an indispensable role. This is why it is very important to have an overview of the challenges in child nutrition, the diseases related to malnutrition, their impact on long-term health and development, and the need for supplementation of some vital nutrition for the prevention or treatment of these deficiency conditions.

There have been rapid advances in the medical knowledge of the epidemiological, biochemical, clinical, immunological, and in particular, the commercial aspects of pediatric nutrition. Supplementation of micronutrients in children is highly efficacious in the prevention and treatment of certain health issues, even in the long term. Recommendations only for such supplements will not solve this issue. This age group is fully dependent on their guardians for dosage, and their selectivity for food and taste makes it very difficult to comply with the standard recommendations. Nutritional supplementation throughout the world, specifically in rural and urban areas, has yet to achieve prime importance with regard to maintaining normal health. Various functional regional or global bodies are engaged in raising awareness about nutrition and its importance in health. This leads to advanced supplementation programs in children including applications of novel delivery techniques, modified dosage forms, and novel methods in taste modifications.

2.2 Child nourishment problems, related diseases, and nutritional supplementation solutions

Infancy and childhood are phases of rapid growth marked by significant physical changes. Thus the dietary and nutritional requirements for infants and children differ from those for adults. Poor diet in children, due to food shortage or eating disorders that result in nutritional deficiencies, place children at higher risk of developing significant acute and/or chronic illness and diseases. Persistent nutritional deficiencies may lead to developmental and psychological health problems and a failure to thrive academically (Hunter et al., 1990; Arts-Rodas and Benoit, 1998; Illingworth and Lister, 1964; Archer et al., 1990). Nutrients are vital to good health and essential for the growth and foundation of future health, which is extremely important in children. Infants and children depend on their parents for supplementation of these nutrients. For parents, the selection of foods that satisfy children's daily requirements may be challenging. Feeding is an important part of the everyday life of infants and children and builds a strong parent–child interaction. Other factors such as tolerance and the children's selectivity for food make this situation more critical. The most common nutritional problems in children result from poor appetite, rejection of variety in foods, allergies, intolerances, and vitamin deficiencies or diseases associated with them.

The needs and nutritional requirements of infants and children differ from those of adults. Nutrition in the initial years of life can affect future health significantly. Problems commonly associated with feeding in infants include slow and extended feeding time, refusal to eat, abdominal discomfort, nausea, and vomiting (Arts-Rodas and Benoit, 1998). These problems are estimated to occur in up to 25–40% of normally developing infants and toddlers (Sisson and Van Hasselt, 1989) and are higher in children with brain developmental abnormalities (Palmer and Horn, 1978). Although some of these challenges are temporary, food selectivity and refusal are major concerns. The parental role is vital in early recognition and management with the support of the family physician (Reau et al., 1996; Dahl, 1987; Dahl and Sundelin, 1992; Dahl et al., 1994; Mitchell et al., 2004; Nelson et al., 1998; Bernard-Bonnin, 2006).

Some vitamins and minerals remain critical for lifelong development and maintenance of health. Every aspect of proper growth is governed by these micro- and macronutrients and they must be included in the regular diet. We discuss below these vitamins and minerals, their deficiencies, chief functions, sources, and the symptoms and diseases resulting from their deficiencies.

2.3 Nutritional deficiencies that affect infants and children

In the rapid growth phase of life certain vitamin and mineral deficiencies are of prime concern. They may lead to long-term abnormalities in the growth and development of infants and children. A healthy, balanced diet together with physical exercise ensure all-round growth and development in infants and children. For example, iron is an integral component of erythropoiesis. Folic acid and vitamin B_{12} are very important erythropoietic factors responsible for proper red blood count (RBC) generation and maturation. Calcium and magnesium, the most abundant minerals in the human body, are major components of bones, imparting strength and rigidity. Deficiency of any of these minerals leads to development of brittle bones and increased risk of fracture. Prolonged deficiency of these and other pro-bone minerals may increase the risk of osteoporosis in old age. We next review some of the important vitamin and mineral deficiencies that are very common in children.

2.3.1 Vitamin D deficiency

Vitamin D is the sunshine vitamin that the body produces from sunlight exposure. Some seafoods and dairy products are natural sources of vitamin D. Studies have confirmed that approximately 40% of toddlers and children have vitamin D deficiency. Similar findings were reported in a 2009 study of American children by Mansbach et al., which reported that over 6 million children under the age of 11 had 50 nmol/L (20 ng/ml) of vitamin D and 24 million children had below 75 nmol/L (30 ng/ml) of vitamin D (Mansbach et al., 2009). According to the American Academy of Pediatrics, most children have lower-than-recommended daily intake of Vitamin D, which negatively affects calcium absorption, the immune system, and bone growth and development. Breastfed infants are also at risk for vitamin D deficiency due to lower concentration of vitamin D in breast milk and its dependence on maternal levels of vitamin D. Breastfed infants without vitamin D supplementation are at high risk to develop rickets. This is commonly observed in the first two years of life.

2.3.2 Iron deficiency

Iron deficiency anemia continues is one of the more common nutritional deficiencies in infants and children in the United States. Children under the age of 3 have increased iron needs and are particularly at risk for iron deficiency anemia. Breastfed babies older than six months who do not receive iron-fortified foods such as cereals are also at high risk. Only about 50% of the iron in breast milk is bioavailable to the infant. A baby is born with a store of iron that he depletes in the first few months of life. Exclusive breastfeeding after four to six months puts infants at risk for iron deficiency. Iron deficiency is also common during adolescence, particularly in teenage girls who have just started their menstrual cycles. Some form of dietary iron supplement that provides 1 mg elemental iron per kg per day is recommended for term infants starting at four to six months of age.

2.3.3 Calcium deficiency

Low amounts of calcium can also contribute to development of rickets. Vitamin D deficiency is the most common underlying cause of rickets, but inadequate calcium intake also contributes to higher instances of bone fractures in children. Adequate calcium intake increases total bone mass and strength and decreases the risk of your child developing osteoporosis later in life. Infants, toddlers, and young children usually get enough calcium from their diets. Most of the research on calcium deficiency has concerned older children and adolescents since most bone formation occurs at this stage (Ross et al., 2011).

2.3.4 Vitamin A deficiency

Vitamin A deficiency is the leading cause of preventable blindness in children. It also increases the risk of disease and death from severe infections. Vitamin A plays a crucial role in maternal and child survival. Adequate vitamin A supplementation in high-risk areas has been shown to reduce mortality significantly. In children, vitamin A deficiency leads to severe visual impairment and blindness. It significantly increases the risk of severe illness, and even death, from common childhood infections such as diarrhea and measles. Approximately 127 million preschool-aged children are vitamin A deficient (Aslam et al., 2017). Health consequences of vitamin A deficiency include mild to severe systemic effects on innate and acquired mechanisms of host resistance to infection and growth, mild to severe (blinding) stages of xerophthalmia, and increased risk of mortality. These health issues associated with deficiency of vitamin A are defined as vitamin A deficiency disorders (VADD). Globally, 4.4 million preschool children have xerophthalmia (West, 2003).

There are many reasons a doctor may recommend supplements. Some examples include children who are underweight, have restricted diets, or have illnesses that put them at risk for deficiency (low level) of a vitamin or mineral. It is recommended that all infants, including those who are exclusively breastfed, have a minimum intake of 400 IU of vitamin D per day beginning in the first few days of life. The last 50 years have witnessed an increased understanding of the biochemistry of vitamins and trace minerals and their role in human nutrition and intermediary metabolism. There also has been a growing public awareness of the sometimes dramatic clinical impact of vitamin and mineral administration in deficiency states. As nutritional needs became more clearly defined, essential vitamins and minerals were incorporated into processed formulas such as infant cereal. Supplemental vitamin and mineral drops or tablets continued to be used by a substantial portion of the population, probably to a greater extent than necessary because they are relatively inexpensive and available without prescription. The widespread consumption of these products is also fostered by a combination of advertising pressure and concern about dietary adequacy. Many individuals regard vitamin and/or mineral supplements as a reliable method of ensuring that real or imagined dietary shortcomings are corrected. Others, on far less rational grounds, have come to regard supplements in a wide range of doses as the philosopher's stone for good health or as treatment for a wide array of ailments ranging from mental retardation to the common cold. As a result, vitamin and mineral supplements are widely abused by the general public. Nevertheless, the desired effects of the supplementation are totally dependent on the compliance and acceptance of the targeted population. The palatability of the supplements can be easily modified by the addition of various flavors.

2.4 Effect of flavor on supplement intake and consumption in children

In early childhood children experience new foods and a variety of tastes and textures. Toddlers and children often refuse to try or eat a variety of foods and are thus labeled "picky eaters" (Cermak et al., 2010). Food intake is largely controlled by sensory reactions to food. Palatability is a prime aspect of sensory experience. Sensory indications are operational well before the actual eating process. They include appearance, odor, and taste of the food, which indicate the basic properties of the food, such as its source, and other factors that determine selection of the specific portion. This phenomenon of sensory mechanisms in food consumption and selectivity has been widely studied. Taste plays a central role in this process (McCrickerd and Forde, 2016; Sorensen et al., 2003).

Taste may be decisive in ensuring acceptable compliance with pediatric oral supplements. It is very important to consider the taste aspects of nutritional supplements. Most often there are no specific symptoms of nutritional deficiency conditions. Long-term deficiency may lead to irreversible problems in development of children. Generally, children have a poor acceptance of unpleasant taste. The use of tasteless or palatable flavors can help improve adherence to the recommended therapy (Bazzano et al., 2009; Matsui, 2007; Cram et al., 2009). Taste preferences usually vary with age. Children are fond of sweet and salty flavors and dislike bitter and peppermint taste. This phenomenon should be considered in taste assessment in the development of pediatric formulations (Albani, 1983; Cram et al., 2009; Mennella and Beauchamp, 2008). The preferred route of administration, toxicity, and taste preferences are well known in children and represent a separate and more heterogeneous group than adults (Breitkreutz and Boos, 2007; Moore, 1998). The major difference between adult and child pharmacokinetics–pharmacodynamics is evident in the initial year and a half of life, when organ functions are developing (Allegaert, 2013; Noel et al., 2012). Their responses to both active ingredients and additives in formulations are often different

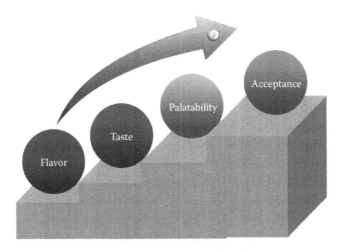

Figure 2.1 The ladder for food and oral formulation acceptance.

compared to adults (Kearns et al., 2003). Childhood is a series of continuous growth and developmental phases characterized by rapid metabolic rate, physiological progress in maturing organ systems, and age-related cognitive development (Ivanovska et al., 2014).

Flavor is the sensory perception of food or dietary substance primarily triggered by the senses of taste and smell and, to some extent, sight. Smell is the key determinant sense of a flavor. The trigeminal senses are part of the sensory system which is involved in perception of chemosensory information in our environment and is majorly responsible for identification of chemical stimuli in the oral cavity. Texture is also important to overall flavor perception. The flavoring of the food or dietary substances can be modified with natural or artificial flavors, which increase acceptance by the target population. A flavoring agent is defined as a substance that imparts its flavoring characteristics to another substance, altering the characteristics of the solute and causing it to become one of the five universally recognized tastes: sweet, sour, bitter, salty, or umami. The flavor of a food can be easily modified by changing its smell or taste. This is very common in artificially flavored candies, soft drinks, and other commercially available food products. Adherence to recommended oral formulation in children depends primarily on acceptances and preferences. These are important factors in achieving the intended outcomes (Figure 2.1). From the commercial point of view, flavoring agents are of primary importance in this process (Schiffman and Dackis, 1975).

2.5 Flavoring agents: Types and applications

Flavoring agents represent the largest group in the category of key food additives. Hundreds of varieties such as fruit, nut, seafood, spice blends, vegetables and wine are used in the food industry to enhance or alter the taste and smell of the foods. The U.S. Code of Federal Regulations defines natural flavorings as "the essential oil, oleoresin, essence or extractive, protein hydrolysate, distillate, or any product of roasting, heating or enzymolysis, which contains the flavoring constituents derived from a spice, fruit or fruit juice, vegetable or vegetable juice, edible yeast, herb, bark, bud, root, leaf or any other edible portions of a plant, meat, seafood, poultry, eggs, dairy products, or fermentation products thereof, whose primary function in food is flavoring rather than nutritional." Their use should not present an unacceptable risk to human health and should not mislead consumers. The quantity added to foods should be the lowest level necessary to achieve the intended flavoring effect. Flavors and flavoring substances should also be of appropriate food grade quality and be prepared and handled in the same way as a food ingredient. It is well established that consumption of food is highly influenced by flavor. More than 1200 flavoring agents are currently available. The majority are of natural origin or nature-identical. Alcohols, esters, aldehydes, ketones, protein hydrolysates and monosodium glutamate (MSG) are examples of flavoring agents.

Synthetic flavoring agents are structurally identical to naturally occurring flavoring agents, but are of greater uniformity in use, cheaper, and are more readily available than their natural counterparts. Some examples of synthetic flavoring agents include amyl acetate (used as banana flavoring), benzaldehyde (used to create cherry or almond flavor), ethyl butyrate (used as pineapple flavor), and methyl anthranilate (used as grape flavor). In addition to natural flavors there are chemical derivatives used as flavors that imitate natural flavors. Some examples of chemical flavoring agents include alcohols, which have a bitter and medicinal taste; esters that impart a fruity taste; ketones and pyrazines used to flavor caramel; phenolics, which have a smoky flavor; and terpenoids, which have a citrus or pine flavor.

Natural flavor enhancers such as MSG are used to intensify the flavor of other compounds in food. They have a taste outside of the basic sweet, sour, salty or bitter tastes. MSG, which was chemically derived from seaweed in the early twentieth century, is now manufactured commercially by the fermentation of starch, molasses, or sugar. Examples of natural and artificial flavors are given in Table 2.1.

Table 2.1 Examples of Natural and Artificial Flavors

Type	Example	Properties
Natural	Peppermint, natural honey, castoreum extract	Less stable
Artificial	Vanilla, chocolate	Highly stable
Natural and artificial	Strawberry, orange, lemon, lime	Effective at low concentrations

Table 2.2 Agents for Masking and Complementing Basic Taste

Basic Taste	Masking Agent
Sweet	Banana, caramel, cream, chocolate, grape, vanilla
Acid	Cherry, lemon, lime, mandarin, orange, strawberry
Bitter	Cherry, chocolate, grapefruit, liquorice, strawberry, peach, raspberry, tutti-frutti
Alkaline	Aniseed, caramel, passion fruit, peach, banana
Salty	Caramel, grapefruit, lemon, orange, vanilla

Source: Adapted from CHMP, EMEA, 2006.

Table 2.3 Examples of Dietary Supplements That Use Sweeteners and Flavoring Agents to Achieve Taste Masking

Drug	Category	Taste	Taste Masking Agent Used
Zinc	Mineral supplement	Bitter	Saccharin sodium
Amino acids and proteins	Diet supplement	Metallic	Sucralose
Iron compounds	Iron supplement	Metallic	Sucralose, sorbitol, xylitol, maltitol, or erythritol
Multi-minerals	Diet supplement	Metallic	Glycyrrhizin, acesulfame potassium
Vegetable crude drug	Diet supplement	Variable	Caramel
Vitamins	Diet supplement	Bitter	Cocoa powder, Stevia extract, Aspartame

Salty and bitter tastes in a number of drug products can be effectively masked by using syrups of cinnamon, orange, citric acid, cherry, cocoa, wild cherry, raspberry, or glycyrrhizin elixir. The cooling effects of some flavors, such as menthol, help reduce after-taste perception. Many mouthwashes and cough syrup formulations use eucalyptus oil as a major constituent. Both impart flavor and odor of their own to products and have a mild anesthetic effect on sensory receptor organs associated with taste (Lachman et al., 1987). Oral liquid vitamin supplements are usually bitter in taste. Their taste is masked by adding sugars, amino acids, and apple flavor. Oral supplements containing vitamin B-complex have a characteristic unpleasant bitter taste. Use of inosinate, orange flavor, or fruit flavors helps improve the taste considerably. The astringent taste of zinc in mouthwashes like Listerine was masked with a combination of sweet note (vanillin–ethyl vanillin), one fruity note (raspberry and lemon), and one spicy note (ginger, clove, anise cinnamon or mixtures), in combination with a taste receptor blocker, which eliminated the burning sensation and astringency associated with eucalyptol and zinc (Stier et al., 2002). Tables 2.2 and 2.3 provide examples of taste masking achieved with the help of sweeteners and flavors (CHMP, EMEA, 2006).

2.6 Regulatory consideration of pediatric flavoring agents

The use of excipients in pediatric medicines is driven by functional requirements and should be justified through a risk-based assessment that takes into account the pediatric age group, frequency of dosing, and duration of treatment. An added challenge for pediatric medicines compared to adult medicines is that excipients in children may lead to adverse reactions that are not experienced or seen to the same extent in adults. Reviews on adverse reactions attributed to excipients show that the currently available data on excipient safety are of limited quantity and variable quality.

In the development of pediatric medicine, the number and levels of excipients in a formulation should be the optimum required to ensure an appropriate product with respect to performance, stability, palatability, microbial control, dose uniformity and other considerations that support product quality. Risks of adverse reactions are mostly associated with excipients used in liquid dosage form. Choice of excipient should include consideration of:

- The safety profile of the excipient for children of the target age groups;
- The route of administration;
- Single and daily dose of the excipient;
- Duration of treatment;
- Acceptability by the intended pediatric population;
- Potential alternatives; and
- Regulatory status in the intended market

Potential alternatives should be sought for excipients that pose a significant risk. Another dosage form or route of administration might be necessary to avoid significant risk. Although well-known excipients with well-defined safety profiles are preferred, new excipients cannot be excluded. Novel excipients should only be used when safety, quality, and appropriateness of use in children have been established. In addition, alternative excipients may need to be considered for different cultural or religious reasons, e.g., the use of gelatin may not be acceptable for all patients. Palatability of oral pediatric medicines often requires the use of sweetening agents such as cariogenic and noncariogenic sweeteners. In addition to the considerations listed above, attention should be paid to:

- Safety of the sweetening agent in relation to specific conditions of the child, e.g., diabetes, fructose intolerance, use of aspartame in patients with phenylketonuria;
- The laxative effect of poorly absorbed or non-digestible sweeteners in high concentrations; and
- The severity of the condition to be treated, i.e., are potential adverse reactions of the sweetening agent secondary to patient compliance?

Taste masking of medicines for oral use or use in the mouth is often needed to improve palatability of the medicine. Children have a well-developed sensory system for detecting tastes, smells, and chemical irritants. They are able to recognize sweetness and saltiness from an early age and the sweet taste in oral liquids as well as the degree of sweetness. Children seem to prefer higher levels of sweetness than adults. The unpleasant taste of an active pharmaceutical ingredient (API), e.g., bitterness or metallic taste, is, therefore, often masked in an oral liquid by using sweetening agents and flavors. Use of coloring agents that match the flavor is discouraged unless it is necessary to cover unpleasant API-related color. Ernest et al. (2007) discussed some successful approaches to taste masking. Children's preference for flavors is determined by individual experience and culture. The target for taste masking need not necessarily be good-tasting medicines, only acceptable-tasting ones that account for cultural differences in taste and provide a taste acceptable to as many countries as possible. An example of a "qualitative evaluation of the taste by a taste panel" for zinc formulations can be found in the UNICEF/WHO publication on production of zinc formulations (WHO, 2007; Cram et al., 2009; WHO Technical Report Series, 2014).

An ideal pediatric supplement should be efficacious, tolerable, affordable in price, and palatable. Modern medical formulations are complex mixtures of active ingredient and excipients, which consist of flavors, sweeteners, coloring agents, and other additives. The excipients are added primarily to mask the unpleasant taste and odor, add attractive color, increase volume, and ensure a uniformity of active ingredient in the final mixture. Thus, formulation of suitable pediatric supplements remains a challenge and restricts their availability in most acceptable formulations such as chewable or liquid formulations (Kumar and Pawar 2002).

In most countries, excipients are not under the same strict regulations as active ingredients. Although excipients are added in fractions and are well tolerated, adverse and idiosyncratic reactions have been reported. The taste masking or flavoring of pediatric formulations is now rigorously supervised by the U.S. FDA and agencies in other countries. In the United States, drug and formulations approved for adults can be labeled for pediatric use only after successful testing in children. The European Commission issued two regulations in October 2012 that clarify the rules for using more than 2100 authorized flavoring substances (Regulations EU/872/2012 and EU/873/2012).

Sweetness and saltiness and their strength or degree can be recognized from a young age. Sucrose is the most commonly used sweetening agent. It is a disaccharide that is readily hydrolyzed in the intestine to the absorbable monosaccharides fructose and glucose. It should be avoided for pediatric patients suffering from hereditary fructose intolerance. Formulations with high amounts of sugar should be avoided in therapy of pediatric patients suffering from diabetes. The use of cyclamate and saccharine is controversial because of the possibility that they are carcinogenic. There is evidence that heavy use of artificial sweeteners is associated with a significantly increased risk of bladder cancer. The U.S. FDA, but not the Canadian Health Protection Agency, banned the use of cyclamate in drugs and food. The reverse is true for saccharin, which is authorized by the U.S. FDA, but not the Canadian Health Protection Agency (Kumar and Pawar 2002).

Market research has revealed standard combinations of specific sweeteners with relevant flavors, which may vary by country and target market (CHMP, EMEA, 2006). National favorites include bubble-gum and grape in the United States, citrus and red berries in Europe, and liquorice in Scandinavia. A bubble-gum or cherry flavor in combination with a high-intensity sweetener may suit the U.S. pediatric market, while a less intense sweetness may be more appropriate for Japan. Children may perceive as unpleasant and reject irritating sensations in the mouth, such as effervescence or peppermint. Peppermint may be described as spicy or hot and rejected in the same way as bitter tastes. When selecting a suitable flavor for a pediatric formulation consideration should be given to the type of flavor (acid, alkaline, bitter, salty or sweet; see Table 2.2) acceptable to the target population.

2.7 Success stories

2.7.1 Advanced demasking of iron supplements

Iron deficiency is the most common nutritional deficiency worldwide. It affects approximately 20% of the world population. Lack of iron may lead to unusual tiredness, shortness of breath, decrease in physical performance, learning problems in children and adults, and increased susceptibility to infection. This deficiency is caused in part by plant-based diets, which contain low levels of poorly bioavailable iron. The most effective technological approach to combating iron deficiency in developing countries is supplementation targeted at high-risk groups (Navarretea et al., 2002).

Albion has developed a form of iron called Taste Free™. A few factors are associated with the less pleasant taste of metals and metal amino acid chelates, such as iron and others. One of those factors involves the metal's coordination number. Iron exists in the primary valence state as the ferrous form (+2 oxidation state), which is easily oxidized to the ferric form (+3 oxidation state), a more stable state for iron. In addition to oxidation state, iron (as well as other metals) has a secondary valence that is referred to as its coordination number. The ferric form can have a coordination number of six (6). To decrease the metal taste of a mineral like iron, the coordination number of the metal must be satisfied by the electron donor groups of the organic ligand (from an amino acid, for instance), which forms an ionic or coordinate bond which is sufficiently stable within the pH environment involved. At least one ligand must be of the polydentate type of the alpha amino acid configuration. Preferably two or more ligands should be of the polydentate type (alpha-amino acid ligands would be ideal) so the overall charge must be balanced (a second taste factor) (www.AlbionMinerals.com).

2.7.2 Masking the fishy odor and taste of omega 3 fatty acids

Nutritional supplements have a market size of approximately US$600 million, with rapid growth of 30% annually. A major reason for this is consumer awareness of health benefits of supplements. The importance of polyunsaturated fatty acid (PUFA) consumption for human health has been established for more than 80 years. The U.S. FDA in 2004 approved the use of ω-3 PUFAs in supplements. The market for ω-3 PUFA ingredients grew by 24.3% as compared to 2013, which confirms their popularity and public awareness of their benefits. PUFAs are essential for normal human growth; however, only minor quantities of the beneficial ω-3 PUFAs eicosapentaenoic acid (EPA) and docosahexaenoic acid (DHA) are synthesized by the human metabolism (Ganesan et al., 2014). Though omega fatty acids go rancid with exposure to oxygen and lead to fishy odor that is considered unacceptable, they are usually delivered through the soft gel technology invented by Robert Pauli Scherer in 1933. The soft gels are now available in chewable form with more palatable and acceptable flavor such as natural lemon flavor.

As an alternative to soft gels, Croda Health Care developed Incromega 3 emulsion DHA, which contains highly concentrated DHA with natural lemon-flavored emulsion in convenient sachet or syrup form. This is ideal for pediatric use because it delivers the recommended daily intake of omega-3 in just one sachet or spoon serving. The use of natural flavor prevents the use of artificial flavors and its delicious taste and lack of fish reflux improves compliance in children (Granato, 2014).

2.7.3 Masking the taste of calcium and magnesium supplements

The salts of divalent cations such as calcium and magnesium are characterized primarily by bitter and salty tastes and to a lesser extent by other basic and metallic tastes and astringent and irritative sensations. The tastes associated with calcium chloride are largely suppressed when calcium is combined with larger organic ions such as lactate, gluconate, or glycerophosphate. However, the combination of these minerals with larger organic ions may reduce the elemental dose per serving. This means multiple dosage is required to achieve the required daily intake, which itself is challenging with children.

To achieve required supplementation dose with more palatable taste, Vitane Pharma has developed a suspension of these bone minerals with delicious chocolate flavor. Elemental calcium higher in calcium carbonate salt was used with the chocolate flavor. The suspension is marketed for and well accepted by children. This is an ideal example of combining a widely accepted flavor like chocolate with important minerals. This combination has proven to supplement optimum minerals and increase compliance.

2.8 Discussion and conclusion

Infants and children are the heterogeneous class in therapeutics. The various challenges of this group require special attention for management of health conditions and needs. Nutrition is fundamental to normal growth and development. Many nutritional deficiencies in this age group affect normal development and have the potential to lead to irreversible abnormalities in the long term. Oral supplementation of vital nutrients serves as the best preventive means not only for this age group but also throughout the human life span. Nevertheless, the efficacy of the supplementation, especially in the case of oral supplements, depends on various aspects such as compliance and adherence to the dosages, acceptance of the supplements by the target population, and quality of the formulations. Adherence and acceptance in infants and children are more problematic to achieve than in adults because children depend on their parents for dosing and children are very selective regarding the taste of food or supplements.

It is important to mention that not all micro- and macronutrients have potential for deficiency. Risk varies for different types of nutrients. As a basic property, each nutrient has a characteristic natural taste and smell. Unfortunately, most nutrients vital to health have a very strong, unacceptable taste and odor. These characteristics are manageable with the addition of certain acceptable flavors and taste enhancers. Table 2.4 summarizes risk of deficiency, deficiency symptoms, and characteristic taste and odor of various nutrients.

The use of oral supplements and fortified foods is important to ensure good health. These formulations are developed using various flavors and taste enhancers. The variety of flavors used are natural, synthetic, or semi-synthetic. Regulatory bodies such as the U.S. FDA have strict regulations for their use in pediatric formulations because some of them pose a risk of toxicity in infants and children. Overall the use of natural flavors has been proved to be much safer and more effective in taste masking of various formulations and they are commonly used at the industrial scale.

Pediatric nutritional supplements made palatable by common, well-accepted flavors have had a great impact on increasing compliance and acceptance in infants and children. Future efforts should focus on the development of nutritional supplements using novel flavors with greater effectiveness with respect to taste masking and their evaluation in larger populations.

Table 2.4 Deficiency Risk and Taste and Odor of Vitamins, Minerals, and Essential Fatty Acids

Vitamin/Mineral	Deficiency Risk	Deficiency Symptoms	Characteristic Taste/Odor
Vitamin A	Moderate	Vision problems, immune dysfunction, fat malabsorption	harsh, bitter, soapy, or piney taste
Vitamin B_1	High	Decreased free radical protection, heart health, cognitive decline, fatigue	obnoxious sour-bitter taste
Vitamin B_2	High	cataracts, poor thyroid function, fatigue	Bitter taste
Vitamin B_3	Low	Cracking, scaling skin, digestive problems, confusion, anxiety, fatigue	Bitter
Pantothenate	Low	Stress tolerance, wound healing, skin problems, fatigue	Bitter
Vitamin B_6	High	Depression, sleep and skin problems, confusion, anxiety, fatigue	Bitter
Biotin	Moderate	Depression, nervous system, premature graying, hair, skin	Metallic
Folate	High	Anemia, immune function, fatigue, insomnia, hair, high homocysteine, cardiovascular disease	Bitter or unpleasant
Vitamin B_{12}	High	Anemia, fatigue, constipation, loss of appetite/weight, numbness and tingling in the hands and feet, depression, dementia, poor memory, oral soreness	Slightly bitter to tasteless
Vitamin C	Moderate	Muscle spasms, muscle cramps and tetany, tooth decay, periodontal disease, depression, possibly hypertension	Strong sour, tingling, sharp, acidic and fruity
Vitamin D	High	Osteoporosis, reduced calcium absorption, thyroid dysfunction	Metallic
Vitamin E	High	Skin, hair, immune dysfunction rupturing of red blood cells, bruising, eczema, psoriasis, wound healing, muscle weakness	Very fatty, repulsive
Vitamin K	Low	Excessive bleeding, a history of bruising, appearance of ruptured capillaries or menorrhagia (heavy menstrual periods)	Unpleasant
Calcium	High	Osteoporosis, osteomalacia, osteoarthritis, muscle cramps, irritability, acute anxiety, colon cancer risk	Chalky
Chromium	Moderate	Metabolic syndrome, insulin resistance, decreased fertility	Tasteless, weak metallic aftertaste
Copper	Low	Osteoporosis, anemia, baldness, diarrhea, general weakness, impaired respiratory function, myelopathy, decreased skin pigment, reduced resistance to infection	Metallic
Iodine	Moderate	Lethargy and tiredness, muscular weakness and constant fatigue, difficulty concentrating, slowed mental processes and poor memory	Unpleasant, metallic

(*Continued*)

Table 2.4 (Continued) Deficiency Risk and Taste and Odor of Vitamins, Minerals, and Essential Fatty Acids

Vitamin/Mineral	Deficiency Risk	Deficiency Symptoms	Characteristic Taste/Odor
Iron	High	Extreme fatigue, weakness, pale skin, headache, dizziness, tongue soreness, brittle nails, unusual cravings for non-nutritive substances, poor appetite	Metallic
Magnesium	High	Appetite, nausea, vomiting, fatigue cramps, numbness, tingling, seizures, heart spasms, personality changes, heart rhythm	Strong alkaline, chalky
Manganese	Moderate	Dermatitis, problems metabolizing carbohydrates, poor memory, nervous irritability, ataxia, fatigue, blood sugar problems,	Strong, undesirable metallic
Phosphorus	Low	Loss of appetite, bone pain, fragile bones, stiff joints, fatigue, irregular breathing, irritability, weakness, and weight change	Slightly bitter taste
Selenium	Moderate	Destruction to heart/pancreas, fragility of red blood cells, immune system	Metallic
Zinc	High	Growth retardation, hair loss, diarrhea, impotence, loss of appetite, taste, weight loss, wound healing, mental lethargy	Strong metallic or unpleasant
Carnitine	Low	Elevated cholesterol, liver function, muscle weakness, reduced energy, impaired glucose control	Sour
N-Acetyl Cysteine, Glutathione	High	Free radical overload, elevated homocysteine, cancer risk, cataracts, macular degeneration, immune function, toxin elimination	Tart/sour
Omega-3-Fatty Acid	Moderate	Diabetic neuropathy, reduced muscle mass, atherosclerosis, Alzheimer's, failure to thrive, brain atrophy, high lactic acid	Unpleasant, fatty with strong fishy odor
Lysine	Moderate	Anemia, apathy, bloodshot eyes, depression, edema, fatigue, fever blisters, hair loss	Bitter
Arginine	Moderate	Skin rash and hair loss	Bitter

References

A Matter of Taste: Taste-Free Iron Chelate (2005). 14(1). Available online at http://www.albionhumannutrition.com. Accessed March 2017.

Albani M, Wernicke I (1983). Oral phenytoin in infancy: Dose requirement, absorption, and elimination. *Pediatr Pharmacol (New York)*. 3(3–4):229–236.

Allegaert K (2013). Neonates need tailored drug formulations. *World J Clin Pediatr*. 2(1):1–5.

Archer LA, Szatmari P (1990). Assessment and treatment of food aversion in a four-year-old boy: A multidimensional approach. *Can J Psychiatr*. 35:501–505.

Arts-Rodas, D, Benoit D (1998). Feeding problems in infancy and early childhood: Identification and management. *Paediatr Child Health*. 3(1):21–27.

Aslam MF, Majeed S, Aslam S, Irfan JA (2017). Vitamins: Key role players in boosting up immune response—A mini review. *Vitam Miner*. 6:1.

Bazzano AT, Mangione-Smith R, Schonlau M, Suttorp MJ, Brook RH (2009). Off-label prescribing to children in the United States outpatient setting. *Acad Pediatr.* 9(2):81–88.
Bernard-Bonnin, A-C (2006). Feeding problems of infants and toddlers. *Can Fam Phys.* 52:1247–1251.
Breitkreutz J, Boos J (2007). Paediatric and geriatric drug delivery. *Expert Opin Drug Deliv.* 4(1):37–45.
Cermak S, Curtin C, Bandini LG (2010). Food selectivity and sensory sensitivity in children with autism spectrum disorders. *J Am Diet Assoc.* 110(2):238–246.
CHMP EMEA (2006). Reflection paper: Formulation of choice for the paediatric population. Reference number EMEA/CHMP/PEG/194810/2005. Published 28/07/2006.
Cram A, Breitkreutz J, Desset-Brèthes S, Nunn T, Tuleu C, European Paediatric Formulation Initiative (EuPFI) (2009). Challenges of developing palatable oral paediatric formulations. *Int J Pharm.* 365(1–2):1–3.
Dahl M (1987). Early feeding problems in an affluent society. III. Follow-up at two years: Natural course, health, behaviour and development. *Acta Paediatr Scand.* 76:872–880.
Dahl M, Sundelin C (1992). Feeding problems in an affluent society. Follow-up at four years of age in children with early refusal to eat. *Acta Paediatr.* 81:575–579.
Dahl M, Rydell AM, Sundelin C (1994). Children with early refusal to eat: Follow-up during primary school. *Acta Paediatr.* 83:54–58.
Ernest TB, Elder DP, Martini LG, Roberts M, Ford JL (2007). Developing paediatric medicines: Identifying the needs and recognizing the challenges. *J Pharm Pharmacol.* 59:1043–1055.
Ganesan B, Brothersen C, McMahon DJ (2014). Fortification of foods with omega-3 polyunsaturated fatty acids. *Crit Rev Food Sci Nutr.* 54(1):98–114.
Granato H (2014). Delivery methods: The next generation of omega 3s takes shape. *Omega-3 Inside.* 3(1):1–16.
Hunter JG (1990). Pediatric feeding dysfunction. In: Semmler CJ, Hunter JG, Eds. *Early Occupational Therapy Intervention: Neonates to Three Years.* Gaithersburg, MD: Aspen Publishers; 124–184.
Illingworth RS, Lister J (1964). The critical or sensitive period, with special reference to certain feeding problems in infants and children. *J Pediatr.* 65:839–848.
Ivanovska V, Rademaker CMA, van Dijk L, Mantel-Teeuwisse AK (2014). Pediatric drug formulations: A review of challenges and progress. *Pediatrics.* 134(2):361–372.
Kearns GL, Abdel-Rahman SM, Alander SW, Blowey DL, Leeder JS, Kauffman RE (2003). Developmental pharmacology—Drug disposition, action, and therapy in infants and children. *N Engl J Med.* 349(12):1157–1167.
Kumar A, Pawar S (2002). Issues in the formulation of drugs for oral use in children. *Pediatr Drugs.* 4(6):371–379
Lachman L, Liberman HA, Kanig JS (1987). *The Theory and Practice of Industrial Pharmacy.* 419–428. 3rd ed. Bombay, India: Varghese Publishing House.
Mansbach JM, Ginde AA, Camargo CA Jr (2009). Serum 25-hydroxyvitamin D levels among US children aged 1 to 11 years: Do children need more vitamin D? *Pediatrics.* 124(5):1404–1410.
Matsui D (2007). Assessing the palatability of medications in children. *Paediatr Perinat Drug Ther.* 8(2):55–60.
McCrickerd K, Forde CG (2016). Sensory influences on food intake control: Moving beyond palatability. *Obes Rev.* 17(1):18–29.
Mennella JA, Beauchamp GK (2008). Optimizing oral medications for children. *Clin Ther.* 30(11):2120–2132.
Mitchell MJ, Powers SW, Byars KC, Dickstein S, Stark LJ (2004). Family functioning in young children with cystic fibrosis: Observations on interactions at mealtime. *J Dev Behav Pediatr.* 25:335–346.
Moore P (1998). Children are not small adults. *Lancet.* 352(9128):630.

Navarretea NM, Camachoa MM, Lahuertab JM, Monzoa JM, Fitoa P (2002). Iron deficiency and iron fortified foods—A review. *Food Res Int.* 35:225–231.

Nelson SP, Chen EH, Syniar GM, Christoffel KK (1998). One-year follow-up of symptoms of gastroesophageal reflux during infancy. *Pediatrics.* 102:e67.

Noel GJ, Van Den Anker JN, Lombardi D, Ward R (2012). Improving drug formulations for neonates: Making a big difference in our smallest patients. *J Pediatr.* 161(5):947–949 pmid:23095694.

Palmer S, Horn S (1978). Feeding problems in children. In: Palmer S, Ekvall S, Eds. *Pediatric Nutrition in Developmental Disorders.* Springfield, IL: Charles C Thomas; 13:107–129.

Reau NR, Senturia YD, Lebailly SA, Christoffel KK (1996). Pediatric Practice Research Group. Infant and toddler feeding patterns and problems: Normative data and a new direction. *J Dev Behav Pediatr.* 17:149–53.

Regulation EU/872/2012 and EU/873/2012; http://eur-lex.europa.eu/JOHtml.do?uri=OJ:L:2012:267:SOM:EN:HTML Accessed July 2017.

Ross AC, Taylor CL, Yaktine AL et al. Eds. (2011). Institute of Medicine (US) Committee to Review Dietary Reference Intakes for Vitamin D and Calcium. Washington, DC: National Academies Press.

Sisson LA, Van Hasselt VB (1989). Feeding disorders. In: Luiselli JK, Ed. *Behavioral Medicine and Developmental Disabilities.* New York: Springer-Verlag; 45–73.

Sorensen LB, Moller P, Flint A, Martens M, Raben A (2003). Effect of sensory perception of foods on appetite and food intake: A review of studies on humans. *Int J Obes (London).* 27:1152–1166.

Stier RE (2002). A taste receptor blocker for oral hygiene compositions. *Cosmetics & Toiletries.* 117(5):63–70.

Schiffman SS, Dackis C (1975). Taste of nutrients: Amino acids, vitamins, and fatty acids. *Percept Psychophys.* 17(2):140–146.

WHO Expert Committee on Specifications for Pharmaceutical Preparations (2012). Development of paediatric medicines: Points to consider in formulation. WHO Expert Committee on Specifications for Pharmaceutical Preparations Forty-sixth Report; 197–225.

WHO Technical Report Series 2014; No. 990. Evaluation of certain food additives.

World Health Organization (2007). Production of zinc tablets and zinc oral solutions: Guidelines for programme managers and pharmaceutical manufacturers, Annex 7. Geneva: World Health Organization.

West KP Jr (2003). Vitamin A deficiency disorders in children and women. *Food Nutr Bull.* 24(4):S78–S90.

3

Flavors in Probiotics and Prebiotics

Deepak Yadav, Kummaripalli Srikanth, and Kiran Yadav

Contents

3.1 Introduction 51
3.2 Flavors in probiotics and prebiotics 55
3.3 Flavor characteristics of probiotic/prebiotic products 56
    3.3.1 Fermented dairy products 56
        3.3.1.1 Yogurt 56
        3.3.1.2 Ayran 58
        3.3.1.3 Kefir 59
        3.3.1.4 Kumiss 61
        3.3.1.5 Fermented milk drinks 61
        3.3.1.6 Miscellaneous dairy-based probiotic beverages 63
    3.3.2 Fermented nondairy probiotic beverages 63
3.4 Flavor/aroma defects in probiotics and prebiotics 65
3.5 Testing of flavor/sensory defects 66
3.6 Conclusion 68
References 68

3.1 Introduction

Functional foods are products that contain added nutrients and other substances that provide health benefits in addition to basic nutritional value. These encompass a broad range of products and often include nutraceuticals. Probiotics and prebiotics are market leaders in the functional foods sector throughout the world. Traditionally, lactic acid bacteria were used in food preservation, but in the last decades there has been a shift to probiotic products. Probiotics are live microorganisms, which, when administered as part of food or pill intake, provide a health benefit to the host. Selected strains of *Lactobacillus* and *Bifidobacterium* have found the widest applications in the production of probiotic food. The main ones used are *L. acidophilus*, *L. casei*, *B. bifidum*, *B. infantis*, and *B. longum* (Nagpal et al. 2007). Within lactic acid bacteria, a large strain-to-strain diversity exists with respect to flavor formation and the products formed. Probiotics are delicate; they are sensitive to heat and stomach acid. There is substantial clinical evidence that probiotics help in the prevention and treatment of a number of diseases of the gastrointestinal, respiratory, and urogenital tracts. There is increasing interest in promoting the consumption of food/milk products containing such beneficial microbes.

In the dairy industry, probiotic products are often sold as yogurt drinks in shot form in small bottles. Beneficial probiotic bacteria can be added to yogurt (Katan 2008), buttermilk (Antunes et al. 2009), cheese (Ong and Shah 2008), cottage cheese (Shah 2007), ice cream (Hagen and Narvhus 1999), and frozen dairy desserts (Shah 2007). Probiotics can take the form of capsules or sachets. Yogurt is a perfect baby food due to its smooth texture and rich tangy flavor. Probiotics have also been introduced into infant formula; confectionery and bakery products such as cereal bars, breads, biscuits, cookies, and muesli-type products; chocolate and tablet candy; and powdered soup, mayonnaise, and fruit and vegetable products. Bacteria are introduced into the filling of confectionery products. This requires careful selection of strains that are tolerant in a low water activity medium in the presence of oxygen and at room temperature because such nondairy products are stored over extended periods without refrigeration in warehouses and on shop shelves. Probiotic fruit and vegetable juices (Yoon et al. 2006) and probiotic fermented meat products (Tyopponen et al. 2003) are also popular.

Probiotics must meet certain criteria. All probiotic strains must have generally recognized as safe (GRAS) status, be nonpathogenic, and cause no adverse health effects to the recipient (Collins et al. 1998, WHO 2001). Strains of human origin are most suitable because some health-promoting benefits may be species specific, and microorganisms may perform optimally in the species from which they were isolated. Probiotic strains should be able to colonize in the intestinal tract at least temporarily. Adherent strains are desirable because they have a greater chance of becoming established in the gastrointestinal tract (GIT), thus enhancing their probiotic effect (Lee and Salminen 1995).

Probiotic microorganisms should be technologically suitable for incorporation into food products so that they retain both viability and efficacy in the food products (on a commercial scale) and following consumption. Probiotics should be capable of surviving industrial applications (e.g., common dairy processing or pharmaceutical manufacturing processes) and be able to grow/survive at high levels in the product until the end of shelf life (Rogelj 1994, Stanton et al. 1998). It is assumed that functional food needs to be consumed with at least 10^6–10^9 cells per day for beneficial effects to develop (Gilliland 1989, Robinson 1991). This generally means that the minimum cell density in the product throughout its entire shelf-life should be 10^5 cells/mL or g (Lee and Salminen 1995). Cultures of probiotic microorganisms are usually produced in the form of frozen concentrates, either freeze dried or spray dried (Holzapfel and Schillinger 2002). Probiotic microorganism biomass is most commonly preserved by freeze drying with a final moisture content of 2–6%. The resulting product is a lyophilized powder, which is easy to use and highly stable (Castro et al. 2000).

Prebiotics are indigestible food components that selectively stimulate the growth and/or activity of one or a limited number of microorganisms in the large intestine, thus improving human health (Shah 2004). Plant products are used increasingly in pharmaceuticals and nutraceuticals (Yadav and Kumar 2014, Yadav et al. 2015a,b, McChesney et al. 2007, Noviendri et al. 2011, Yang et al. 2015, Wang et al. 2008a, Femenia and Waldron 2007). Prebiotics are a special type of plant fiber added to foods that beneficially nourish the good bacteria already in the large bowel or colon. These prebiotics are only digestible by some probiotics so they provide a selective environment for them (Kaplan and Hutkins 2003). While probiotics introduce good bacteria into the gut, prebiotics facilitate survival and proliferation of good bacteria already present there. The body itself does not digest these plant fibers. Instead, it uses these fibers to promote the growth of many of the good bacteria in the gut.

Consumption of prebiotics with the diet provides similar benefits to the consumer as the consumption of probiotics. Prebiotics were first introduced into the market in Japan. Prebiotics must meet certain criteria. They should be nondigestible by human digestive enzymes in the GIT. They should stimulate growth of selected groups of microorganisms beneficial for human health, especially bacteria such as lactic acid bacteria and *Bifidobacteria* in adult humans, and have an indirect regulatory effect on the microbial equilibrium in the alimentary tract. Their metabolism should have a beneficial effect, including the production of short chain fatty acids and organic acids, reducing the pH of intestinal contents. They should be safe for human health (Roberfroid 2001, Gibson and Roberfroid 1995). Unlike probiotics, prebiotics are not destroyed when ingested in the body. They are not affected by heat or bacteria. Prebiotic fiber is found in many fruits and vegetables, such as the skin of apples, bananas, onions, garlic, Jerusalem artichokes, chicory root, and beans. Raw chicory root has the highest percentage (64.6%) of prebiotic fiber per gram.

Most prebiotics are oligosaccharides. They include certain polysaccharides, protein, peptides, and fats. The most thoroughly investigated oligosaccharides are fructo-oligosaccharides, malto-oligosaccharides, galacto-oligosaccharides, isomalto-oligosaccharides, palatinose oligosaccharides, glucosylsucrose, soy oligosaccharides, arabinogalactan, xylo-oligosaccharides, and stachyose. Their metabolism in the large intestine leads to their transformation to volatile fatty acids (acetic acid, lactic acid, propionic acid, and butyric acid) and gaseous products (carbon dioxide, hydrogen, and methane), and to the formation of bacterial biomass. Their presence also contributes to a reduction in triacylglycerol, phospholipid, and cholesterol levels. Inulin is the most important source of fructo-oligosaccharides. It is broken down by the enzyme inulinase to oligosaccharides with a degree of polymerization between two to six. Its degradation products strongly stimulate the growth of bifidobacteria, which inhibit the growth of *Clostridium, Fusobacterium, Salmonella,* and *Escherichia.*

Inulin, the most popular prebiotic in the world, is is the most important source of fructo-oligosaccharides. It is found in chicory, artichokes, onions, garlic, leeks, asparagus, tomatoes, wheat germ, barley, and bananas. Various commercial preparations are produced from chicory by controlled enzymatic hydrolysis, containing fructo-oligosaccharides with different degrees of polymerization. Different plant products are sources of other oligosaccharides. We get galactanes from lupin seeds and soy oligosaccharides from soybeans (Martínez-Villaluenga et al. 2006). Mannanoligosaccharides are isolated from yeast cell walls (Martínez-Villaluenga et al. 2006). Enzymatic synthesis is also used for production of oligosaccharides. For example, oligofructosides are synthesized from sucrose and galacto-oligosaccharides from lactose (Johnson 1999).

Raftilose® is a well-known prebiotic preparation containing inulin. Raftiline HP® is another popular prebiotic that contains oligofructose. Both are used as additives in dairy products. Other commercial prebiotic products include Prebiotin™, NutraFlora®, Actilight®, Neosugar, and Meioligo (Figure 3.1). Prebiotics are stable, not as sweet as fructose, and practically imperceptible in food products. For these reasons they can be introduced into a wide range of products such as cookies, bread, soups, ready-to-eat dinner dishes, puff snacks, chocolate products, and food concentrates (Shah 2007). The French company Vivis has marketed Actilight, which is manufactured from beets and added to products such as cookies and soups. The Japanese company Beghin Meiji Industries produces milk enriched with a soluble fraction of dietary fiber. Bauer (Germany) produces the fermented product Probiotic Plus Oligofructose, which contains two probiotic bacterial strains and the prebiotic Raftilose.

Synbiotics are a combination of probiotics (the live bacteria) and prebiotics (the food components they live on), A synbiotic is used primarily because a true probiotic without its prebiotic food source does not survive well in the digestive system (Panesar et al. 2009). Simply put, prebiotics are "food" for probiotics. Probiotics digest prebiotics and use them as a source of energy. Probiotics and prebiotics act synergistically to promote gut health. A typical example includes fermented dairy drinks with the addition of fruits, such as fruit-flavored yogurts. These added preparations usually contain sucrose, sucrose syrup, inverted sugar, oligofructose, and taste and aroma additives such as peach, strawberry, or blueberry flavor. Generally, oligosaccharides are introduced into the matrices of foodstuffs and drinks. Synbiotic preparations are available in tablet form, usually two-thirds bacteria and one-third fructo-oligosaccharides. The tablet formulation usually includes calcium carbonate, microcrystalline cellulose, fructo-oligosaccharides, fructose, powdered

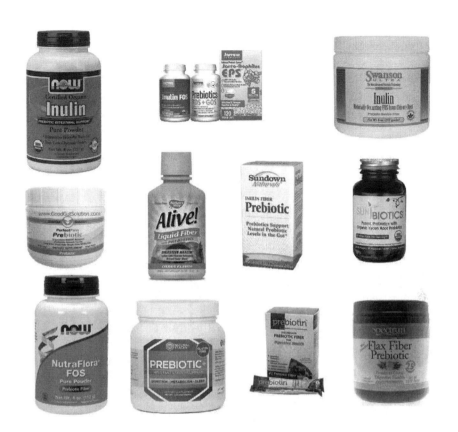

Figure 3.1 Prebiotic products available on the market.

yogurt, encapsulated culture of probiotic bacteria, calcium stearate, ascorbic acid, filling agent, and natural vanilla flavor. Each tablet contains more than 10^9 cells, and their stability is more than one year (Sip and Grajek 2010).

The consumption of functional foods has increased considerably in recent years. Globally this sector is expected to rise in terms of sales to US$ 253 billion by the year 2024 (Statista 2018). Fermented milks are the main vehicle for the incorporation of probiotic cultures or prebiotic ingredients. From a technological standpoint, cheese and ice cream also may be used (Cruz et al. 2009, da Cruz et al. 2009).

3.2 Flavors in probiotics and prebiotics

Prebiotic ingredients generally influence texture and aroma, whereas probiotic products have a greater effect on flavor and aroma. A prebiotic ingredient alters the already existing bonding between different components of the food and usually substitutes the fat, which is directly responsible for the softness and creaminess of the food product. A probiotic culture can result in the production of components that may contribute negatively to the aroma and taste of the product, resulting in probiotic off-flavor. This is why fermented dairy drinks with added probiotic bacteria have a modified taste, i.e., they are milder in taste and less aromatic. Thus, taste and aroma additives are frequently introduced to such products. *Bifidobacterium* produces acetic acid as a product of its metabolism, which can impart a vinegary flavor to the product (Tamime et al. 1995).

Formulation and technical interventions such as microencapsulation (Talwalkar and Kailasapathy 2003) and an increase in inoculums of the probiotic culture (Olson and Aryana 2008) can ensure protection of the probiotic strain only to the extent that it survives in a viable form throughout the shelf life of the product and during its passage through the intestinal tract in amounts sufficient to provide clinical benefit to the host. The choice and selection of flavors are important considerations for these functional food products, particularly for probiotics.

3.3 Flavor characteristics of probiotic/prebiotic products

3.3.1 Fermented dairy products

3.3.1.1 Yogurt

Yogurt is a semisolid fermented milk product that is popular worldwide. Yogurt is made by adding specific bacterial strains to milk, which is subsequently fermented under controlled temperatures (42–43°C) and environmental conditions. The bacteria consume milk sugar lactose and release lactic acid as a waste product. As a result of increased acidity, the milk proteins are coagulated into a solid mass or curd. This process is called denaturation (Robinson and Tamime 1986). The increased acidity (pH = 4–5) also prevents the proliferation of potentially pathogenic bacteria.

The bacterial species used for yogurt production are *Streptococcus salivarius* subsp. *thermophilus* and *Lactobacillus delbrueckii* subsp. *bulgaricus*. These two are often cocultured with other lactic acid bacteria for better taste or health effects. These include *L. acidophilus, L. casei,* and *Bifidobacterium* spp. In the United States and European Union a product may be called yogurt only if live bacteria are present in the final product; therefore nonpasteurized yogurt is marketed there containing live active culture (Chandan et al. 2013). When yogurt is pasteurized, it kills both harmful and essential bacteria. Pasteurized products that have no living bacteria may be called fermented milk products.

Although the consistency, flavor, and aroma may vary from one region to another, the basic ingredients and manufacturing are consistent. Important parameters in yogurt manufacturing include ingredients, starter culture, and manufacturing methods. Sweeteners such as glucose or sucrose and high-intensity sweeteners such as aspartame are used. Commonly used stabilizers include gelatin, carboxymethyl cellulose, locust bean guar, alginates, carrageenans, and whey protein concentrate. Fruit preparations, including natural and artificial flavoring, are added as flavors (Vinderola et al. 2002).

3.3.1.1.1 Starter culture characteristics for yogurt manufacture and flavor The starter culture for yogurt production is a symbiotic blend of *S. salivarius* subsp. *thermophilus* and *L. delbrueckii* subsp. *bulgaricus*. The rate of lactic acid production is much higher when these two microorganisms are used together than if either is grown individually. These microorganisms are ultimately responsible for the formation of typical yogurt flavor and texture. The yogurt mixture coagulates due to the pH drop in the fermentation medium. Fermentation products such as lactic acid, acetaldehyde, acetic acid, and diacetyl contribute to yogurt flavor (Vinderola et al. 2002). Acetaldehyde is the main flavoring component of yogurt (at levels of up to 40 mg/kg). Its desirable acceptable level in yogurt is 25 ppm (Ray and Bhunia 2007, de Oliveira 2014). During yogurt fermentation acetaldehyde is produced from lactic acid bacteria in different pathways from glucose, thymidine, or threonine. Acetic acid also contributes to flavor of yogurt. It is produced from pyruvate by *S. thermophilus*. Bifidobacteria are heterofermentative. They produce acetic acid in addition to lactic acid. The ratio of lactic acid to acetic acid is 2:3 (Hassan and Frank 2001). Acetic acid causes harshness in the product and in high levels it imparts a "vinegary" taste that decreases consumer acceptability of the product (De Vuyst 2000). Diacetyl is a yellow/green liquid with an intensely buttery flavor. Diacetyl is also a flavor compound with 0.5 ppm level in yogurt (Gurakan and Altay 2010).

3.3.1.1.2 Yogurt types and flavors A number of yogurt types are available on the market. Set yogurt is a solid set where the yogurt forms in a consumer container and is not disturbed. Stirred yogurt, the most popular commercial product, is made in a large container and then spooned or dispensed into secondary serving containers. It is less firm than set yogurt. Sweet drinking yogurt is stirred yogurt to which additional milk and flavors are added. Fruit or fruit syrups are added as flavoring agents (De Ramesh et al. 2006). In fruit yogurt, fruits, fruit syrups, or pie filling can be added on the top or bottom or stirred into the yogurt (Robinson and Tamime 1986). Yogurt cheese is a fresh cheese made by draining overnight and separating the whey. Its flavor is similar to that of sour cream and it has the texture of soft cream cheese (Keceli et al. 1999). The popular East Indian dessert mishti-dahi, a variation of traditional dahi, is thicker, more custard-like in consistency, and usually sweeter than Western yogurt (Pantaleao et al. 1990). Tarator and cacik/tzatziki are popular cold summer soups in Turkey, Bulgaria, and Macedonia. They are made with ayran (a cold yogurt beverage mixed with salt; see Section 3.3.1.2), cucumber, dill, salt, olive oil, and sometimes garlic and ground walnuts (Prakash and Urbanska 2007). Jameed is a salted dried yogurt popular in Jordan. Raita is a yogurt-based South Asian/Indian condiment used as a sauce or dip. The yogurt is seasoned with coriander, cumin, mint, pepper, and other herbs and spices. Vegetables such as cucumber and onions or fruits are sometimes mixed in. Raita is served chilled (Keceli et al. 1999).

Yogurt drinks with fruit and added sweeteners, such as honey, are widely available in the United States and United Kingdom. These are marketed as drinking/drinkable yogurt (Dong et al. 2015; Taillie et al. 2017; Shah 2017). Activia is a low-fat probiotic available commercially in more than 30 countries, both as a semisolid yogurt and a yogurt drink. It contains the probiotic bacterium *B. animalis*. Activia is available plain or in a range of varieties including strawberry, raspberry, peach, mango, oatmeal, pear, walnut, coconut, vanilla, blueberry, prune, fig, pineapple, aloe vera, fibers, fruit of the forest, kiwi cereals, and rhubarb (Coudeyras et al. 2008).

3.3.1.2 Ayran

Ayran is a salty, yogurt-based drink popular in the Middle East. It is the most popular fermented drinkable product in Turkey. It is made by mixing yogurt, made primarily with cow's milk, with water and salt. The same drink is known by different names in different countries. Plain ayran with its characteristic acidic taste is produced in Turkey. Different colors and flavors such as mango and coconut are added in Europe and United States. Carbonated ayran was produced to increase ayran's market competitiveness with carbonated nonalcoholic beverages. A similar drink, doogh, is also popular in parts of the Middle East. It differs from ayran in the addition of herbs, usually mint, and it is carbonated, usually with mineral water. Ayran can also be made with cucumber juice in place of some or all of the water, and can be flavored with garlic. It may be seasoned with black pepper. Another recipe popular in some regions includes finely chopped mint leaves mixed into the ayran (Kucukoner et al. 2006).

3.3.1.2.1 Starter culture characteristics for ayran manufacturing and flavor The bacteria *S. thermophilus* and *L. delbrueckii* subsp. *bulgaricus* are used for fermentation in ayran production. Selection of starter bacteria is crucial because the resulting flavor and texture are strongly dependent on this bacterial activity. Starter culture-producing companies such as Chr. Hansen, Denmark, have focused on natural solutions such as cultures for yogurt, including ayran. Low-viscosity culture strains and slow fermenting starter cultures are preferred, which yield a stable product (Nilsson et al. 2006). The pH of ayran at the end of fermentation is also important (Ozdemir and Kilic 2004). The incubation temperature and starter culture activity should be optimized as they directly affect the final pH. In addition to cow's milk, ayran can also be manufactured from sheep's or goat's milk, or a mixture of the two. Ayran from the Turkish Saanen breed of goat's milk has a more "goaty" flavor than others made from Turkish Hair and Maltese goat breeds (Uysal Pala et al. 2006).

A numbers of stabilizers such as carboxymethyl cellulose (CMC), pectin, gelatin, carrageenan, and locust bean gum. are commonly used in concentrations of 0.3–1%. Pectin is most widely used to provide textural stabilization and reduce serum separation in stirred yogurts and drinking yogurts (Nilsson et al. 2006, Foley and Mulcahy 1989). Pectin molecules interact with casein through calcium ions and prevent their aggregation, sedimentation, and hence serum separation by ionic and steric stabilization in these products (Lucey et al. 1999). Pectin can be used at 0.2–0.8% in aryan without causing off-flavor (Atamer et al. 1999). Stabilizers should be used at the optimum level and should be added to the product at the right step in the process. Excess and misuse of stabilizers may mask flavor and cause a sandy structure. For example, pectin should be added to the fermented product prior to final heat treatment to prevent sedimentation and sandy mouthfeel. Pectin functions best in the pH range 3.7–4.3 (Nilsson et al. 2006). Homogenization is also an important factor affecting ayran stability. It improves the mouthfeel of ayran and prevents the formation of a creamy layer on top of the product. Flavors used commercially in ayran production include mint, basil, vanilla, honey, and various fruit flavors such as strawberry, cherry, blueberry, blackberry, raspberry, mango, and peach. Spices such as mint and basil and various fruits are often included.

3.3.1.3 Kefir

Kefir is a traditional fermented dairy product that originated with the tribes of the Northern Caucasus mountain region in Russia. It is produced by fermentation of kefir grains when added to milk (Figure 3.2). Cow's, goat's, or sheep's milk is used. Kefir grains are small, cauliflower-shaped, semihard

Figure 3.2 *Dairy-based probiotics and their commercial products.*

granules that contain a symbiotic culture of bacteria and yeast. Kefir is a self-carbonated refreshing fermented milk drink with a unique flavor due to a mixture of lactic acid, pyruvic acid, acetic acid, diacetyl, and acetoin (the latter two of which impart a "buttery" flavor), citric acid, acetaldehyde, and amino acids resulting from protein breakdown. Depending on the process, ethanol concentration can be as high as 1–2%, with the kefir having a bubbly appearance and carbonated taste. A number of American companies also offer noncarbonated, alcohol-free kefir with fruit flavors.

3.3.1.3.1 Starter culture characteristics for kefir manufacture and flavor Kefir grains are added to previously pasteurized and cooled milk and incubated with stirring for approximately 24 h at 25°C. After complete fermentation, the kefir grains start floating due to carbon dioxide release. The kefir grains are recovered after fermentation for future use in subsequent kefir fermentations. The metabolic activity of a variety of lactic acid bacteria and yeast imparts a unique flavor to kefir. The most frequently reported bacterial genera in kefir grains are homofermentative and heterofermentative *Lactobacillus*, *Lactococcus*, *Leuconostoc*, and acetic acid bacteria (Simova et al. 2002, Rea et al. 1996, Yuksekdag et al. 2004a,b, Toba et al. 1987, Duitschaever et al. 1988, Koroleva 1988, Angulo et al. 1993, Santos et al. 2003).

Industrial production of kefir using kefir grains is difficult due to postfermentation separation requirements. As a result, much of the kefir produced in industrial practice is not considered authentic kefir because it is not incubated with the grains. Commercial production involves use of lyophilized starter cultures containing lactic acid bacteria and yeast. Using this method, activated starter culture is added to homogenized and pasteurized milk. After fermentation at 25°C, the pH gradually drops to 4.6 and fermentation is completed in approximately 20–24 h, which is sufficient time for the formation of taste and aroma substances (Farnworth and Mainville 2003, Simova et al. 2002). The typical final kefir product has lactic acid, ethanol, carbon dioxide, and other flavor products such as acetaldehyde, diacetyl, and acetoin (Wood and Hodge 1985). The combined yeast and lactic acid fermentation impart a refreshing taste to traditional kefir. Some companies use starter cultures without yeast culture because package swelling may occur due to carbon dioxide production by the yeast. The resulting product lacks the authentic flavor of kefir because of the lack of alcohol fermentation and little-to-no production of carbon dioxide.

Kefir cultures have been used experimentally as starter cultures for cheese manufacture. Freeze-dried kefir culture and freeze-dried kefir immobilized on casein have been used in preparation of hard-type cheese (Katechaki et al. 2008). It yielded open cheese texture due to carbon dioxide generation. Use of 1 g/L freeze-dried kefir culture had better sensory results such as texture, color, and flavor than the control (Katechaki et al. 2008). Cheese made with kefir culture had higher populations of total aerobic bacteria, yeast, lactococci, and lactobacilli than the control cheese. Forty-seven different flavor substances were identified. Kefir culture cheese had higher levels of flavor compounds, and free fatty acids, benzaldehyde, and butanal were significantly higher in kefir cheeses. Kefir co-culture was reported to suppress spoilage and pathogenic bacteria during ripening of white soft cheese as well as provide good flavor due to increased amounts of esters, free fatty acids, alcohols, and carbonyl compounds (Kourkoutas et al. 2006). Flavors added can be spicy, sweet, or savory. Fruits, syrups, or jams can also be added, for example, fresh bananas and strawberries, a little maple syrup and pecans, some stevia, cinnamon, and dried apple pieces, or some orange marmalade. Kefir is also available in the United States in peach, pineapple and raspberry flavors.

3.3.1.4 Kumiss

Kumiss is a traditional fermented milk drink (from mare's milk) that originated with the nomadic tribes of Central Asia and Russia. It is especially popular in Kazakhstan and Kyrgyzstan (Wang et al. 2008b). It has not been globally commercialized (Malacarne et al. 2002). Kumiss is similar to kefir but produced from a liquid starter culture, in contrast to the solid kefir grains. Also, Koumiss has a slightly higher alcohol content because mare's milk contains more sugar than the cow's or goat's milk fermented into kefir. Starter cultures used include a variety of lactic acid bacteria and yeasts. Like kefir, kumiss is produced by both lactic acid and alcohol fermentations. Traditionally, the natural starter culture is the previously fermented koumiss (Yaygin 1992). Depending on the lactic acid content, there are three types of kumiss: strong (pH 3.3–3.6), moderate (pH 3.9–4.5), and light (pH 4.5–5) (Danova et al. 2005). Compared to cow's milk, mare's milk has higher lactose and lower fat and protein content (Pagliarini et al. 1993). Mare's milk is translucent, is less white, and is sweeter than either cow's milk or goat's milk because of higher lactose content. Mare's milk pH is approximately 7. Mare's milk contains less casein but a higher amount of immunoglobulins than cow's milk (Malacarne et al. 2002). It contains higher levels of polyunsaturated fatty acids and phospholipids than cow's milk (Malacarne et al. 2002, Iametti et al. 2001). Commercial flavors used include mostly fruit flavors such as strawberry, mango, guava, and peach.

3.3.1.5 Fermented milk drinks

Fermented milk drink products can be classified into three groups: viscous products, diluted or beverage products, and carbonated products (Nilsson et al. 2006). Most of these products contain sugar, fruit syrup, sweeteners, aroma compounds, and coloring to attract consumers.

3.3.1.5.1 Acidophilus milk Acidophilus milk is milk fermented with *L. acidophilus*. The warm product is rapidly cooled to less than 7°C before agitation and pumped to a filler where it is filled into bottles or cartons (Goodarzi et al. 2017; Gilliland 2018). Stabilizers may be added to reduce the risk of separation in the carton and to improve the mouthfeel of the product (Tamime and

Robinson 1988). Acidophilus milk has higher free amino acids than milk. Fermented acidophilus milk is a bit tangier and thicker than regular milk and is usually sold unflavored.

3.3.1.5.2 Sweet acidophilus milk

Acidophilus milk is limited in popularity in Western countries due to its undesirable sour milk flavor. In response to this, sweet acidophilus milk has been developed, which is more palatable and appealing to consumers (Salji 1992).

3.3.1.5.3 Acidophilus-yeast milk

Acidophilus-yeast milk is very popular in the former Soviet Union. Whole or skimmed milk is heat treated at 90–95°C for 10–15 min, then cooled down to 35°C. It is inoculated with a mix culture of *L. acidophilus* and *Saccharomyces lactis* until 0.8% lactic acid and 0.5% ethanol is produced. The product is viscous, slightly acidic, with a yeasty taste (Surono and Hosono 2003). Acidophilin, a sweetened milk beverage, is another fermented dairy drink that is similar to acidophilus-yeast milk. It is available in vanilla or banana flavors.

3.3.1.5.4 Bifidus milk and acidophilus–bifidus milk (A/B milk)

Bifidus and acidophilus–bifidus (A/B) milk are both prepared in a way similar to that of acidophilus milk. In both products, the milk is standardized to the desired protein and fat levels. For the manufacture of bifidus milk, the milk is then heat-treated at 80–120°C for 5–30 min and rapidly cooled to 37°C. The heat-treated milk is inoculated with frozen culture of *B. bifidum* and *B. longum* at a level of 10% and left to ferment until a pH of 4.5 is obtained. When the fermentation is stopped, the product is cooled to <10°C and packaged. The final product has a slightly acidic flavor, and the ratio of lactic acid to acetic acid is 2:3. A/B milk has a characteristic aroma and slightly acidic flavor. The viscosity of the product is high. In some countries this product is produced in set form. These milks are available commercially in vanilla, coffee, and mocha flavors.

3.3.1.5.5 Mil-mil

Mil-mil is a fermented milk drink similar to A/B milk. It is very popular in Japan. Pasteurized milk is inoculated with a mix culture of *L. acidophilus*, *B. bifidum*, and *B. breve*. Glucose or fructose is added to balance the taste of the product, and carrot juice is added as a colorant (Kurmann et al. 1992). Mil-mil is available in strawberry, banana, mango, melon, mocha, almond, and honey flavors.

3.3.1.5.6 Yakult

Yakult is a sweetened, therapeutic milk beverage made from skim milk or skim milk powder, with sugar, dextrose, and water added. Yakult has been known since 1935. It is very popular in Japan. Live *L. paracasei* subsp. *paracasei* Shirota strain is added to the fermentation tank and the solution is allowed to ferment. High heat treatment of skim milk triggers the Maillard reaction between glucose and skim milk and milk proteins, resulting in the characteristic light coffee color of Yakult (Akuzawa and Surono 2002). The product is usually flavored with parsley, tomato, celery, carrot, cabbage, or other vegetable juices. Yakult can be added to cereals, smoothies, milkshakes, cheesecake, and other cold foods.

3.3.1.6 Miscellaneous dairy-based probiotic beverages

3.3.1.6.1 Lassi Lassi is a yogurt-based beverage that originated on the Indian subcontinent. It is a staple of the Punjab region. It is usually slightly salty or sweet. The sweet version may be commercially flavored with rosewater, mango, or other fruit juices to create a totally different drink. Salty lassi is usually flavored with roasted ground cumin and red chillies. This salty variation may also use buttermilk. It is variously called ghol (Bangladesh), mattha (Bihar), tak (Maharashtra), or chhaas (Gujarat) (Keceli et al. 1999).

3.3.1.6.2 Cacik Cacik is a Turkish dish of seasoned, diluted yogurt, eaten throughout Bulgaria, Serbia, Cyprus, Greece, and the Middle East. It is made of yogurt, salt, crushed garlic, chopped cucumber, dill, and mint, diluted with water to a somewhat soupy consistency. Olive oil, lime juice, and sumac are sometimes added. Dill and mint (fresh or dried) may be used. A similar side dish prepared in India is known as raita. It consists of cucumbers, onions, tomatoes, and quite often, grated carrots. Unlike other versions, lime juice is not used in raita. The Indian dish include heated oil, mustard seeds, or other vegetables.

3.3.1.6.3 Buttermilk Buttermilk is a fermented dairy product with a characteristic sour taste. It is produced from cow's milk. The product is made in one of two ways. Originally buttermilk was the liquid left over from churning butter from cream. In India buttermilk (chhaas) is the liquid left over after extracting butter from churned curd (dahi). Today this is called traditional buttermilk. Cultured buttermilk is made by adding lactic acid bacteria *S. lactis* to milk. It is much thicker than traditional buttermilk. Whether traditional or cultured, the tartness of buttermilk is due to the presence of lactic acid. As lactic acid is produced by the bacteria, the pH of the milk decreases, and milk protein casein precipitates, causing the curdling or clabbering of milk. This process makes buttermilk thicker than plain milk (Mistry et al. 1996). Cultured buttermilk was originally the fermented byproduct of butter manufacture, but today it is more common to produce cultured buttermilk from skim or whole milk. Milk is usually heated to 95°C and cooled to 20–25°C before the addition of the starter culture at 1–2%. The fermentation is allowed to proceed for 16–20 h, to an acidity of 0.9% lactic acid. Buttermilk is frequently used in the baking industry and sold in retail. Buttermilk is available in flavors such as banana, maple, vanilla, strawberry, blueberry, apple, lemon, and mint.

3.3.2 Fermented nondairy probiotic beverages

Boza is a traditional fermented product native to Turkey, Romania, Albania, and Bulgaria. It is made from maize, wheat, rye, millet, and other cereals.

Lactic acid bacteria and yeast are used for fermentation in a ratio of 2:4 (Gotcheva et al. 2000). The product has a light to dark beige color and a slightly sharp to slightly sour/sweet taste (Prado et al. 2008). Fermented cassava flour is made using *L. plantarum* A6. The product contains cassava flour at about 20%. Incubation is carried out at 35°C for 16 h. The level of probiotic bacteria is 2.3×10^9 cfu mL^{-1}. The resulting product can be stored for 28–30 days. Probiotic oat drink is made using a fermentation of *L. plantarum B28* for about 8 h. Aspartame, cyclamate, and saccharine are added. The shelf life of the product is 21 days at 4°C. Fermented soybean drink is made by using *Bifidobacterium* spp. (Chou and Hou 2000). The raw material is usually heated at 65°C for 10 min or over 100°C (Wang et al. 2002).

Fruit juice–based probiotics are also available (Figure 3.3). Fruit juices are added to milk containing *Bifidbacterium* spp. and/or *L. acidophilus*. They impart flavor and sensory properties to the product pleasing to all age groups. Fermentation of different raw materials yields probiotic products of different flavors such as probiotic fermented beet juice made by inoculating

Figure 3.3 *Probiotic fruit juices and other nondairy fermented probiotic products.*

L. acidophilus for about 48 h. Probiotic tomato juice was made by *Lactobacillus* fermentation carried out at 30°C for 72 h (Yoon et al. 2004). Probiotic cabbage juice is produced by using 24-h-old *Lactobacillus* culture and fermenting at 30°C for 48 h. Probiotic-fermented carrot and beetroot juice is made using yeast and lactic acid bacteria (Rakin et al. 2007). Another product is probiotic blackcurrant juice. Probiotic orange juice has medicinal (an iron-like aroma associated with cough syrups) and dairy (a condensed sweet milk taste) flavors that result in an entirely different taste from that of conventional orange juice (Luckow and Delahunty 2004a).

A number of other nondairy beverage products are available. Grainfields Whole Grain Probiotic is a dairy-free effervescent product that contains pearl barley, linseed, maize, rice, alfalfa seed, mung beans, rye grains, wheat, and millet with no added sugar. Vita Biosa (Live Superfoods, USA) is another sugar-free product made by fermenting aromatic herbs and other plants by lactic acid bacteria. It also contains high amounts of antioxidants (Prado et al. 2008). ProViva (Skane Dairy, Sweden) is the first probiotic food that does not contain milk/milk constituents. Its active component is oatmeal gruel fermented by lactic acid bacteria. Malted barley is added to enhance liquefaction (ref-citation). Biola (Tine BA, Norway) is a probiotic fruit juice containing 95% fruit juice with no added sugar. It is available in orange, mango, and apple-pear flavors.

3.4 Flavor/aroma defects in probiotics and prebiotics

Different terminology is used for commercial flavor and organoleptic evaluation of probiotic drinks based on odor, taste, aroma, and texture (Muir and Hunter 1991). Lack of the characteristic aroma/flavor profile is among the most common organoleptic defects in yogurt. The carbonyl compounds produced by starter bacteria such as acetaldehyde, acetone, acetoin, and diacetyl are responsible for the characteristic aroma/flavor of yogurt. Any factor interfering with the metabolic activity of the starter culture likely causes aroma/flavor defects in the end product. The processing variables that affect starter culture activity are incubation temperature and period, presence of inhibitor substances in raw milk, and bacteriophage activity. *Lactobacillus delbrueckii* is mainly responsible for the production of these carbonyl compounds and *S. thermophillus* is responsible for development of acidity during fermentation. Any factor affecting the growth or metabolic activity of these bacteria leads eventually to lack of characteristic aroma/flavor in the final product. The incubation temperature and inoculation level of starter culture must be selected properly. Because the yogurt starter bacteria are thermophilic, the incubation temperature of fermenting milk should be set at 41–43°C. Optimum inoculation levels are 2.5–3.0%. Too high or too low starter inoculation levels may cause flavor/aroma defects in yogurt. At higher inoculation levels, depending on the increased lactic acid concentration, the characteristic flavor/aroma compounds are masked. In contrast, at lower inoculation levels, insufficient growth of starter bacteria may disturb flavor/aroma. Too rapid cooling after fermentation may also result in flavor/aroma defects in fermented milks.

Excessive sourness is another major flavor problem in fermented milks. Although improper storage conditions (e.g., high storage temperature) and high level of starter inoculation are the main causes of sourness, other factors including too slow cooling after fermentation and low levels of fat and protein in yogurt are also responsible for development of excess sour taste in yogurt. Table 3.1 lists common flavor defects in probiotic beverages and the preventive measures that can be taken when formulating and processing these products. Cooked flavor is developed in yogurt as a result of the heat treatment at high temperatures. The heating temperature of yogurt milk should not be higher than 90–95°C for 5–10 min. Kefir has a characteristic acidic, milky taste, a slight yeasty flavor, and uniform texture (Wszolek et al. 2006). The common defects in kefir are attributed to taste and aroma (Wszolek et al. 2006). In particular, *Saccharomyces cerevisiae* causes a vinegar- or solvent-like aroma in kefir (Seiler 2003). Similarly, *Acetobacter* spp. leads to development of acetic acid aroma. Molds (e.g., *Geotrichum candidum*) and some yeasts present in kefir can cause bitterness in the end product (Seiler 2003). In koumiss, the levels of acidity and alcohol affect the sensory quality of the final product.

3.5 Testing of flavor/sensory defects

Although analytical tests are extremely important, they may not detect the presence of certain chemical compounds that may be responsible for disagreeable off-flavors. For example, disagreeable flavors or tastes may be transferred from a plastic container to the beverage contained in it, as a consequence of inadequate storage conditions with respect to temperature. Different discriminative, affective, and descriptive tests are used in industry for flavor and other sensory evaluation of probiotic and prebiotic products.

Devereux and others investigated the sensory acceptance of different foods such as cookies, carrot cake, chocolate cake, ice creams, and frankfurters with partial substitution of fat by oligofructose and inulin. Good acceptance was observed for all the products, with no significant difference between these products and the controls (i.e., products containing no prebiotics) (Devereux et al. 2003).

Aryana and McGrew studied the effect of prebiotic inulin with three different-sized chain lengths—short, medium, and long—on the sensory acceptance of the yogurt products with added *L. casei*. The yogurt sample with *L. casei* and no addition of inulin (control) and the sample with *L. casei* plus the shortest chain inulin achieved the best flavor score compared to the samples that contained longer-chain prebiotics. These results show that inulin can be an alternative fat substitute in yogurts (Aryana and McGrew 2007).

Conventional currant syrup and a sample with added *L. plantarum* were analyzed. A descriptive analysis revealed the presence of nontypical aromas and

Table 3.1 Common Flavor/Aroma Defects in Fermented Milk Products: Causes and Preventive Measures

Defect	Possible Causes	Preventive Measures
Insipidness	– Insufficient incubation	– Cease incubation at pH < 4.6
	– Presence of bacteriophages	– Evaluate sanitation conditions and culture rotation
	– Insufficient growth of *Lactobacillus delbrueckii* subsp. *bulgaricus*	– Set incubation temperatures at 42–43°C and monitor the presence of inhibitor substances such as antibiotics
	– Insufficient inoculation	– Add starter culture at the level of 2.0–2.5%
	– Too rapid cooling after fermentation	– Apply two-phase cooling after incubation
Sourness	– Too high starter culture inoculation level	– Add starter culture at the level of 2.0–2.5%
	– High ratio of *L. delbrueckii* subsp. *bulgaricus* in starter culture	– Inoculate milk with fresh culture
	– High incubation temperature	– Set incubation temperatures at 42–43°C
	– Insufficient cooling after fermentation	– Apply two-phase cooling after incubation (first phase at 20–22°C, second phase at <7°C)
	– Too long incubation period	– Cease incubation at pH 4.6
	– Improper storage temperature	– Set storage temperature at 4°C
Cooked flavor	– Reactivation of sulfhydryl groups as a result of high heat treatment	– Heat treatment at 80–85°C for 20–30 min or at 90–95°C for 5–10 min
Metallic flavor	– Metal contamination of yogurt or yogurt milk	– Use stainless steel equipment
Malty/yeasty flavor	– Contamination of yeast and/or mold in yogurt	– Provide proper sanitation conditions throughout manufacturing process
Feed flavor	– Feeding animals with rotten feeds	– Revise silage production method and select proper fermentation conditions for silage-making
	– Feeding animals with too aromatic feeds	– Remove feeds with bad smell/odor from feeding program

Source: Ozer, B. & H. A. Kirmaci. (2010). Quality attributes of yogurt and functional dairy products. In *Development* and *Manufacture of Yogurt and Other Functional Dairy Products*, Ed. F. Yildiz. Boca Raton, FL: CRC Press.

tastes in the syrup with added probiotic microorganisms, such as perfumed and lactic aroma and sour and salty tastes. In a study carried out with orange juice, the authors also found unusual aromas and flavors, such as lactic, medicinal, and salty taste, among others, reinforcing the fact that the addition of probiotic microorganisms alters the sensory profile of the final product (Luckow and Delahunty 2004b). Studies are underway to find techniques to mask these undesirable flavors. The use of tropical fruits has been successful in masking medicinal flavor and improving the sensory quality and acceptance of probiotic juices. However, the sensory impact of prebiotic or probiotic cultures on foods or beverages to which they are added has not been studied as much as it should have been, although it is understood that products to which these functional ingredients have been added can create different flavor profiles when compared to the conventional products (Mattila-Sandholm et al. 1999).

3.6 Conclusion

The commercial success of a probiotic/prebiotic product in the consumer market implies consumer acceptance of its sensory characteristics, safety characteristics for consumption, and nutritional qualities. Studies have shown that flavor is the first indicator with respect to the choice of a food, followed by health considerations (Tepper and Trail 1998, Tuorila and Cardello 2002). These studies also indicated that consumers are not interested in a product if the added ingredients contribute disagreeable flavors to the product, even if this results in health advantages (Cruz et al. 2010). Consumers who purchase functional foods motivated by a beneficial "healthy" effect may not consume the required amount to reach the necessary level to obtain the desired physiological effect if they do not experience the expected flavor (Tuorila et al. 1994). This means that flavor is related not only to intrinsic sensory properties of the product such as overall acceptability but also to purchase intent. Companies that market probiotic/prebiotic products focus in particular on the flavor, aroma, and textural attributes of the final products. High nutritive value and probiotic effects have to be combined with sensory attractiveness. A number of flavors are currently used in probiotic/prebiotic products, but in view of the expanding market for functional foods there is enormous scope for development of new flavors to mask the unwanted flavors and for enhancement of manufacturing technologies to prevent the development of off-flavor compounds.

References

Akuzawa R, Surono IS (2002). Fermented milks of Asia. In *Encyclopedia of Dairy Sciences*, 2nd ed., Eds. Fuquay JW, Fox PF, McSweeney PLH, 1045–1048. London: Academic Press.

Angulo L, Lopez E, Lema C (1993). Microflora present in kefir grains of the Galician region (North-West of Spain). *J Dairy Res.* 60:263–267.

Antunes AEC, Silva ERA, Van Dender AGF, Marasca ETG, Moreno I, Faria EV, Padula M, Lerayer ALS (2009). Probiotic buttermilk-like fermented milk product development in a semiindustrial scale: Physicochemical, microbiological and sensory acceptability. *Int Dairy Technol.* 62:556–563.

Aryana KJ, McGrew P (2007). Quality attributes of yogurt with *Lactobacillus casei* and various prebiotics. *LWT - Food Sci and Technol.* 40:1808–1814.

Atamer M, Gursel A, Tamucay B, Gencer N, Yildirim G, Odabasi S, Karademir E, Senel E, Kirdar S (1999). A study on the utilization of pectin in manufacture of longlife ayran. *Gida.* 14:119–126.

Castro AG, Bredholt H, Strom AR, Tunnacliffe A (2000). Anhydrobiotic engineering of gram-negative bacteria. *Appl Environ Microbiol.* 66:4142–4144.

Chandan RC, White CH, Kilara A, Hui YH (Eds.) (2013). *Manufacturing Yogurt and Fermented Milks*, 2nd Ed. Ames, IA: Blackwell Publishing Professional.

Chou C-C, Hou J-W (2000). Growth of bifidobacteria in soymilk and their survival in the fermented soymilk drink during storage. *Int J Food Microbiol.* 56:113–121.

Collins J, Thornton G, Sullivan G (1998). Selection of probiotic strains for human applications. *Int Dairy J.* 8: 487–490.

Coudeyras S, Marchandin H, Fajon C, Forestier C (2008). Taxonomic and strain-specific identification of the probiotic strain *Lactobacillus rhamnosus* 35 within the *Lactobacillus casei* group. *Appl Environ Microbiol.* 74:2679–2689.

Cruz AG, Antunes AE, Sousa ALO, Faria JA, Saad SM (2009) Ice-cream as a probiotic food carrier. *Food Res Int.* 42:1233–1239.

Cruz AG, Cadena RS, Walter EHM, Mortazavian AM, Granato D, Faria JAF, Bolini HMA (2010). Sensory analysis: Relevance for prebiotic, probiotic, and symbiotic product development. *Comp Rev Food Sci Food Saf.* 9:358–373.

da Cruz AG, Buriti FCA, de Souza CHB, Faria JAF, Saad SMI (2009). Probiotic cheese: Health benefits, technological and stability aspects. *Trends Food Sci Technol.* 20:344–354.

Danova S, Petrov K, Pavlov P, Petrova P (2005). Isolation and characterization of *Lactobacillus* strains involved in koumiss fermentation. *Int J Dairy Technol.* 58:100–105.

de Oliveira MN (2014). FERMENTED MILKS | Fermented Milks and Yogurt A2 - Batt, Carl A. In *Encyclopedia of Food Microbiology*, 2nd ed., Ed. Tortorello ML, 908–922. Oxford: Academic Press.

De Ramesh C, White C, Kilara A, Hui Y (2006). *Manufacturing Yogurt and Fermented Milks*. Hoboken: Blackwell Publishing.

De Vuyst L (2000). Technology aspects related to the application of functional starter cultures. *Food Technol Biotechnol.* 38:105–112.

Devereux HM, Jones GP, McCormack L, Hunter WC (2003). Consumer acceptability of low fat foods containing inulin and oligofructose. *Food Sci.* 68:850–1854.

Dong D, Bilger M, van Dam RM, Finkelstein EA (2015). Consumption of specific foods and beverages and excess weight gain among children and adolescents. *Health Affairs.* 34(11):1940–1948.

Duitschaever CL, Kemp N, Emmons D (1988). Comparative evaluation of five procedures for making kefir. *Milchwissenschaft.* 43:343–345.

Farnworth ER, Mainville I (2003). Kefir: A fermented milk product. *Handbook of Fermented Functional Foods.* 2:89–127.

Femenia A, Waldron K (2007). *High-Value Co-Products from Plant Foods: Cosmetics and Pharmaceuticals*. Cambridge, U.K.: Woodhead Publishing Limited.

Foley J, Mulcahy AJ (1989). Hydrocolloid stabilisation and heat treatment for prolonging shelf life of drinking yoghurt and cultured buttermilk. *Irish J Food Sci Technol.*13:43–50.

Gibson GR, Roberfroid MB (1995). Dietary modulation of the human colonic microbiota: Introducing the concept of prebiotics. *J Nutr.* 125:1401.

Gilliland SE (1989). Acidophilus milk products: A review of potential benefits to consumers. *J Dairy Sci.* 72:2483–2494.

Gilliland SE (2018). The lactobacilli: Milk products. In *Bacterial Starter Cultures for Food*. Boca Raton, FL: CRC Press. 41–56.

Goodarzi A, Hovhannisyan H, Grigoryan G, Barseghyan A (2017). Acidophilus milk shelf-life prolongation by the use of cold sensitive mutants of Lactobacillus acidophilus MDC 9626. *Appl Food Biotechnol.* 4(4):211–218.

Gotcheva V, Pandiella SS, Angelov A, Roshkova ZG, Webb C (2000). Microflora identification of the Bulgarian cereal-based fermented beverage boza. *Process Biochem.* 36: 127–130.

Gurakan G, Altay N (2010). Yogurt microbiology and biochemistry. In *Development and Manufacture of Yogurt and Other Functional Dairy Products*. Ed. Yildiz F. Boca Raton, FL: CRC Press. 98–116.

Hagen M, Narvhus JA (1999). Production of ice cream containing probiotic bacteria. *Milchwissenschaft.* 54:265-268.
Hassan AN, Frank JF (2001). Starter cultures and their use. In *Applied Dairy Microbiology*, 2nd ed., Eds. Marth EH, Steele JL. New York: Marcel Dekker. 151-206.
Holzapfel WH, Schillinger U (2002). Introduction to pre-and probiotics. *Food Res Int.* 35:109-116.
Iametti S, Tedeschi G, Oungre E, Bonomi F (2001). Primary structure of I°-casein isolated from mares' milk. *J Dairy Res.* 68:53-61.
Johnson KF (1999). Synthesis of oligosaccharides by bacterial enzymes. *Glycoconj J.* 16:141-146.
Kaplan H, Hutkins RW (2003). Metabolism of fructooligosaccharides by *Lactobacillus paracasei* 1195. *Appl Environ Microbiol.* 69:2217-2222.
Katan MB (2008). The probiotic yogurt Activia shortens intestinal transit, but has not been shown to promote defecation. *Ned Tijdschr Geneeskd.* 152:727-730.
Katechaki E, Panas P, Rapti K, Kandilogiannakis L, Koutinas AA (2008). Production of hard-type cheese using free or immobilized freeze-dried kefir cells as a starter culture. *J Agric Food Chem.* 56:5316-5323.
Keceli T, Robinson RK, Gordon MH (1999). The role of olive oil in the preservation of yogurt cheese (labneh anbaris). *Int J Dairy Technol.* 52:68-72.
Koroleva NS (1988). Technology of kefir and kumys. *Bulletin of the International Dairy Federation (Belgium)*, No. 227, Federation Internationale de Laiterie.
Kourkoutas Y, Kandylis P, Panas P, Dooley JSG, Nigam P, Koutinas AA (2006). Evaluation of freeze-dried kefir coculture as starter in feta-type cheese production. *Appl Environ Microbiol.* 72:6124-6135.
Kucukoner E, Tarakci Z, Sagdic O (2006). Physicochemical and microbiological characteristics and mineral content of herby cacik, a traditional Turkish dairy product. *J Sci Food Agric.* 86:333-338.
Kurmann JA, Rasic JL, Kroger M (1992). Encyclopedia of fermented fresh milk products.
Lee Y-K, Salminen S (1995). The coming of age of probiotics. *Trends Food Sci Technol.* 6:241-245.
Lucey JA, Tamehana M, Singh H, Munro PA (1999). Stability of model acid milk beverage: Effect of pectin concentration, storage temperature and milk heat treatment. *J Texture Stud.* 30:305-318.
Luckow T, Delahunty C (2004a). Consumer acceptance of orange juice containing functional ingredients. *Food Res Int.* 37:805-814.
Luckow T, Delahunty C (2004b). Which juice is "healthier"? A consumer study of probiotic non-dairy juice drinks. *Food Qual Prefer.* 15:751-759.
Malacarne M, Martuzzi F, Summer A, Mariani P (2002). Protein and fat composition of mare's milk: Some nutritional remarks with reference to human and cow's milk. *Int Dairy J.* 12:869-877.
Martínez-Villaluenga C, Frías J, Vidal-Valverde C (2006). Functional lupin seeds (*Lupinus albus* L. and *Lupinus luteus* L.) after extraction of α-galactosides. *Food Chem.* 98:291-299.
Mattila-Sandholm T, Blum S, Collins JK, Crittenden R, De Vos W, Dunne C, Fonden R, Grenov G, Isolauri E, Kiely B (1999). Probiotics: Towards demonstrating efficacy. *Trends Food Sci Technol.* 10:393-399.
McChesney JD, Venkataraman SK, Henri JT (2007). Plant natural products: Back to the future or into extinction? *Phytochemistry.* 68:2015-2022.
Mistry VV, Metzger LE, Maubois JL (1996). Use of ultrafiltered sweet buttermilk in the manufacture of reduced fat cheddar cheese. *J Dairy Sci.* 79:1137-1145.
Muir DD, Hunter, EA (1991). Sensory evaluation of cheddar cheese: Order of tasting and carryover effects. *Food Qual Prefer.* 3:141-145.
Nagpal R, Yadav H, Puniya AK, Singh K, Jain S., Marotta F. (2007). Potential of probiotics and prebiotics for synbiotic functional dairy foods: An overview. *Int J Probiotics Prebiotics.* 2:75.

Nilsson LE, Lyck S, Tamime AY (2006). Production of drinking products. In *Fermented Milks*, Ed. A. Y. Tamime, Oxford: Blackwell Publishing. 95–127.

Noviendri D, Hasrini RF, Octavianti F (2011). Carotenoids: Sources, medicinal properties and their application in food and nutraceutical industry. *J Med Plants Res.* 5:7119–7131.

Olson D, Aryana K (2008). An excessively high *Lactobacillus acidophilus* inoculation level in yogurt lowers product quality during storage. *LWT - Food Sci Technol.* 41:911–918.

Ong L, Shah NP (2008). Influence of probiotic *Lactobacillus acidophilus* and *L. helveticus* on proteolysis, organic acid profiles, and ACE-inhibitory activity of cheddar cheeses ripened at 4, 8, and 12°C. *J Food Sci.* 73:M111–M120.

Ozdemir U, Kilic M (2004). Influence of fermentation conditions on rheological properties and serum separation of ayran. *J Text Stud.* 35:415–428.

Ozer B, Kirmaci HA (2010). Quality attributes of yogurt and functional dairy products. In *Development and Manufacture of Yogurt and Other Functional Dairy Products.* Ed. F. Yildiz. Boca Raton, FL: CRC Press.

Pagliarini E, Solaroli G, Peri C (1993). Chemical and physical characteristics of mare's milk. *Ital J Food Sci.* 323–332.

Panesar PS, Kaur G, Panesar R, Bera MB (2009). Synbiotics: Potential dietary supplements in functional foods. *FST Bulletin.* Food Science Central, IFIS Publishing, U.K.

Pantaleao A, Moens E, O'Connor C (1990). The technology of traditional milk products in developing countries. *FAO Animal Production and Health Paper.* 85:333.

Prado FvC, Parada JL, Pandey A, Soccol CR (2008). Trends in non-dairy probiotic beverages. *Food Res Int.* 41:111–123.

Prakash S, Urbanska AM (2007). Fermented milk product and use thereof. US-PTO. United States, McGill University.

Rakin M, Vukasinovic M, Siler-Marinkovic S, Maksimovic M (2007). Contribution of lactic acid fermentation to improved nutritive quality vegetable juices enriched with brewer's yeast autolysate. *Food Chem.* 100:599–602.

Ray B, Bhunia A (2007). *Fundamental Food Microbiology.* Boca Raton, FL: CRC Press.

Rea MC, Lennartsson T, Dillon P, Drinan FD, Reville WJ, Heapes M, Cogan TM (1996). Irish kefir-like grains: Their structure, microbial composition and fermentation kinetics. *J Appl Bacteriol.* 81:83–94.

Roberfroid MB (2001). Prebiotics: Preferential substrates for specific germs? *American J Clin Nutr.* 73:406s–409s.

Robinson R, Tamime A. (1986). Recent developments in yoghurt manufacture. In *Modern Dairy Technology.* Vol. 2. Ed. Robinson RK. London: Elsevier Applied Science Publishers. 1–34.

Robinson RK (1991). *Therapeutic Properties of Fermented Milks.* London: Elsevier Applied Science Publishers.

Rogelj I (1994). Lactic acid bacteria as probiotics. *Mljekarstvo.* 44:277–284.

Salji J (1992). Acidophilus milk products: Foods with a third dimension. *Food Sci Technol Today.* 6:142–147.

Santos A, San Mauro M, Sanchez A, Torres JM, Marquina D (2003). The antimicrobial properties of different strains of *Lactobacillus* spp. isolated from kefir. *Syst Appl Microbiol.* 26:434–437.

Seiler H (2003). A review: Yeasts in kefir and kumiss. *Milchwissenschaft.* 58:392–396.

Shah N (2004). Probiotics and prebiotics. *Agro Food Ind Hi-Tech.* 15:13–16.

Shah NP (2007). Functional cultures and health benefits. *Int Dairy J.* 17:1262–1277.

Shah NP (2017). *Yogurt in Health and Disease Prevention.* Academic Press.

Simova E, Beshkova D, Angelov A, Hristozova T, Frengova G, Spasov Z (2002). Lactic acid bacteria and yeasts in kefir grains and kefir made from them. *J Ind Microbiol Biotechnol.* 28:1–6.

Sip A, Grajek W (2010). Probiotics and prebiotics. *Funct Food Prod Dev.* 2:146.

Stanton C, Gardiner G, Lynch P, Collins J, Fitzgerald G, Ross R (1998). Probiotic cheese. *Int Dairy J.* 8:491–496.

Statista (2018). Sales of functional foods worldwide from 2015 to 2024, by region (in billion U.S. dollars). Retrieved 14/04/2018, from https://www.statista.com/statistics/253083/global-functional-food-sales-by-country/.

Surono IS, Hosono A (2003). Fermented milks: Types and standards of identity. In *Encyclopedia of Dairy Sciences.* Vol. 2. Eds. Roginsi H, Fuquay JW, Fox PF. London: Academic Press. 1018–1023.

Taillie LS, Ng SW, Xue Y, Busey E, Harding M (2017). No fat, no sugar, no salt... No problem? Prevalence of "low-content" nutrient claims and their associations with the nutritional profile of food and beverage purchases in the United States. *J Acad Nutr Diet.* 117(9):1366–1374. e1366.

Talwalkar A, Kailasapathy K (2003). Effect of microencapsulation on oxygen toxicity in probiotic bacteria. *Aust J Dairy Technol.* 58:36.

Tamime AY, Marshall VM, Robinson RK (1995). Microbiological and technological aspects of milks fermented by bifidobacteria. *J Dairy Res.* 62:151–187.

Tamime AY, Robinson RK (1988). Fermented milks and their future trends. Part II. Technological aspects. *J Dairy Res.* 55:281–307.

Tepper BJ, Trail AC (1998). Taste or health: A study on consumer acceptance of corn chips. *Food Qual Prefer.* 9:267–272.

Toba T, Abe S, Adachi S (1987). Modification of KPL medium for polysaccharide production by *Lactobacillus* sp. isolated from kefir grain. *Jpn J Zootech Sci.* 58:987–990.

Tuorila H, Cardello AV (2002). Consumer responses to an off-flavor in juice in the presence of specific health claims. *Food Qual Prefer.* 13:561–569.

Tuorila H, Cardello AV, Lesher LL (1994). Antecedents and consequences of expectations related to fat-free and regular-fat foods. *Appetite.* 23:247–263.

Tyopponen S, Petaja E, Mattila-Sandholm T (2003). Bioprotectives and probiotics for dry sausages. *Int J Food Microbiol.* 83:233–244.

Uysal Pala C, Karagul Yuceer Y, Pala A, Savas T (2006). Sensory properties of drinkable yogurt made from milk of different goat breeds. *J Sens Stud.* Manufacturing Yogurt and Fermented Milks Blackwell Publishing. 21:520–533.

Vinderola C, Mocchiutti P, Reinheimer J (2002). Interactions among lactic acid starter and probiotic bacteria used for fermented dairy products. *J Dairy Sci.* 85:721–729.

Wang J-F, Wei D-Q, Chou K-C (2008a). Drug candidates from traditional chinese medicines. *Curr Top Med Chem.* 8:1656–1665.

Wang J, Chen X, Liu W, Yang M, Zhang H (2008b). Identification of Lactobacillus from koumiss by conventional and molecular methods. *Eur Food Res Technol.* 227:1555–1561.

Wang Y-C, Yu R-C, Chou C-C (2002). Growth and survival of bifidobacteria and lactic acid bacteria during the fermentation and storage of cultured soymilk drinks. *Food Microbiol.* 19:501–508.

WHO (2001). Evaluation of health and nutritional properties of powder milk and live lactic acid bacteria. *Food and Agriculture Organization of the United Nations and World Health Organization Expert Consultation Report.* 1–34.

Wood BJB, Hodge MM (1985). Yeast-lactic acid bacteria interactions and their contribution to fermented foodstuffs. In *Microbiology of Fermented Foods.* Vol. 1. Ed. Wood BJB, London: Elsevier Applied Science Publishers. 263–293.

Wszolek M, Kupiec-Teahan B, Guldager HS, Tamine AY (2006). Production of kefir, koumiss and other related products. In *Fermented Milks*, Ed. Tamime AY, Oxford: Blackwell. 174–216.

Yadav D, Kumar N (2014). Nanonization of curcumin by antisolvent precipitation: Process development, characterization, freeze drying and stability performance. *Int J Pharm.* 477:564–577.

Yadav K, Yadav D, Yadav M, Kumar S (2015a). Noscapine-loaded PLA nanoparticles: Systematic study of effect of formulation and process variables on particle size, drug loading and entrapment efficiency. *Pharm Nanotechnol.* 3:134–147.

Yadav K, Yadav D, Yadav M, Kumar S (2015b). Noscapine loaded PLGA nanoparticles prepared using oil-in-water emulsion solvent evaporation method. *J Nanopharm Drug Deliv.* 3:97–105.

Yang, X, Jiang Y, Yang J, He J, Sun J, Chen F, Zhang M, Yang B (2015). Prenylated flavonoids, promising nutraceuticals with impressive biological activities. *Trends Food Sci Technol.* 44:93–104.

Yaygin H (1992). *Who and What.* Antalya, Turkey: New Printing House.

Yoon KY, Woodams EE, Hang YD (2004). Probiotication of tomato juice by lactic acid bacteria. *J Microbiol (Seoul, Korea).* 42:315–318.

Yoon KY, Woodams EE, Hang YD (2006). Production of probiotic cabbage juice by lactic acid bacteria. *Bioresour Technol.* 97:1427–1430.

Yuksekdag ZN, Beyath Y, Aslim B (2004a). Metabolic activities of *Lactobacillus* spp. strains isolated from kefir. *Mol Nutr Food Res.* 48:218–220.

Yuksekdag ZN, Beyatli Y, Aslim B (2004b). Determination of some characteristics coccoid forms of lactic acid bacteria isolated from Turkish kefirs with natural probiotic. *LWT - Food Sci Technol.* 37:663–667.

4

Natural Ingredients/ Botanical Extracts for the Nutraceutical Industry

Rahul Maheshwari, Kaushik N. Kuche, Ashika Advankar,
Namrata Soni, Nidhi Raval, Piyoosh A. Sharma,
Muktika Tekade, and Rakesh Kumar Tekade

Contents

4.1 Introduction 76
4.2 The global nutraceutical industry..77
4.3 Nutraceutical markets for plant extracts 77
    4.3.1 The European market 77
    4.3.2 The U.S. market...78
4.4 Classification of nutraceuticals based on natural sources 78
    4.4.1 Plants 78
    4.4.2 Minerals 80
        4.4.2.1 Iodine 80
        4.4.2.2 Iron...81
        4.4.2.3 Zinc 81
        4.4.2.4 Manganese 82
    4.4.3 Microbes 82
    4.4.4 Marine..84
        4.4.4.1 Carbohydrates 84
        4.4.4.2 Proteins 88
        4.4.4.3 Peptides 88
        4.4.4.4 Fatty acids 88
        4.4.4.5 Phenolic compounds and prebiotics 89
        4.4.4.6 Enzymes, vitamins, and minerals 89
4.5 Methods for extracting herbal products..90
    4.5.1 Maceration 90
    4.5.2 Percolation 91
        4.5.2.1 Modifications made to the percolation process..............92
        4.5.2.2 Types of percolators 92
    4.5.3 Infusion 93
    4.5.4 Decoction..95

4.5.5 Continuous hot extraction 95
    4.5.5.1 Description 95
    4.5.5.2 Process 96
    4.5.5.3 Advantages and disadvantages 97
4.5.6 Aqueous alcoholic extraction 97
4.5.7 Countercurrent extraction 98
    4.5.7.1 Applications 98
4.5.8 Distillation 98
    4.5.8.1 Types of distillation for extraction 99
4.5.9 Supercritical fluid extraction 101
    4.5.9.1 Components 101
    4.5.9.2 Applications 103
4.5.10 Microextraction techniques 103
    4.5.10.1 Solid-phase microextraction 103
    4.5.10.2 Stir bar sportive extraction (SBSE) 104
    4.5.10.3 Liquid-phase microextraction 105
4.6 Common plant/herbal extracts and their therapeutic applications 106
4.7 Ongoing clinical trials on natural ingredients/botanical extracts 106
4.8 Future prospects 106
4.9 Acknowledgments 112
References 113

4.1 Introduction

The term nutraceutical is a combination of two words, nutrition/nutrients (a nourishing food component) and pharmaceutical (medicine or a substance used as a medication). The term was coined in 1989 by Stephen DeFelice, founder and chairman of the Foundation for Innovation in Medicine (Wildman 2016). Nutraceuticals have health and medical benefits, usually derived from plant components, that extend beyond basic nutrition.

The philosophy behind nutraceuticals is to describe and use nutrient products derived from food sources and food ingredients. The focus is on prevention. The Greek physician Hippocrates, known as the father of medicine, said, "Let food be your medicine." The modern world is again embracing the concept of healthy living and healthy nutrition using food from natural sources with pharmaceutical properties and biochemical compounds of the food products to maintain good health and prevent and treat disease (Kim and Wijesekara 2013).

Nutraceuticals are products derived from foods that are generally marketed in medicinal forms. Nutraceuticals are a way to minimize use of expensive, high-tech medical treatments currently employed in developed countries. Dietary supplements consist of nutrients derived from food and often utilized in liquid, capsule, powder, or pill form. The regulation of dietary supplements differs from the regulation of drugs and other food products but the U.S.

Food and Drug Administration (FDA) oversees the regulation of all of these products (Ronis et al. 2017).

Food products and ingredients of animal and vegetable origin are not governed by the same regulatory accountability as pharmaceuticals. There is nominal regulation over the words used on product labels. There are two major nutraceutical product types: functional foods, and vitamins, minerals, and herbal supplements (Eaglstein 2014).

Major issues with nutraceuticals are their low bioavailability and poor solubility in the digestive tract, which leads to their incomplete absorption from the gut. In recent years nanotechnology-based approaches have addressed these problems and developed approaches that modified the physicochemical properties of the active substances in many drug formulations and nutraceutical-based formulations (Soni et al. 2016, 2017; Maheshwari et al. 2015b; Maheshwari et al. 2012). Researchers have developed nanosystems to meet the desired release profiles from the novel carrier systems with the capability to transport the drugs and herbs (Maheshwari et al. 2015a; Kumar Tekade et al. 2015; Sharma et al. 2015; Lalu et al. 2017).

4.2 The global nutraceutical industry

In recent years there has been a growing appreciation of the role of nutraceuticals and dietary supplements in preventing health risks and improving overall health. The nutraceutical and functional food industry is now a multi-billion-dollar industry. It experienced tremendous growth in the last decade. It is estimated that it will grow up to $578.2 billion worldwide by 2025 (da Costa 2017).

The industrial aspect of nutraceuticals began in the early 1990s. From 1999 to 2002 the industry advanced at an annual average growth rate of 7.3%, which then doubled to 14.7%. It is estimated that it will rise to $278.96 billion by 2021 (Shao 2017).

Personalization and customization are current trends in the development of nutraceuticals, especially in developed markets across the globe. Main strategies for the industry include R&D ventures to procure innovative approaches, confirmation of health claims of nutraceutical products, and market research (Shao 2014).

4.3 Nutraceutical markets for plant extracts

4.3.1 The European market

The European market accounted for $35 billion in neutraceutical revenue in 2010. Research studies estimated that the worldwide market, which was

$151 billion in 2011, would grow at a rate of 6.5% by 2020 (da Costa 2017). This growth rate is perhaps exaggerated for the European market, but it has been estimated to increase by 5.8% per annum over the same period in the third-largest global market, the United States. Subsequently, the market value for Europe will continue to grow up to $65 billion by 2020. The global market will have matured to $207 billion by 2016 (Roccato 2016).

The European nutraceutical industry is quite mature. It depends on the amalgamation of food and pharmaceuticals. Even large companies are investing in the future of worldwide dietary supplements. Undeniably, large food companies are channeling their energy in a seemingly pharma way. Well-known players such as Nestlé and Danone have broadened their portfolios to include advanced functional food products. These companies have sales teams tasked with selling products to physicians and sponsoring studies and startups, side stepping pharmaceuticals altogether (Williams and Nestlé 2015).

4.3.2 The U.S. market

The United States is the largest nutraceutical market in the world. Nutraceutical companies are turning increasingly to natural components in their products. This is the result of a push from U.S. consumers, who are extremely concerned about their health and demand specific natural ingredients in the products they consume. The United States is the market's big gun in terms of both share and maturity. The U.S. market share was 39% in 2017 (Wildman 2016). It is assumed the United States will continue to maintain its position,

The United States, like other regions, is likely to experience a surge due to shifts in product offerings and marketing in response to consumer needs. The rise in U.S. healthcare cost has caused increased anxiety, and consumers are seeking alternative pathways to address health concerns. Nutraceuticals are consumed in great quantities by elder populations. The market opportunity in the United States now lies with the younger generations, who have different need perceptions of healthy living (Chauhan et al. 2013).

4.4 Classification of nutraceuticals based on natural sources
4.4.1 Plants

Lifestyles worldwide have transformed in the last century due to rise in income, limited physical activity, and consumption of junk foods. Different health problems related to improper selection, planning, and consumption of foods in the regular diet as well as lack of time have increased interest in the preparation of nutrient-dense food products (Lang and Heasman 2015). As a result, there is a growth in diseases such as obesity, cardiovascular disease, diabetes mellitus, and rheumatoid arthritis, and a similar increase in health awareness and interest in herbal alternatives worldwide. Herbal nutraceuticals offer efficient resources for disease prevention (Jafari 2017). They are the earliest source of flavorings, aromatic compounds, and medicines. Nutraceuticals play a dual role as food and therapeutic agent in disease prevention and treatment. They are natural and have virtually no side effects. Nutraceuticals obtained from plants may vary from single isolated nutrients, dietary supplements, and secondary metabolites to genetically engineered designer foods. Soybeans have received great attention as a source of protein and oil. Soybean proteins are gaining in importance as a vegetable source of protein-containing products with essential amino acids (Carbonaro et al. 2015).

The importance of natural health and the use of herbs and herb products have increased since 1960. Herbal bioactives, a significant type of nutraceuticals with vitamins, minerals, and other active components, have health-promoting medicinal properties (Williams et al. 2015). Phytochemicals are natural bioactive molecules found in plants—vegetables, fruits, medicinal plants, flowers, leaves, and roots—which act as a protective system to fight disease with nutrients and fiber (Prakash and Sharma 2014).

The effects of such mixtures may be due to the synergistic action of various phytochemicals. Phytochemical-based foods have important relevance in modern medicine as well as a strong historical background. These biomolecules are applied in medicinal and commercial industries in nutraceuticals and food additives. The plant-based diet phytoconstituents are macro- and micronutrients, e.g., proteins, fats, carbohydrates, vitamins, and minerals, which are essential for regular metabolism and may also play a compelling role in health enhancement (Girdhar et al. 2017).

Carotenoids (isoprenoids) are biomolecules in different fruits and vegetables that may have anticancer properties, enhance natural killer immune cells, and protect the cornea from UV light (Cardoso et al. 2017). Legumes, such as chickpeas and soybeans, grains, and palm oil, contain noncarotenoid substances and are anticarcinogenic. They also reduce cholesterol. Flavonoid and polyphenolic substances are found in berries, fruits, vegetables, and legumes. They have potent antioxidant and diuretic properties, prevent breast and prostate cancer, and control diabetes. For example, flavonoids such as apiol and psoralen, found in parsley (*Petroselinum cripsum*), have diuretic, carminative, and antipyretic effects (Prameela et al. 2017). Nonflavonoid polyphenolics are found in dark grapes, raisins, berries, peanuts, and turmeric root. They may have significant anti-inflammatory and antioxidant effects as well as effective anticlotting and hypolipidemic properties (Ananga et al. 2017).

Tannin found in lavender (*Lavandula angustifolia*) is useful in treating depression, hypertension, stress, colds, cough, and asthma. Proanthocyanidin in cranberries (*Vaccinium erythrocarpum*) has been used in ulcer, urinary tract infection, and cancer treatments. Menthol derived from peppermint (*Mentha piperita*) is effective in treating colds, cough, and asthma (Kumar and Zandi 2014).

Glucosamine from ginseng, omega-3 fatty acids from linseed, *Epigallocatechin gallate* from green tea, and lycopene from tomato are accepted phytonutraceuticals with multiple therapeutic benefits. The targets of the flourishing plant biotechnology industry are an enhancement of the edible nutritional values of fruits, vegetables, and other crops, and augmentations of the bioactive components in herbal plants (Farzaneh and Carvalho 2015). Various nutraceuticals are important components of medicinal herbal plants, dietary nutrients linked with disease prevention. Bioactive molecules are usually concentrated in all plant cells as secondary metabolites; however, concentrations of secondary metabolites differ according to plant parts. The leaf, for example, has the highest accumulation of these biomolecules. Some bioactive molecules inhibit microbial growth that causes disease. They either act alone or in combination with other components (Maheshwari et al. 2015b).

4.4.2 Minerals

Minerals found in plant, animal, and dairy products help prevent osteoporosis and anemia. They build strong bones, teeth, and muscles, and improve nerve impulses and heart rhythm (Palacios et al. 2017). Minerals are structural components and elements necessary for important functions in the body and cell transport, and they serve as various catalytic metalloenzyme cofactors in a wide range of metabolic processes. Trace elements help build the binding sites of metalloenzymes, where every element has a precise role in the body and most have many helpful functions (Ananga et al. 2017). Trace elements are found in seaweed.

4.4.2.1 Iodine

Dietary iodine is important for the management of thyroid hormones, thyroxine and triiodothyronine, which balance many important physiological processes in humans (Suleria et al. 2015b). More than 1.9 billion individuals worldwide have poor iodine nutrition. Iodine deficiency is lowest in America and highest in Europe. Iodine deficiency affects growth and development due to inadequate production of thyroid hormones. Consequences of iodine deficiency include goiter, a rise in hypothyroidism in moderate-to-severe iodine deficiency or decreased hypothyroidism in mild iodine deficiency, and increased susceptibility of the thyroid gland to nuclear radiation (Aronson 2017). Abortion, stillbirth, congenital anomalies, perinatal and infant mortality, or endemic cretinism may occur in neonates. Iodine deficiency during childhood and adolescence may cause a delay in physical development and impairment of mental function or iodine-induced hyperthyroidism in adults as well. In severe iodine deficiency, hypothyroidism and developmental brain damage are the dominant disorders. Excess iodine may result in thyrotoxicosis and be connected with hyperthyroidism, euthyroidism, hypothyroidism, or autoimmune thyroid disease. The thyroid has adaptation mechanisms that regulate thyroid hormone synthesis and secretion and protect from thyrotoxicosis (Bhurtyal 2015).

4.4.2.2 Iron

Iron is an essential element for humans because of its role in fundamental cell functions. Iron is the most abundant transition metal in the body. It takes part in the utilization of oxygen. As a component of numerous enzymes it alters many critical metabolic processes, including DNA synthesis and oxygen and electron transport (Raab et al. 2016). Approximately 60–70% of iron is bound to hemoglobin in circulating erythrocytes; 10% is present in the form of myoglobins, cytochromes, and iron-containing enzymes; and approximately 25% is stored as ferritins and hemosiderins. Iron deficiency is the most common nutritional disorder worldwide. This results mainly from excessive bleeding, but it may also be induced by plant-based vegan diets, which contain less bioavailable iron. Iron deficiency adversely affects cognitive performance, behavior, physical growth, immune status, and death from infections in all age groups (Ciccolini et al. 2017).

Iron-deficient humans have defective gastrointestinal functions, which can also alter patterns of hormone production and metabolism. Homeostatic mechanisms are essential for the prevention of aggregation of excess iron that is believed to generate oxidative stress by catalysis of a variety of chemical reactions involving free radicals, which could result in cell damage. Excess iron aggregation may cause cancer and increase cardiovascular risk. An increase in iron can be observed in some cases, including excessive dietary iron intake, inherited diseases such as idiopathic hemochromatosis and congenital atransferrinemia, or the medical treatment of thalassemia (Zheng et al. 2016).

4.4.2.3 Zinc

Zinc is one of the most important essential elements. It is present in zinc metalloenzymes and in protein domains such as zinc fingers (Sarris 2017). Zinc is vital for growth and development. It is a structural unit of biological membranes and plays a role in gene expression and endocrine function, DNA synthesis, RNA synthesis, and cell division. Zinc interacts with important hormones involved in bone growth and enhances the effects of vitamin D on bone metabolism.

Most zinc in the body (85%) is deposited in muscles and bones, 11% is in the skin and liver, and the remainder is in other tissues. Zinc is present in the brain. Zinc homeostasis disturbances are associated with diseases including diabetes mellitus. The alteration of zinc homeostasis in the brain may be involved in the manifestation of epileptic seizures. Zinc deficiency is seen in those whose diets contain an increased amount of phytate, a powerful chelator, and low protein. Zinc deficiency negatively affects the epidermal, central nervous, immune, gastrointestinal, skeletal, and reproductive systems (Sarris 2017).

Exposure to mounting levels of zinc and zinc-containing compounds may lead to adverse effects in the human gastrointestinal, hematological, and respiratory systems together with changes in the cardiovascular and neurological systems. Increased zinc intake leads to diarrhea, vomiting, and headache. Chronic zinc toxicity manifests as functional impairment in immunological response, reduced copper status, altered iron function, or cholesterol metabolism (Mullin et al. 2014; Tekade et al. 2018).

4.4.2.4 Manganese

Manganese is an essential trace element required for biological processes. The highest manganese levels are stored in tissues requiring more energy, such as the brain, and in the retina and the skin, which contains a large amount of melanin (Suleria et al. 2015b).

The liver, pancreas, kidneys, and bone contain high manganese concentration. Manganese is a factor in the metabolism of proteins, lipids, and carbohydrates, and performs as various enzyme cofactors. Manganese is required for maintaining normal immune function, blood sugar and cellular energy, reproduction, digestion, and bone growth. It also aids in defense mechanisms against free radicals, and together with vitamin K, it helps blood clotting and haemostasis. It is essential for the development and functioning of the brain (Prasad et al. 2017). A large portion of manganese is bound to manganese metalloproteins. Approximately 3–5% of ingested manganese is absorbed, and it is cleared from the blood by the liver and excreted in bile. Manganese absorption is influenced by the presence of other trace elements, phytate, and ascorbic acid. Manganese deficiency can lead to several conditions, including osteoporosis, epilepsy, impaired growth, poor bone formation and skeletal defects, abnormal glucose tolerance, and altered lipid and carbohydrate metabolism (Carrasco-González et al. 2017).

Manganese toxicity is linked with damaged ganglia structures. It leads to neuropsychiatric symptoms and behavioral dysfunction reminiscent of Parkinson's disease, the most common form of parkinsonism, caused by neurodegenerative disease, drugs, toxicants, and infections. High liver manganese content has been reported in alcoholic liver disease, and it may affect hepatic fibrogenesis (Suárez-Martínez et al. 2016).

4.4.3 Microbes

Microbial production by metabolic engineering has resulted in ecofriendly strategies aimed at nutraceutical manufacture of food products from simple carbon sources. In context, *Saccharomyces cerevisiae* microbial platforms are most extensively engineered for the manufacture of assorted and multi-form value-added chemicals such as prebiotics, phytochemicals, polyamino acids, and polysaccharides. Several health benefits are associated with this. When healthy bacteria are consumed in the gut, it also helps maintain the immune system (Krivoruchko and Nielsen 2015).

Microorganisms have long been manipulated to harvest food ingredients (Joana Gil–Chávez et al. 2013). Microorganisms have great potential for the treatment and prevention of diseases such as diabetes, atopic dermatitis, Crohn's disease, obesity, anemia, diarrhea, and cancer. Such microorganisms are possible sources of essential enzymes, antibiotics, vitamins, colors, immunosuppressants, hypocholesterolemic agents, enzyme inhibitors, and natural antioxidants (Liu et al. 2017).

The highly nutritional and ecofriendly Spirulina (*Arthrospira platensis*) has functional components such as phenolics, phycocyanins, and polysaccharides, with antioxidant, anti-inflammatory, immunostimulating, hypolipidemic, hypoglycemic, and antihypertensive properties. Spirulina is safe in healthy subjects, but eating habits likely affect the acceptability of foods that contain Spirulina. The combination of Spirulina and probiotics may constitute a novel approach to improving the growth of beneficial intestinal microbiota suggested by microbial-modulating in vitro (Ríos-Hoyo et al. 2017).

Lactic acid bacteria (LAB) *Streptococcus thermophilus* and *Lactococcus lactic* are employed all over the world in nutraceuticals as well as food products. LAB can strengthen fermented dairy products with folate by natural means, i.e., by avoiding external supplements (Mann et al. 2017). The probiotic preparation is nothing more than the live microorganisms that are beneficial to the host consuming it if taken in an appropriate dose. There has been an increase in the application of LAB in the fermentation of raw foods such as meat and vegetables on an industrial scale. The preparation of several nutraceuticals, including sugars such as tagatose, sorbitol, and mannitol, are possible due to metabolic engineering of LAB (Li and Shah 2017).

Bacterial nutraceuticals have a recognized biological advantage in human health. Nutraceuticals have been proven to diminish the risk of certain chronic diseases such as immune-regulatory diseases, diabetes, and hypertension (Kumar and Kumar 2015). Antioxidants, the important nutrient constituents in bacteria, are effective in treating and preventing a variety of chronic disorders. Polyphenolics such as tocopherols and tocotrienols are antioxidants thought to be involved in battling oxidative stress in individuals, a process allied with certain neurodegenerative ailments and some cardiovascular disorders. There are numerous forms of nutraceutical preparations, depending on the intended application, e.g., water-in-oil or oil-in-water suspensions, colloidal dispersions, solutions, gels, foams, creams, lotions, powders, suppositorie, and mousses (Del Ben et al. 2017).

The most critical cancer chemotherapeutic agents are microbial metabolites, including anthracyclines such as doxorubicin, epirubicin, daunorubicin, valrubicin, and pirirubicin. Bleomycin, actinomycin D, anthracenones (mithramycin, streptozotocin, and pentostatin), mitosanes (mitomycin C), enediynes (calcheamycin), taxol, and epothilones are also listed as acceptable products (Kamal-Eldin and Budilarto 2015). Taxol (paclitaxel), a successful nonactinomycete molecule, is synthesized and isolated from the endophytic fungi *Nodulisporium sylviforme* and *Taxomyces andreanae*. This compound constrains the rapid division of mammalian cancer cells by endorsing tubulin polymerization and altering the normal microtubule. Various microbial compounds such as tacrolimus (FK506) and *Cyclosporin sirolimus* (rapamycin), which are antifungal peptides, have shown activity in diminishing the immune response and are thus employed in liver, kidney, and heart transplants (Kamal-Eldin and Budilarto 2015).

4.4.4 Marine

Marine nutraceuticals are in high demand worldwide for their part in nutraceutical supplementation that reduces the frequency of routine illness. The investigation of marine-originated biomolecules has resulted in novel bioactive composites such as omega-3 fatty acids, taurine, carotenoids, siphonaxanthin, oligosaccharides of chitin, glucosamine, collagen, vitamins, enzymes, and minerals (selenium and iodine) as well as fucoidan. The use of marine nutraceuticals is growing rapidly in food and supplement industries due to their extensive beneficial properties (Kim 2013).

Hippocrates, the father of modern medicine, noted the beneficial effects of several marine invertebrates and their components on human health. The marine species represent a variety of bioactive molecules such as minerals, polysaccharides, fatty acids, vitamins, polyphenols, enzymes, probiotics, proteins, and peptides with widespread relevance as nutraceuticals in the food and supplement industries (Table 4.1).

4.4.4.1 Carbohydrates

Marine carbohydrates are considered vital organic components of marine molecules synthesized by photosynthetic organisms. They are energy sources for heterotrophic organisms (Pallela and Park 2014). Carbohydrates exist mainly in monosaccharide, disaccharide, and polysaccharide forms in the marine system. The polysaccharides are some of the richest bioactive biomolecules in marine organisms. Marine polysaccharides such as carrageenan, chitosan, fucoidan, chitin, and alginate have many pharmaceutical properties, such as immuno-stimulatory, anticoagulant, antioxidative, anticancer, antiviral, and antibacterial effects. They participate in cell proliferation as well as cell cycle, and modify various other metabolic pathways, which have been industrialized into efficient nutraceuticals. Due to atmospheric variations such as high

Table 4.1 Nutraceuticals Obtained from Marine Organisms

Marine Organisms	Examples	Bioactive Molecules Isolated	Applications	References
Marine algae	Microalgae	β-Carotene; vitamins H, C, A, E, B_6, B_2, B_1, and B_{12}; astaxanthin; polysaccharides; and polyunsaturated fatty acids	Food additives; incorporated into infant milk preparations and dietary supplements	Enzing et al. 2014
	Macroalgae/seaweed (red and brown seaweed)	Polyunsaturated fatty acids; proteins; L-α kainic acid; phenolics; pigments; furanone; phlorotannins; hydrocolloids (carrageenan and agar) and minerals; fucan; alginate; laminaran; carrageenan; and agar	Agarose production; peptides; improves protein digestibility; antihypertensive; hydrocolloids; food and cosmetics industries	Milledge et al. 2016
Marine fish	Salmon; cod; flounder; tuna; mullet; and anchovy	Proteins; unsaturated essential fatty acids; minerals such as calcium, iron, selenium, and zinc; fish collagen; and vitamins A, B_3, B_6, B_{12}, E, and D	Severe and chronic diseases such as viral infections, hypertension, cancer, and Alzheimer's disease; bone treatment as a substitute for mammalian collagen, which is known to be immunogenic	Khora 2013
Marine invertebrates	Sponges; molluscs; echinoderms; and crustaceans	Terpenoids; peptides; steroids; strigolactones, phenols; ether; and alkaloids	Antibacterial; antiviral; anthelmintic; antifungal; antihypertensive; anticancer; and immune modulatory	Suleria et al. 2015b
Sponges	Spicules and spongia	Sterols; terpenoids; macrolides; polyketides; polyphenolic compounds; peptides; and alkaloids	Cure various diseases such as Alzheimer's and several different types of cancers	Monaco and Quinlan 2014
Molluscs; echinoderms; sand crustaceans	Crabs; prawns; and shrimps	Vitamin A; specific proteins; eicosapentaenoic acid; omega-3 fatty acids	Improve value-added health food products	Kim and Himaya 2013
Sea cucumber	*Holothuria scabra*; *H. leucospilota*; *H. excellens*; and *H. atra*	Flavonoids; terpenoids; phenol; saponin; alkaloid; anthraquinone; and glycoside	Antioxidant	Santos et al. 2016

pressure or low temperature, the biomolecules extracted from marine organisms represent a huge unexploited pool of bioactive molecules that offer additional nutritional value to foods, preservatives, pigments, and flavors (Chen et al. 2016).

Various significant species of algae of nutritional interest include *Phaeophyceae*, brown algae: *Ascophyllum nodosum, Ecklonia cava, E. kurome, Laminaria digitata, Lessonia flavicans, Saccharina japonica, Sargassum horneri,* and *Undaria pinnatifida; Cholorophyta,* green algae: *Caulerpa racemosa, Codium fragile, C. pugniforme, Gayralia oxysperma, Monostroma latissimum, Ulva australis, U. conglobata,* and *U. lactuca; Rhodophyta,* red algae: *Cryptonemia crenulata, Grateloupia indica, Gigartina skottsbergii, Nemalion elminthoides, Nothogenia fastigiata, Pyropia haitanensis,* and *Schizymenia binderi.* These edible algae may be consumed directly as a part of the diet or indirectly in the form of nutraceutical and functional food supplement extracts (Claeys et al. 2014).

Several carbohydrates obtained from seaweed have anticoagulant properties. They also hinder thrombin by activating anti-thrombin III, thereby increasing clotting time equally in both intrinsic and extrinsic pathways. In fucans and fucoidans, the anticoagulant properties are due to the presence of sulfate, disulfate, or fucose. The greater the molecular weight, the more linear the backbone, which will further induce stronger anticoagulant activity. The marine polysaccharide laminaran shows its anticoagulant activity only when it is structurally modified by reduction, sulfation, or oxidation (Ruocco et al. 2016).

Spatoglossum schroederi is an S-galactofucan (brown seaweed) that shows good antithrombotic activity during in vivo study. Antithrombogenic property is also shown by the interfering blood coagulation-fibrinolytic system observed in the case of spirulan (Júnior et al. 2015).

Sulfated galactans and fucans obtained from *Ulva fasciata* are used as functional foods and nutraceuticals due to their anticoagulant effect. Marine waste materials such as shellfish waste from cockles (*Cerastoderma edule, Clinocardium nuttalli*), scallops (*Chlamys hastate*), whelks (*Buccinum undatum*), clams and mussels (*Mercenaria mercenaria, Mytilus galloprovincialis, M. edulis*), crustaceans [*Cancer pagurus* (crab), *Nephrops norvegicus* and *Homarus americanus* (lobster), and *Crangon crangon* (shrimp)], and oysters (*Crassostrea gigas, C. gryphoides*) can be manipulated for the removal of carbohydrate molecules, which can be utilized as nutritional constituents and animal feed (Ngo and Kim 2013).

Most marine algal polysaccharides are characterized as both potential dietary supplements and nutraceuticals because of the presence of antioxidant properties. Edible seaweed is considered dietary fiber and tends to reduce plasma LDL cholesterol, cholesterol, and triacylglycerol (TAG), which makes it a significant food for human consumption, in addition to nutraceutical and medicinal applications. Exopolysaccharides from cyanobacteria and marine carbohydrates such as algins can be employed for stabilizing emulsions and as bioflocculants in food. The marine polysaccharides, which are extracted from crustaceans, algae, and additional marine extracts such as hydrocolloids, fucans/fucanoids, carrageenans, and glycosaminoglycans have many physiological roles such as anti-inflammatory, antiproliferative, antiviral, anticoagulant, and antithrombotic activity (Ruocco et al. 2016).

Carrageenan and furcellaran are used to stabilize milk protein due to their reactivity to proteins. Gelidium, Gracilaria, Hypnea and Gigartina are red algae used as a main source of agar. Agar E406 has been used as a gelling agent and additive in the food industry. A recent study suggested that chitooligosaccharide (COS) has significant nutraceutical properties. It is used in food additives and dietary supplements because of its hypocholesterolemic, antidiabetic, and adipogenesis inhibition properties (Pallela and Park 2014).

Fucoidan, which is extracted from the alga *E. cava,* has been proven to reduce the levels of prostaglandin E2, nitric oxide, and cyclooxygenase-2. The green seaweed *Ulva rigid* and the marine dinoflagellate *Gymanganese odiniumimpudicum* synthesized polysaccharides further to trigger the synthesis of nitric oxide and immunostimulate the formation of cytokines within the macrophages (Ruocco et al. 2016).

Other biomolecules obtained from *Chlorella stigmatophora* (Chlorophyta), *Porphyridium* (Rhodophyta), *U. pinnatifida* (Phaeophyceae), and Phaeodactylum (Bacillariophyta) have shown immune-suppression activity in both in vitro and in vivo studies by hindering Th2 action. Tabarsa et al. demonstrated that the carbohydrate (polymer) extracted from *Codium fragile* (Chlorophyta) can help release nitric oxide upon its binding with protein moiety by triggering NF-κB and MAPK pathways (Tabarsa et al. 2013). Many algal polysaccharides play a vital role in generating an innate immune response because they can bind to toll-like receptor-4 or pattern recognition receptors. Chitin administered in the vascular system augments the release of cytokines by macrophages and thus enhances Th1 immunity and diminishes Th2 immunity (Gomes et al. 2017).

Marine polysaccharides such as DAEB polysaccharide, which is extracted from the green alga *U. intestinalis* (Chlorophyta), contains glucose, xylose, and galactose. In tests on mice it reportedly prevented carcinogenesis. Brown seaweed such as fucoidan fractions restrain leukemia growth but not the sarcoma developed in mice and unfractionated fucoidan from the algae. A nodosum showed the property of causing apoptosis in the HCT116 cell line, which are human colon tumor cells via initiation of caspases 3 and 9 and the PARP cleavage that encouraged a modification on mitochondrial membrane and its permeability (Foley et al. 2011).

Fucoxanthin was also found to encourage apoptosis in several cell lines such as LNCaP, PC-3, and DU 145, which are human prostate cancer cell lines; and apoptosis in the human leukemia cell line (HL-60) and human colon cancer cell lines such as DLD-1, Caco-2, and HT-29. Siphonaxanthinis restrains cell viability and encourages apoptosis in human leukemia along with colon cells, isolated from the marine green alga *Codiumfragile* (Maeda 2015).

4.4.4.2 Proteins

Several sources of marine origin such as fish (haddock, herring, tuna, cod, trout, hake, and pollock), molluscs, crustaceans, and extremophiles, such as seaweed and Dunaliella, are proteins with exclusive antioxidant, antimicrobial activity, gel-formation, film and foaming capacity, and anticoagulant properties. Collagen, gelatin, and albumin are considered universal marine proteins used in nutraceuticals. They are also hydrolyzed enzymatically for the production of bioactive peptides, which has great potential for applications in nutraceuticals. The marine protein protamine is used as a natural preservative in the food industry (Li-Chan 2015).

4.4.4.3 Peptides

Bioactive peptides consist of protein fragments 2–20 amino acids long. Such residues can be produced from initial parent protein by the process of digestion or other processing. The development of marine-derived pharmaceutical peptide composites mainly for antihypertensive and ACE inhibition function has led to an increase in focused research. Several proteins isolated from molluscs, fish, and crustaceans are considered among the wealthiest sources of marine bioactive molecules. A study done by Bourseau et al. (2009) demonstrated some novel peptides such as fish protein hydrolysatesin action on osteoporosis and Paget's disease by adhering to cell receptors, thereby augmenting calcium absorption (Šližytė et al. 2009). Collagen in pork and beef is widely utilized in the biomedical, food, cosmetic, and pharmaceutical industries (Bourseau et al. 2009).

4.4.4.4 Fatty acids

Algae and marine fishes are sources of polyunsaturated fatty acids consisting of ω-3 or ω-6 fatty acids. They are widely used in the food industry as nutraceuticals (Koskela et al. 2016). The main reason why marine-based nutraceuticals are attracting great attention is because of many unique features that are not found in formulations and perpetrations obtained from terrestrial resources. Fish (salmon, tuna, sardines, and herring), fungi (Phycomycetes), extremophiles, microalgae, macroalgae (Bryophyta, Rhodophyta), and krill are most common sources of marine oils. They offer various visual and neurodevelopmental health features, including treatment of arthritis and hypertension and reduced risk of cardiovascular problems (Abdullah et al. 2017).

4.4.4.5 Phenolic compounds and prebiotics

Phenolic compounds are recognized mainly as mechanisms of adaptation to oxidative stress found in marine algae. The most abundantly found polyphenols, phlorotannins with antioxidant activity found in marine brown algae, are used as active ingredients in nutraceuticals, whereas flavonoids contribute to the total phenolic content in green algae. The brown algal phlorotannin profile mainly consists of phloroglucinol, eckol, and dieckol (Wang et al. 2016).

Polyphenols and carotenoids can also be synthesized. This displays their antioxidant properties by increasing their use as nutraceuticals. Carotenoids consist of 40 carbon structures, which make them lipid soluble. They are used as natural pigments (Liu et al. 2016). Carotenoids such as fucoxanthin, β-carotene, and astaxanthin are used as nutraceuticals because of their high antioxidant capacity, which is synthesized within marine organisms. Antioxidants shield against additional reactive oxygen species and oxidative rancidity and peroxidation products such as hydroxyl radicals, superoxide anions, and hydrogen peroxide (H_2O_2), which are responsible for food spoilage. β-Carotene and astaxanthin preparations, which use *Haematococcus* and *Dunaliella* species, respectively, are on the market (Sweazea et al. 2017).

Siphonaxanthin demonstrates significant anti-angiogenic activity by suppressing endothelial cell proliferation in addition to human umbilical vein endothelial cell (HUVEC) tube formation. Ganesan et al. (2011) reported that siphonaxanthin induces sapoptosisin HL-60 cells through caspase-3 activation. This has been related to the enhancement of DR5 and GADD45α expression levels as well as the depression of Bcl-2 expression (Ganesan et al. 2011). GADD45α is considered a DR5 death receptor and a significant apoptosis regulator that enables cell cycle detention.

4.4.4.6 Enzymes, vitamins, and minerals

Enzymes used in the nutraceutical industry can modify other molecules in food ingredients that can reduce spoilage and improve shelf life, storage, processing, and food safety (Mcneil et al. 2013). Enzymes such as polyphenol oxidase (phendase, polyphenolase, catecholase, tyrosinase, catechol oxidase, and cresolase), lipase, chitinolytic enzymes and transglutamase, and red algal enzymes actively participate in starch degradation pathways (e.g, α-1,4-glucanase) that are isolated from marine origin. They have exceptional physical, chemical, and catalytic properties compared with their terrestrial counterparts. They can become inactivated at moderate temperatures and provide high catalytic activity at low temperatures. These enzymes are employed in food processing and food ingredients. They have high salt tolerance and specificity and varied properties and enhanced activity at mild pH (Aditya et al. 2017).

Several minerals and vitamins, including zinc, iron, manganese, and iodine, are essential for proper functioning of the cells and therefore play an important role in maintaining the body's homeostasis. They also act as cofactors in various metabolic pathways. Iodine can also be extracted from seaweed for the same use (Suárez-Martínez et al. 2016).

4.5 Methods for extracting herbal products

Every plant contains multiple bioactive constituents, with widespread applications in the fields of pharmaceuticals, cosmetics, flavors, fragrances, and pigments (Aulton and Taylor 2017). Processes been developed to extract the active constituents and utilize them at an industrial level. The following sections discuss the most widely used methods employed in extracting herbal products.

4.5.1 Maceration

The term maceration means softening. This process is employed in the formation of extracts, tinctures, and concentrated infusions. The drug is kept in contact with the menstruum (solvent) until the cellular structure becomes soft, and solvent can penetrate inside and dissolve the active constituent into it and diffuse out. This is a well-established method that has been used for years. It is an official method described in the Indian Pharmacopeia (Indian Pharmacopoeia, 2014).

This methodology places the material in contact with the suitable solvent, termed menstruum, in a closed vessel—to avoid menstruum evaporation and to prevent batch-to-batch variation—for a week with intermediate shaking. Then the menstruum is strained to collect the liquid. The residue, called marc, is further collected and pressed to remove the soaked menstruum (Pandey and Tripathi 2014). The expressed liquid obtained is then mixed with the strained liquid and clarified by filtration. The liquid should be well pulverized to avoid clarification. In the case of animal tissue andvegetables the vessel is kept closed for long periods so that menstruum can penetrate inside the cell wall and then solubilize the active constituents present within the cells and diffuse them out (Yalçınçıray and Anlı 2015).

The reason for intermediate or occasional shaking is to bring intra- and extracellular fluids into rapid equilibrium, which brings new menstruum onto the surface for additional extraction. An important factor in this process is the ratio of drug to menstruum, which ideally should be 1:10. There are three types of maceration: simple, modified, and multiple.

Simple maceration is the process used to extract components from roots, stems, and leaves. The final volume is not predetermined because regardless of the method used to press the marc—hydraulic press or manual press—the concentration remains same. Only the net yield will be affected. The volume and the concentration of drug will be the same. If the volume is adjusted equally then the concentration of product will vary (Pandey and Tripathi 2014).

In modified maceration, the drug is kept at a 4:5 ratio of menstruum and kept closed for 2–7 days with infrequent shaking. Then the volume is adjusted by filtering 1:5 of menstruum (Damre 2015).

Maximum extraction is done in the multiple maceration process. Menstruum is segregated in various segments so that volume used in each segment will be equal in each maceration. Often a single process of maceration is not enough, so the process is repeated two or three times so that extraction is effective. The drug is classified as double or triple maceration, depending on how many times the drug is macerated. In the case of double maceration, menstruum is segregated in two volumes (Hegde and Kesaria, 2013):

$$V_1 = \frac{V_t - V_r}{2} + V_r$$

where
V_1 = volume of menstruum required for the first maceration;
V_t = total volume of menstruum; and
V_r = volume of menstruum retained by drug or marc.

The volume needed for the second maceration is calculated by determining the difference between V_t and V_1, whereas V_r is calculated by performing a trial run with the known weight of the drug. In triple maceration, the volume of menstruum is divided into three equal parts:

$$\text{volume for 1st maceration} = \frac{\text{total vol of menstruum} - \text{vol retained by drug}}{3} + (\text{vol retained by drug})$$

The same formula is applied for the second run, but the volume of menstruum and volume retained is divided by 2 instead of 3. The vacuum maceration process involves a specially designed maceration assembly that is connected to the vacuum pipe. This assembly allows the permeability of menstruum deep inside the cells, thereby facilitating the extraction process (Tsai et al. 2016).

4.5.2 Percolation

Percolation refers to the traveling of solvent or any fluid through a porous medium. Percolation is employed in the process of extraction, enabling the menstruum to descend down through the powdered drug to absorb the active constituents from it and then drip down where it is collected in a container and then filtered and processed (Kapure et al. 2015).

In the process of percolation, the drug is pulverized into a suitably sized powder and moistened uniformly with menstruum for at least 4 h in a separate vessel. This is known as imbibition. Then paper is placed over the packed drug. Next clean sand is put on the surface so that the drug at the top surface is not disturbed. Menstruum is then added to saturate the drug, opening the knob at the bottom to permit the gas to pass and then closing it. A small layer of menstruum may cover the drug, which is then left undisturbed for 24 h. The knob at the bottom is opened, and menstruum is allowed to percolate. The same volume of fresh menstruum is added until 75% volume of the product is collected. The marc is then pressed, which increases the percolate level up to 78–92% of the final volume. The extract is then subjected to evaporation or concentration to obtain the finished product (Wenzel et al. 2017).

4.5.2.1 Modifications made to the percolation process

The percolation process can lead to several issues. As in the case of thermolabile drug constituents, evaporation of such a large volume of percolate results in the loss of its active constituents (Kataoka et al. 2017). Another issue arises with the alcohol–water mixture. When it is subjected to evaporation this causes preferential evaporation of alcohol, which disturbs the alcohol–water mixture equilibrium required for maintaining the drug in soluble form, thus leading to precipitation (Myszka and Trzaskowski 2017).

In addition to the normal percolation method, in the reversed percolation process evaporation completely removes the traces of water and then dissolves the mixture again in the reverse portion. This is acidic in nature so there is no chance of precipitation.

There is also a method known as the "cover and run-down" method. This includes a combination of maceration and percolation, but it cannot be applied to a system containing volatile or thermolabile constituents. It is used to separate methylated spirit that can be reused, thus increasing profit. In this process, after imbibition has been completed and the drug has been loaded inside the percolator, maceration takes place for 3–4 h, using alcoholic spirit to run off the liquid and obtain more menstruum. Excess alcohol is removed using evaporation under reduced pressure. Finally, the concentrate is diluted with ethanol and water to produce a precise concentration of active component (Myszka and Trzaskowski 2017).

4.5.2.2 Types of percolators

Percolators are classified based on the scale of percolation to be carried out (Nanjo et al. 2014).

4.5.2.2.1 Small-scale percolators for laboratory use
At this level, the processes involve manufacturing of concentrated preparations by using both maceration and percolation for extraction, which is then subjected to evaporation of solvents. Continuous extraction involves a combination of both processes (Maccoss et al. 2016).

4.5.2.2.2 Soxhlet apparatus In a small scale or laboratory-scale operation, a soxhlet is used in the continuous extraction process. The soxhlet apparatus consists of a flask, a soxhlet extractor, and a reflux condenser. The crude products are packed inside a thimble constructed of filter paper and then placed inside the extractor tube. Alternatively, the drug could be packed into the central extractor tube directly after completion of imbibition with the menstruum in a thimble, ensuring that the bottom outlet for the extract is not blocked. The solvent is placed in a flask and heated to boiling point. The vapors formed are allowed to pass through the condenser (Oladoja 2016).

The vapors are condensed into hot liquid, which drips down onto the drug. As it percolates down, the increased temperature of the solvent allows for easy extraction of constituents from the drug. As solvent level reaches the top of the siphon tube the whole percolate siphons over the back into the flask. This process continues until all the drug is extracted from the flask. This extraction process is a short series of macerations. This setup is discussed in depth in Section 4.5.5 on continuous hot extraction (Oladoja 2016).

4.5.2.2.3 Extractor method (British and Indian Pharmacopeias) The official extractor method is mentioned in monographs in both the British and Indian Pharmacopeias. This method is based on a continuous extraction procedure wherein the vapor formed by heating rises from the extraction chamber, passes through the drug container, is condensed in the reflux condenser, and drips back onto the drug, thereby extracting all the soluble active constituents along with it into the flask.

This method has several limitations. It is not applicable for thermolabile constituents. It can be used only with pure solvents or with solvent mixtures forming azeotropes. If ordinary binary mixture menstruum were used it would form a vapor composition different from what is actually present in liquid form (British Pharmacopoeia 2016; Indian Pharmacopoeia 2007).

4.5.2.2.3.1 Large-scale extractor In the large-scale extractor method, the drug is maintained on a perforated metal plate enclosed with a layer of sacking or straw, as depicted in Figure 4.1. The extractor has a removable lid with apertures for witnessing the flow of solvent, for stuffing the drug, and for running in the solvent. The percolator outlet has a tap and pipeline to eliminate the percolate for subsequent processing or to reuse it as a menstruum for the second percolator if the system is arranged in series. Copper percolators were used in small-scale opeations, but they have been replaced with stainless-steel or glass percolators (ElSohly et al. 1990).

4.5.3 Infusion

An infusion is a dilute solution containing the soluble active constituents from the crude drug. Infusions are generally prepared by macerating the drug

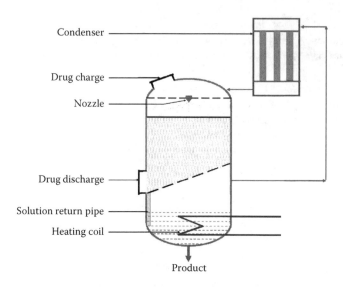

Figure 4.1 *A typical industrial continuous extraction system. Drug is placed in the container of a large-scale plant and solvent is evenly distributed using the nozzle. The solvent is heated, and as it percolates, the drug extract with solvent drips down and is again heated for concentration and collected. The vaporized solvent is condensed back and recirculated for continuous extraction.*

for a specified period with hot or cold water, depending upon the required specifications. Infusions are used to extract significant vitamins and volatile constituents from soft ingredients such as citrus peelings, flowers, and leaves. Infusions are made by diluting one part of a concentrated infusion to ten parts of water, based on volume. They are also prepared by employing modified versions of the percolation or maceration process followed by dilution with water. These will have the same potency and aroma as a fresh infusion. Infusions have high water content that supports rapid bacterial and fungal growth. They should be dispensed or consumed within 12 h of preparation (Lane et al. 2013).

Fresh infusions are subject to fungal and bacterial growth so it is recommended that they be made extemporaneously. A fresh infusion requires roughly 50 g of ground or coarse drug powder, which is then moistened with 50 ml of water in a suitable container with a cover. Approximately 900 ml of boiling water is added. The lid of the container is closed and kept undisturbed for approximately 30 min. The mixture is then strained. Sufficient water is added to make the volume 1 L (Azwanida 2015).

To prepare a concentrated infusion, one can refer to official monographs in the British and Indian Pharmacopoeias. A concentrated infusion contains almost 25% alcohol, which is added throughout the process. Such concentrated infusions contain active and necessary constituents from the crude drug that have the similar solubility in water as in the menstruum employed for infusions and concentrate (British Pharmacopoeia 2016; Indian Pharmacopoeia 2007).

4.5.4 Decoction

Infusion and decoction are often confused. We have already explained infusion. Decoction is employed to extract certain mineral salts or other bitter constituents of plants from the hard parts of plants like wood, seeds, roots, rhizomes, and bark. These hard parts are kept in contact with boiling water for at least 10 min and then stirred continuously for several hours (Martins et al. 2015).

In decoction, the hard drug is thoroughly mashed, then boiled in water to extract essential oils or other required volatile compounds. Decoction is used in the preparation of ayurvedic extracts called Quath or Kawath. For these preparations the ratio of drug to water is 1:4 if the drug is comparatively soft (leaf, stem, flower). The ratio is increased to 1:8 for moderately hard drugs (roots, rhizomes) and to 1:16 for a very hard drug (wood, bark). It is boiled until the quantity is reduced to 25% for soft drugs and 12.5% for moderately hard drugs (Cheung et al. 2017).

Quantity can also be determined based on the weight. For herbs less than 4 tolas (1 tola = 12 g) 16 times the water is required, whereas for herbs that weigh 5–16 tolas almost 8 times the water is required. If the herb weighs more than 16 tolas then 4 times the water is required. Thus, the ratio of the drug is a critical factor to consider while preparing a decoction because it decides the concentration of the final product formed (Cheung et al. 2017).

4.5.5 Continuous hot extraction

In continuous hot extraction the soxhlet system uses menstruum or the solvent for extraction. This results in a lower menstruum requirement and a higher amount of yield. This apparatus, invented by Franz von Soxhlet in 1987, was originally designed for extraction of lipid from solid material. This apparatus is used when the compound is subjected to low solubility in the selected menstruum (Plaza and Turner 2015).

4.5.5.1 Description

The continuous hot extraction process consists of a solvent/menstruum container, also called a still pot, which is a round bottom flask onto which the soxhlet assembly is attached. It contains a side arm that provides the path for the generated vapor of the heated menstruum so that it can move toward the condenser, which is attached above the assembly, as depicted in Figure 4.2. Below the condenser is a thimble into which the sample is to be placed. The thimble may be built in. If there is no thimble, the sample is packed in filter paper and placed in the soxhlet extractor. A syphon tube attached at the

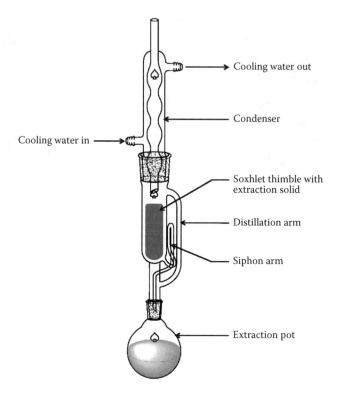

Figure 4.2 *Schematic drawing of the soxhlet apparatus.*

bottom of the soxhlet tube helps drain out the saturated menstruum back into the solvent container (Ramluckan et al. 2014).

4.5.5.2 Process

The process is simple. The sample/herb drug is kept in the thimble tube or packed inside the filter paper in the soxhlet extractor, ensuring that the opening to the syphon tube is not blocked. Once the assembly is set the solvent container is heated. Solvent vapor is formed which passes through the side tube and moves toward the condenser, which condenses the hot vapor via heat transfer and causes entropy loss, thus converting it back into the liquid state (Cheung et al. 2017).

The hot liquid drips down into the thimble. The hot condensed solvent continues to drip down into the thimble and comes in contact with the drug, extracting the soluble material out of the drug. As the process continues the solvent volume inside the soxhlet tube increases. As the level crosses the critical point of return the solvent drains back into the solvent container. The entire process is repeated until the extraction is complete. This is verified by identifying the concentration present in the solvent and weighing the material added in the soxhlet (Schmidt et al. 2014).

4.5.5.3 Advantages and disadvantages

One advantage of this process is that the sample is continuously exposed to fresh portions of extractant, which leads to the rapid establishment of transfer equilibrium. The additional step of filtration is avoided, resulting in an enhanced extraction rate and volume and reduced time and cost. The methodology involved in the soxhlet apparatus is simple and does not require additional training. The apparatus is capable of extracting more sample mass compared to other substitutions present (Subramanian et al. 2016).

There are several advantages associated with this method, but there are also drawbacks. Sample preparation requires a lot of time and the extraction process requires a long time to complete. Any attempt to do this in less time would likely result in unextracted waste, which could increase the overall cost. The samples, which are extracted continuously at the boiling point of the solvent for a considerable time, can eventually degrade the constituent. This method is not suitable for thermolabile compounds. Ultimately, the soxhlet technique is limited by its extractant and problems arising from process automation (Subramanian et al. 2016).

Such limitations associated with the soxhlet technique led to modifications to enhance the output from the extraction process. Most of the work has focused on tilting the soxhlation process and automating the assembly (Gunawan et al. 2008).

4.5.6 Aqueous alcoholic extraction

The aqueous alcoholic extraction process involves soaking crude drugs in the form of powder or a decoction for a specified period. This phase of incubation facilitates the in situ alcohol generation that extracts active constituents from the drug. This technique is employed in formulating various ayurvedic preparations such as asava and arishta, which utilize this fermentation technique to generate in situ alcohol for extraction. The alcohol generated can also act as a preservative (Caldeira et al. 2004). Earthen pots are usually used for fermentation. New earthen pots are not recommended. New pots must first be used to boil water, then they can be used for fermentation. Metal vessels, wooden vats, and porcelain jars are used in industry. Several ayurvedic formulations based on the same principle such as kanakasava, karpurasava, and dasmularista are available.

The method used for extraction is not standardized. Typically a drug in finely powdered form (in the case of asava) or a decoction drug (in the case of arishta) can be made and used. A substrate for fermentation such as jaggery or sugar is added to the powdered drug or filtered decoction. It is then allowed to ferment to produce alcohol. The vessel used should be cleaned to avoid contamination. The setup is kept in clean premises at a uniform temperature (Salem et al. 2016).

Some space must be left in the container to allow for gas formation from the fermentation. The lid should be kept open slightly so that gas formed can escape and to avoid an explosion. Once the process is done, the mixture must be filtered using a clean cloth, and the remaining particles should be discarded (Singh et al. 2016).

4.5.7 Countercurrent extraction

The countercurrent extraction process uses a slurry of drug that is properly ground in a disintegrator. The material to be extracted moves in one direction in the form of slurry and the extraction solvent moves in the opposite direction. When the extraction solvent comes in contact with the slurry, the solvent moves ahead as a more concentrated solvent extract, enabling a complete extraction if the ratio of solvent and its flow rate are optimized. This process is more efficient, less time consuming, and poses less risk at elevated temperature. Eventually, the extracts come out from one end. The marc has no visible signs of solvent, which was expelled from the opposite end (Rao and Arnold 1956).

4.5.7.1 Applications

The countercurrent method of extraction is employed for extracting soyabean oil using hexane as a solvent. It is also involved in the separation and purification of several biochemical compounds. Aside from the extraction of active constituents in inorganic chemistry it has been limited to separating radionuclides and for determining a new technique for tracing geological materials (Modolo et al. 2008).

4.5.8 Distillation

Distillation is a process in which the mixture of substances called feed is segregated into individual components using heat as the extrication agent. Distillation is used for extraction of essential oils, volatile oils, and sometimes cologne liquids (Katiyar 2017). This process of separation is based on differences in the boiling points of the constituents that are to be separated. The constituents with low boiling point tend to concentrate in vapor state, which is then separated and collected via condensers by providing a cold surface. The constituents with higher boiling points remain in the feed solvent itself, which leads to the separation of the required constituent. Work is underway to reduce operating costs and energy consumption to improve the efficiency of separation (Adnan et al. 2012).

Distillation may be a batch operation or a continuous operation. In a batch operation a limited quantity of feed is charged into columanganese. When the required amount of separation or extraction is done the columanganese is removed and cleaned. Then there is a further addition of feed into it. Each successive start for a new operational cycle is called a batch. In continuous operations the feed stream moves uninterruptedly through the columanganese and products are collected without any break in the process (Chen et al. 2014).

The type of columanganese selected depends upon the number of components to be separated. If two constituents are to be separated or extracted then a binary columanganese is selected. If more than two constituents are to be separataed, then a multi column is used (Katiyar 2017).

Columanganese is classified into two types, depending upon the ability of the columanganese to promote the phase contact: tray columanganese and packed columanganese. In tray columanganese, the liquid vapor equilibrium occurs by bubbling vapors arising from the liquid layer that are present on the lower tray inside the columanganese and that are specifically designed to promote the liquid vapor equilibrium. In the case of packed columanganese the liquid vapor equilibrium occurs throughout the columanganese. This is possible due to specifically constructed small bumps inside the columanganese. This results in the production of a greater number of theoretical plates by removing plates, thus ensuring better and purer extraction (Yuan et al. 2015).

4.5.8.1 Types of distillation for extraction

4.5.8.1.1 Hydrodiffusion Hydrodiffusion works on the principle of diffusion, which allows the generated steam to enter into the plant material. The essential oils or volatile oils are diffused out by osmosis and collected. A low-pressure steam passes from the boiler to the plant material, which causes the diffusion. It then passes through the condenser, which is tubular in construction,, cooling the oil and water, which are collected in a collector or in an oil separator. This process yields an efficient extraction but also has many co-extractions of other nonvolatile compounds, the separation of which complicates the process (Yusoff et al. 2016).

4.5.8.1.2 Water/hydrodistillation The water/hydrodistillation technique is the oldest and simplest of all techniques. It is mostly used by small-scale producers for extracting essential or volatile oil from plants/herbs. In hydrodistillation, the plant is completely immersed in water. Then a suspension is placed on a furnace that converts the water into vapor, rich in essential oil, which is carried toward the condenser and collected. This technique is best suited for extraction from powdered spices and herbs (properly pulverized). The best example of this method is Deg Bhabka, used in India, which employs copper stills for the process (Reda et al. 2017).

There are several drawbacks. The process is slow and thus requires more fuel. The rate of distillation is variable. It is not possible to predict whether the extraction is complete or not, but most of the time it is not. Heat is applied directly to the suspension, so chances of damaging or destroying the plant are high, leading to charring of plant tissues. Most important, the process takes a long time and it is not suitable for large-scale production (Reda et al. 2017).

4.5.8.1.3 Steam and water distillation Hydrodistillation is the most primitive distillation technique. It has several drawbacks. Certain modifications were made, which lead to a new system called wet steam distillation. In this technique, the plant is put on a fenestrated grid that is placed over boiling water. The steam that is generated comes in contact with the plants and extracts the essential oil out of the plants. This arrangement prevents direct contact of the plants with the heat source, thus preventing char formation. This modification resulted in advantages such as faster yield, lower fuel requirements, and reproducible yield (Katiyar 2017).

4.5.8.1.4 Direct steam distillation Direct steam distillation is also called dry steam distillation. The plant is again placed on the grid. Steam generated from an outside source such as a boiler is passed through it. The boiler is fitted with a steam coil that generates vapors that accumulate in the boiler, thus generating pressure. Figure 4.3 depicts this process. The advantage of this system is the high-pressure steam that raises the temperature to 100°C, which yields an efficient distillation more rapidly (Chemat and Boutekedjiret 2016). Fuel costs are low compared to traditional methods.

One advantage to this method is batches of 1–3 tons are possible. The steam can be controlled. No thermal degradation has been observed. It is suitable for large-scale applications.

4.5.8.1.5 Distillation with cohobation Several essential oils have limited solubility in water. Some oils such as rose oil are highly solubile in water. A great

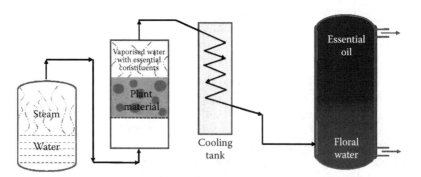

Figure 4.3 *Diagrammatic representation of the direct steam distillation process. Steam generated from the external boiler passes into the extraction tank containing herbal material. The vapors containing the volatile extract then pass through the cooling chamber and are collected.*

quantity of oil can be lost in water. This issue can be resolved by recirculating the condensed water from the separator back to the still. This same process of recirculation cannot be done for steam distillation. In that case, the water level would continue to increase due to continuous steam injection. In this assembly, the condenser is attached above the still. The water that is condensed reverts and drips into the still, limiting the total water content in the operation cycle. This enables us to increase the yield of extraction. Distillation with cohobation is widespread (Chemat and Boutekedjiret 2016).

4.5.9 Supercritical fluid extraction

Supercritical fluid extraction (SCF) is a process that uses fluid in the supercritical state for extracting the active constituent. The supercritical state is achieved when temperature and pressure applied are beyond the critical point such that the system exists in an equilibrium between liquid and vapor form. The extraction is generally made from a solid material, but sometimes even lipids can be extracted. This method may be employed in several applications such as analytical sample preparation or to extract essential or volatile oils from a natural resource on a large scale. For this CO_2 in the supercritical state is the most preferred gas. The extraction state for CO_2 is above 31°C and critical pressure is above 74 bars. Depending on the extraction specifications, the addition of certain modifiers can alter these parameters (Mchugh and Krukonis 2013).

The material to be extracted (solid feed) is added into the extractor or into the modified columanganese either cocurrently or countercurrently. The next step is solvent/mobile phase preparation. The mobile phase is prepared by mixing suitable solutes with supercritical solvents. Then the run parameters such as the mobile phase pressure are set to 50–500 atm. The temperature is set to 300°C which is above the critical point to improve the extraction capabilities of the solvent by enhancing the solubility of essential constituents in it. The widely used solvent for extraction is CO_2, which is compressed into liquid form and then used for extraction (Mendiola et al. 2013).

4.5.9.1 Components

There are many components to SCF. The basis is a cylinder that can withstand high pressure with pressure monitoring systems attached to pressure gauge and controlling knobs for CO_2 in its supercritical state. Pumps attached could be either a reciprocating pump or syringe type pump. The reciprocating type of pump has a pulsed flow. The syringe pump gives a pulse free flow with increased flow rates. The extraction columanganese is constructed of stainless steel which can withstand a pressure of 300–600 atm. For liquids, an open tubular capillary columanganese or packed columanganese is employed. For the extraction to be successful, the extraction solvent must be kept in the supercritical state. To do so, the pressure needs to be regulated well, hence restrictors are used (De Melo et al. 2014).

Restrictors used can be fixed or variable. Fixed restrictors used include linear restrictors, metal restrictors, ceramic frit restrictors, and integral restrictors. Variable restrictors have variable nozzles and back pressure regulators. The system is attached to the collector or trapping system with detectors to identify the extractant. Sometimes a CO_2 extract separator along with condenser is also connected to the system to recirculate and to re-use the solvent again, thereby yield a more efficient extraction process (De Melo et al. 2014). This process is shown in Figure 4.4.

Modes used are classified according to the way the supercritical state solvent is used for the extraction of constituent. In the static extraction mode the sample is soaked in a fixed amount of supercritical fluid, just like a tea bag is steeped in a cup of water (Sharif et al. 2014).

Another method of extraction is the dynamic extraction mode. This is a nonsteady process in which the supercritical fluid continuously passes through the sample mixture and results in extraction. An analogy for this process is how a coffee maker works. Certain parameters related to the material that can be used for extraction require particular attention to result in effective extraction. Other parameters that can affect the solubility of the essential

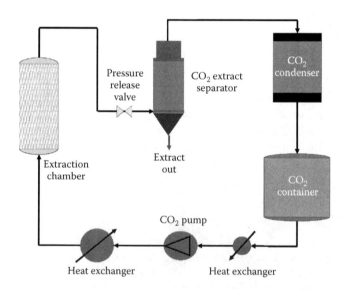

Figure 4.4 *Schematic representation of components involved in the supercritical extraction process. After the extraction using supercritical fluid the material is separated from the extract and collected. The extractant is then condensed and collected in a container and passed through the heat exchanger. It is then again brought into supercritical form and reused, thus increasing the extraction efficiency.*

constituent in the liquid are the vapor pressure of the component, interaction with a supercritical fluid, temperature, density, pressure, and other additives (Zhang et al. 2015).

4.5.9.2 Applications

This technique can be applied by using either single- or multiple-stage extraction. Dried ginger was extracted first with CO_2 at 30°C and 79 bar pressures. The second extraction at an elevated condition of 40°C and 246 bar yielded ginger flavor extract. Similarly, flavor extraction from coriander seeds used CO_2 initially at 40°C and a pressure of 250 bar as the first extraction, followed by a mild condition like 20°C and pressure of 70 bar to yield an efficient extract. Lipid extraction from egg yolk solvent used CO_2 along with 5% ethanol at 45°C and 414 bar for both extraction stages (Sharif et al. 2014).

4.5.10 Microextraction techniques

Extraction techniques are being developed at an increasingly rapid pace because they provide reduced analysis cost. Microextraction techniques are not exhaustive techniques. This means there is not complete extraction of the analyte from the compound. It depends instead on the equilibrium it attains with the extractant phase (Asfaram et al. 2016). The following sections discuss several microextraction techniques.

4.5.10.1 Solid-phase microextraction

The solid-phase microextraction (SPME) technique is based on the principle that a silica rod coated with apolymeric film (or fiber) acts as an extraction medium. It was developed by Arthur and Pawliszyn in 1990 (Arthur and Pawliszyn 1990). The working principle behind SPME is the partitioning of the analyte or constituents between the sample and the head space, which is above the sample. The sample is first allowed to get equilibrated between the sample and fiber. Then it is desorbed from fiber to the injector of the determination system (either gas chromatography (GC) or liquid chromatography (LC)) (Souza-Silva et al. 2015).

SPME can be performed via two methods: head-space solid-phase microextraction (HS-SPME), which is employed for volatile compound extraction, and direct immersion solid-phase microextraction (DI-SPME), which can be employed for extraction of non-volatile oil compounds. Figure 4.5 shows both processes. In the case of HS-SPME the fiber is not in contact with the sample, thus it has a longer shelf life compared to DI-SPME. After the extraction from fiber the sample is injected into the injection system of GC via thermal desorption, or into the desorption chamber for SPME-LC.

The widely used solvent for desorption is polydimethylsiloxanes (PDMS). It is thermally stable and commercially available in different grades depending

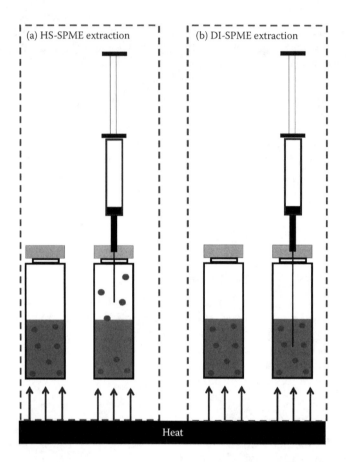

Figure 4.5 *Schematic views of HS-SPME extraction and DI-SPME extraction. (a) The HS-SPME extraction shows the transfer of the material from the head space using injection to transfer it to the detection system. (b) The DI-SPME extraction shows the extraction of sample by directly immersing it into the matrix and transferring it into the detection system.*

on viscosity. Other compounds such as carboxen, carbowax (the commercial name for polyethylene glycol), and divinylbenzene (DVB) have been employed for the same purpose. The advantage of this system is that it does not involve organic solvent and can be automated easily; however, it may take long hours for the equilibrium to occur. Other issues are great likelihood of cross-contamination and problems with reproducibility and repeatability (Souza-Silva et al. 2015).

4.5.10.2 Stir bar sportive extraction (SBSE)

The stir bar sportive extraction (SBSE) technique was initially developed by Baltussen et al. (Baltussen et al. 1999). In this technique, there is a stir bar coated with polydimethylsiloxane (PDMS), hence the name. The working principle for SBSE is similar to that of SPME, but the latter technique is more beneficial because the amount of extraction phase is 100 times higher. Thus, enhanced sensitivity can be expected but the equilibrium time can be slow.

In SBSE the material that is separated is absorbed on a magnetic rod and then coated with PDMS. Then the material or analyte is transported into the injection port of GC via thermal desorption. It may also be desorbed via liquid desorption. However, thermal desorption parameters depend completely on the volatility of the compound that is to be extracted. That should range from 150°C to 300°C. Similarly, in the case of liquid desorption the solvents used are acetonitrile or methanol. Desorption can be brought about by sonicating the complete mixture at room temperature for 5–10 min (He et al. 2014).

There are several advantages to this process that are related to reduction insolvent utility, which is 1 mL per sample or even less. Automation of the process is simple. The same bar can be reused subsequent to the systematic cleaning procedure. Methods are quite robust and reproducible and can thus be employed in the extraction of many compounds from both liquid samples and matrices (e.g., fruit pulp). Disadvantages include handling errors in liquid desorption techniques and long extraction times because the time required to achieve equilibrium is long. PDMS can only be used to extract nonpolar compounds thus restricting use to nonpolar compounds (Nogueira 2015).

4.5.10.3 Liquid-phase microextraction

Liquid-phase microextraction (LPME) is a miniaturized approach to liquid–liquid extraction (LLE) developed in 1996. This single-step extraction and enrichment technique uses only a few microliters of organic solvent and is thus an economic approach. It is considered a miniaturized version of the conventional LLE (Spietelun et al. 2014).

Single-drop microextraction (SDME) was the first technique developed using this approach. A drop of organic solvent known as acceptor is injected directly through a syringe into the liquid sample. Diffusion of analyte then begins in the drop, after which the solvent is extracted out and later injected into the chromatographic system. Here, the sensitivity is greater for the headspace mode compared to direct immersion although the efficiency of the procedure varies, based on the type of the solvent, the volume of the drop, the total extraction time, the rate of stirring, and the amount of salt added (Stanisz et al. 2014). SDME is mostly used for the extraction of volatile and semivolatile compounds because it allows speedy extraction, with an enrichment factor of more than 100. The enrichment factor can be defined as the ratio of the total analytical concentration of a solute in the extract (regardless of its chemical form) to its total analytical concentration in the other phase. Advantages of this technique include speed, cost effectiveness, and low solvent consumption. It has several issues related to its instability and poor reproducibility (Wu et al. 2013).

Hollow fiber liquid-phase microextraction (HF-LPME), developed by Pedersen-Bjergaard, uses porous membrane or fiber to immobilize solvent. Use of fibers allows easy agitation as well as selectivity compared to SDME. It can be a two- or three-phase system, depending on the solvent used. In the two-phase system extraction of the analytes is done through immobilization in the pores of the hollow fibers using the same organic solvent, whereas in the three-phase system immobilization uses another aqueous system. The two-phase system can be used for the analysis of hydrophobic compounds, whereas the three-phase system can be used for analysis of ionizable compounds (Wang et al. 2014). This type of technique can be used for extraction of pesticides from water sources and in the extraction of some drugs, hormones, and personal care products.

4.6 Common plant/herbal extracts and their therapeutic applications

As evident from the previous sections a single plant has multiple activities. A single dose of these plant extracts and nutraceuticals may improve well-being enormously. These activities have been studied and reported in the ethno-pharmacological literature, thus substantiating the accuracy of the claims of traditional medicine. The activities listed in Table 4.2 are based on both in vivo and in vitro studies performed by groups worldwide. We summarize nutraceutically important plants and herbs and their reported activities.

4.7 Ongoing clinical trials on natural ingredients/botanical extracts

Lack of clinical data is the major factor preventing the mix of traditional and herbal medicinal knowledge and modern medicine. This scenario has changed in recent decades with an increased number of clinical trials being conducted on herbal extracts and nutraceuticals. These clinical data help us establish the safety and efficacy of the extracts and optimize their use. Table 4.3 summarizes some recent representative clinical trials on natural ingredients/botanical extracts.

4.8 Future prospects

Despite an increase in the use of synthetic drugs and biologicals, plants are still the biggest source of novel molecules with varied structures. The nutraceutical industry has made its impact on most strata of society and has fueled a revolution in the healthcare industry. Many of us consume nutraceuticals as a part of our daily diets. Marketing campaigns promising good health have been very effective.

Table 4.2 Nutraceutically Important Plants and Herbs

Nutraceutical	Scientific Name	Active Constituent	Uses	References
St. John's wort	*Hypericum perforatum*	Hypericin, hyperoside, hyperforin	Antidepressant, cognitive enhancer, anticancer, anti-inflammatory, anti-arthritis, antibacterial, antiviral	Forsdike and Pirotta 2017; Ben-Eliezer and Yechiam 2016
Turmeric	*Curcuma longa*	Curcumin, demethoxy curcumin, bisdemethoxycurcumin	Antiviral, antibacterial, antitumor, anti-inflammatory, antioxidant, antiulcer, antiseptic, anticancer, antiallergic, analgesic	Afrose et al. 2015; Illuri et al. 2015
Ginger	*Zingiber officinale*	Volatile oil that contains zingiberene and gingerol	Antistress, anti-inflammatory, against dysmenorrhea, gastroprotective, analgesic, antibacterial, antidiabetic, rheumatoid arthritis, renal toxicity, anthelmintic, hepatoprotective	Moon et al. 2017; Shirvani et al. 2017
Garlic	*Allium sativum*	Alliin, allicin, diallyl disulphide, allyl propyl disulphide	Colon cancer, common cold, antihypertensive, atherosclerosis, hypercholesterolemia, anti-microbial, vaginitis, breast cancer, antifungal, leukemia, antidiabetic, immunomodulatory,	Bayan et al. 2014
Liquorice	*Glycyrrhiza glabra*	Triterpenoids such as glycyrrhetic acid, and glycoside glycyrrhizinic acid and flavonoids such as liquiritin and isoliquiritin	Stomach ulcers, hepatoprotective, antiviral, memory enhancer, antimicrobial, antidepressant, antioxidant	Dastagir and Rizvi 2016

(*Continued*)

Table 4.2 (Continued) Nutraceutically Important Plants and Herbs

Nutraceutical	Scientific Name	Active Constituent	Uses	References
Ginseng	*Panax ginseng*	Triterpenoidal saponin glycosides are grouped as ginsenosides, panaxosides, and chikusetsuasaponin; panaxosides yield oleanolic acid, panaxdiol, and panaxatriol; ginsenoside Rg2	Antidiabetic, memory enhancer, hepatoprotective, immunomodulatory, anticancer, antihyperlipidemic	Lee and Kim 2014
Onion	*Allium cepa*	Alliin, allicin, 2,5-dimethyl thiophene, benzyl thioisocyanate	Anticancer, antispasmodic, respiratory problems, asthma, cardiovascular diseases, hyperglycemia	(Suleria et al., 2015a)
Ginkgo, maidenhair tree	*Ginkgo biloba*	Ginkgetin, isoginkgetin, ginkgolides A, B, C, and J	Migraine, Alzheimer's, hypertension, anxiety, Schizophrenia, tinnitus, vitiligo, attention deficit disorder (ADD), macular degeneration, allergic conjunctivitis	Karsch-Volk et al. 2014
Valerian	*Valeriana officinalis*	Alkaloids such as valerine, chatinine; valerenic acid and sesquiterpenes such as valepotriates and valtrates	Manganese, anxiety	(Hadley and Petry, 2003)
Aloe	*Aloe vera*	Aloin is the mixture of glucosides that contain barbaloin, isobarbaloin, and β-barbaloin; resin present is aloesin; it also contains aloetic acid, aloesone, saponins, and choline	Wound healing, anti-inflammatory, antioxidant, psoriasis, antidiabetic, bacteriostatic, peptic ulcer	Hashemi et al. 2015
Senna	*Cassia angustifolia* (Indian), *C. acutifolia* (Alexandrian)	Anthroquinone glycosides and sennosides A, B, C, and D	Antioxidant, antimicrobial, anticancer	Ulbricht et al. 2011

(Continued)

Table 4.2 (Continued) Nutraceutically Important Plants and Herbs

Nutraceutical	Scientific Name	Active Constituent	Uses	References
Asafoetida	*Ferula asafoetida*	Ferulic acid, umbellic acid, and umbelliferone	Asthma, bronchitis, anticancer, dysmenorrhoea, antimicrobial, anthelminthic, antibacterial, anticonvulsant, antitumor, antispasmodic, hypotensive	Sgarbossa et al. 2015
Bael	*Aegle marmelos*	Marmelosine A, B, and C along with vitamins C and A	Antidiabetic, antimicrobial, anti-inflammatory, antifungal, hypoglycemic, antipyretic, hyperthyroidism, antidiarrheal, antidyslipidemic, hepatoprotective, analgesic	Baliga et al. 2010
Brahmi	*Bacopa monnieri*	Alkaloids brahmine and herpestine along with bacosides A and B	Antistress, cognitive enhancer, neuroprotection, anti-parkinsonian, antidiabetic, antimicrobial, hepatoprotective, antimicrobial, hepatoprotective, anticancer, spasmolytic	Shinomol et al. 2011
Green tea	*Camellia sinensis*	Caffeine, myricetin, quercetin	Antioxidant, anticarcinogenic, ultraviolet protection, antiviral, antibacterial, neuroprotection	Baladia et al. 2014
Kava	*Piper methysticum*	Kawain, dihydrokawain, methysticin, dihydromethysticin, yangonin	Neuroprotection, anxiety, antioxidant, colon cancer, analgesic, prostate cancer	Sarris et al. 2011

(*Continued*)

Table 4.2 (Continued) Nutraceutically Important Plants and Herbs

Nutraceutical	Scientific Name	Active Constituent	Uses	References
Echinacea, purple coneflowers	*Echinacea angustifolia, E. purpurea, E. serotina, E. simulate, E. pallida, E. atrorubens*	Fucogalactoxyglucans, arabinogalactans, echinacosides, echinacin, polyacetylene	Anti-inflammatory, angiogenic, immunostimulant, antifungal, anticancer, antiviral, common cold	Karsch-Volk et al. 2014
Spiruline	*Arthrospira platensis, A. maxima*	Gamma linolenic acid, amino acids, and vitamins	Antiviral, anemia, antihyperglycemic, anti-inflammatory	Halidou Doudou et al. 2008
Psyllium or ispaghula	*Plantago ovate*	Pentosane, aldobionic acid	Ulcerative colitis, anti-inflammatory, hypertension, hyperglycemia, colon cancer	Mehmood et al. 2011
Guarana	*Paullinia cupana*	Caffeine, theophylline, theobromine	Antibacterial, antioxidant, cognitive enhancer	Schimpl et al. 2013
Hawthorn	*Crataegus monogyna*	Rutin, quercetin, vitexin, crataegolic acid, ursolic acid	Congestive heart failure, antioxidant, ischemia	Zhang et al. 2014
Arnica	*Arnica montana*	Sesquiterpene-type lactones and helenanolides including 11,13-dehydrohelenalin, epoxyhelenalin, and arnifolin	Antimicrobial, osteoarthritis, anti-inflammatory, wound healing	Ho et al. 2016
Belladonna	*Atropa belladonna*	Alkaloids such as l-hyoscyamine, atropine, belladonine, and scopoletin	Hallucinations, asthma, analgesic,	Glatstein et al. 2014
Burdock	*Genus Arctium*	Matairesinol, arctigenin, arctiin, lappaol	Constipation, hypertension, antimicrobial	Segueni et al. 2016
Capsicum	*Capsicum annuum* and other species	Capsaicin and capsanthin	Antioxidant, antimicrobial, cancer	Kehie et al. 2015
Cocillana	*Guarea species*	α-Santalene, α-copalene, sapthulenol, kaurene	Expectorant, anti-inflammatory, antiviral	George and Topaz 2013
Damiana	*Turnera diffusa*	Pinocembrin, acacetin, Luteolin-8-C-E-propenoic acid, luteolin 8-C-β-[6-deoxy-2-O-(α-L-rhamanganese opyranosyl) xylo-hexopyranose-3-uloside]	Peptic ulcer, gastroprotective, hypoglycemic, antioxidant	Szewczyk and Zidorn 2014

(*Continued*)

Table 4.2 (Continued) Nutraceutically Important Plants and Herbs

Nutraceutical	Scientific Name	Active Constituent	Uses	References
Dandelion	*Taraxacum officinale*	Tetrahydroridentin B, taraxacolide β-D-glucoside, taraxinic acid β-D-glucoside, ainslioside	Diuretic, hepatitis, antioxidant, antihyperglycemic, anticoagulant	Gonzalez-Castejon et al. 2012
Fenugreek	*Trigonella foenum-graecum*	4-Hydroxyisoleucine, diosgenin	Antihyperglycemic, antioxidant, immunomodulatory, anti-inflammatory, antipyretic, gastroprotective, antihyperlipidemic	Neelakantan et al. 2014
Ipecacuanha	*Cephalis ipecacuanha*	Isoquinoline alkaloids emetine, cephaeline, psychotrine	Expectorant, dysentery, diarrhea	Nomura and Kutchan 2010
Senega	*Polygala senega*	Saponin glycosides senegin and polygalic acid	Anti-angiogenic, anti-inflammatory, anti-inflammatory, hypoglycemic, anticancer, antioxidant	Kako et al. 1996
Garcinia	*Garcinia indica*	Garcinol, hydroxy citric acid, anthocyanins, cyanidin-3-glucoside, cyanidin-3-sambubioside	Antiglycation, anti-ulcer, hepatoprotective, anti-inflammatory, laxative	Semwal et al. 2015
Amla	*Emblica officinalis*	Phlyllembin, vitamin C	Antioxidant, antitumor, antipyretic, dyslipidemia, analgesic, anti-ulcer, hepatoprotective, bitter tonic	Baliga and Dsouza 2011
Kalmegh	*Andrographis paniculata*	Andrographolide, neoandrographolide	Anthelmentic, hepatoprotective, immunostimulant, anticancer, antihyperglycemic, antimicrobial, cardiotonic	Roy et al. 2016
Indian squill	*Urginea indica*	Glucoscillaren A, scillaren A and scillaren B, proscillaridin A	Diuretic, bronchodilator	Shenoy et al. 2006

Table 4.3 Recent Clinical Trials Pertaining to Medicinally Important Extracts/Plant Parts

Study	Extracts Used	Trial Phase	Principal Investigator	Verification
Use of snail slime, calendula extract and propolis extract to treat diabetic foot ulcer and to determine the safety of the use of formulation	Calendula extract and propolis extract	Phase 1	Luis Quinones Sepulveda, University of Chile	May 2017
Study of supplement (Mind Master) containing natural extract	Aloe barbadensis miller gel, grape juice, polygonum cuspidatum extract, and green tea extract		Elizabeth Fragopoulou, Harokopio University	July 2016
Effect of quercetin on green tea polyphenol uptake in prostate tissue from patients with prostate cancer undergoing surgery	Green tea polyphenols	Phase 1	Jonsson Comprehensive Cancer Center	September 2016
Cannabinoid therapy for pediatric epilepsy	Cannabis extract	Phase 1	Blathnaid McCoy, The Hospital for Sick Children	May 2017
Study to evaluate safety and efficacy of SR-T gel in patients with actinic keratosis	*Solanum udatum* plant extract	Phase 2	G & E Herbal Biotechnology Co. Ltd.	June 2017

Nutraceuticals are key to overcoming problems of malnutrition and immunodeficiency in developing countries. Establishing markets in these countries is vital. Maintaining product quality is an important factor in providing better healthcare. Standardization of extracts can help detect impurities and improve efficacy. The unwarranted concomitant use of herb and drugs or herb and food or herb and drug may lead to potentially harmful, synergistic, or ineffective interactions. Studies to prevent unwanted effects are essential. A nutraceutical might show some activity, but understanding the proper mechanism might be necessary to prevent last-stage failure in clinical trials and aid drug development. Developing a pharmacokinetic profile of a plant may help us understand its mechanism.

4.9 Acknowledgments

The authors would like to acknowledge the Science and Engineering Research Board (statutory body established through an Act of Parliament: SERB Act 2008), the Department of Science and Technology (DST), and the Government of India for the grant (#ECR/2016/001964) allocated to Dr. Tekade for research work on drug and gene delivery. The authors also acknowledge DST-SERB for N-PDF funding (PDF/2016/003329) to Dr. Rahul Maheshwari in Dr. Tekade's lab for work on targeted cancer therapy. The authors would also like to acknowledge the Department of Pharmaceuticals, Ministry of Chemicals and Fertilizers, India, for supporting research on cancer and diabetes at NIPER-Ahmedabad.

References

Abdullah MM, Jew S, Jones PJ (2017). Author's reply: Impact on health and healthcare costs if monounsaturated fatty acids were substituted for conventional dietary oils in the United States. *Nutr Rev.*

Aditya N, Espinosa YG, Norton IT (2017). Encapsulation systems for the delivery of hydrophilic nutraceuticals: Food application. *Biotechnol Adv.* 35(4):450–457.

Adnan S, Hoang M, Wang H, Xie Z (2012). Commercial PTFE membranes for membrane distillation application: Effect of microstructure and support material. *Desalination.* 284:297–308.

Afrose R, Saha SK, Banu LA, Ahmed AU, Shahidullah AS, Gani A, Sultana S, Kabir MR, Ali MY (2015). Antibacterial effect of Curcuma longa (turmeric) against Staphylococcus aureus and Escherichia coli. *Mymensingh Med J.* 24:506–515.

Ananga A, Obuya J, Ochieng J, Tsolova V (2017). Grape seed nutraceuticals for disease prevention: Current status and future prospects. *Phenolic Compounds-Biological Activity.* InTech.

Aronson JK (2017). Defining 'nutraceuticals': Neither nutritious nor pharmaceutical. *Br J Clin Pharmacol.* 83:8–19.

Arthur CL, Pawliszyn J (1990). Solid phase microextraction with thermal desorption using fused silica optical fibers. *Anal Chem.* 62:2145–2148.

Asfaram A, Ghaedi M, Goudarzi A (2016). Optimization of ultrasound-assisted dispersive solid-phase microextraction based on nanoparticles followed by spectrophotometry for the simultaneous determination of dyes using experimental design. *Ultrason Sonochem.* 32:407–417.

Aulton ME, Taylor KM (2017). *Aulton's Pharmaceutics E-Book: The Design and Manufacture of Medicines*, Elsevier Health Sciences.

Azwanida N (2015). A review on the extraction methods use in medicinal plants, principle, strength and limitation. *Med Aromat Plants.* 4:2167-0412.1000196.

Baladia E, Basulto J, Manera M, Martinez R, Calbet D (2014). Effect of green tea or green tea extract consumption on body weight and body composition; systematic review and meta-analysis. *Nutr Hosp.* 29:479–490.

Baliga MS, Bhat HP, Pereira MM, Mathias N, Venkatesh P (2010). Radioprotective effects of Aegle marmelos (L.) Correa (Bael): A concise review. *J Altern Complement Med.* 16:1109–1116.

Baliga MS, Dsouza JJ (2011). Amla (Emblica officinalis Gaertn), a wonder berry in the treatment and prevention of cancer. *Eur J Cancer Prev.* 20:225–239.

Baltussen E, Sandra P, David F, Cramers C (1999). Stir bar sorptive extraction (SBSE), a novel extraction technique for aqueous samples: Theory and principles. *J Microcolumn Sep.* 11:737–747.

Bayan L, Koulivand PH, Gorji A (2014). Garlic: A review of potential therapeutic effects. *Avicenna J Phytomed.* 4:1–14.

Ben-Eliezer D, Yechiam E (2016). Hypericum perforatum as a cognitive enhancer in rodents: A meta-analysis. *Sci Rep.* 6:35700.

Bhurtyal A (2015). The new vivious circle of malnutrition and nutraceuticals. *BMJ.* 345.

Bourseau P, Vandanjon L, Jaouen P, Chaplain-Derouiniot M, Masse A, Guérard F, Chabeaud A, Fouchereau-Peron M, Le Gal Y, Ravallec-Plé R (2009). Fractionation of fish protein hydrolysates by ultrafiltration and nanofiltration: Impact on peptidic populations. *Desalination.* 244:303–320.

British Pharmacopoeia. (2016). British pharmacopoeia.

Caldeira I, Pereira R, Clímaco MC, Belchior A, De Sousa RB (2004). Improved method for extraction of aroma compounds in aged brandies and aqueous alcoholic wood extracts using ultrasound. *Anal Chim Acta.* 513:125–134.

Carbonaro M, Maselli P, Nucara A (2015). Structural aspects of legume proteins and nutraceutical properties. *Food Res Int.* 76:19–30.

Cardoso LA, Karp SG, Vendruscolo F, Kanno KY, Zoz LI, Carvalho JC (2017). Biotechnological production of carotenoids and their applications in food and pharmaceutical products. *Carotenoids.* pp. 125–147, InTech.

Carrasco-González JA, Serna-Saldívar SO, Gutiérrez-Uribe JA (2017). Nutritional composition and nutraceutical properties of the Pleurotus fruiting bodies: Potencial use as food ingredient. *J Food Compost Anal.* 58:69–81.

Chauhan B, Kumar G, Kalam N, Ansari SH (2013). Current concepts and prospects of herbal nutraceutical: A review. *J Adv Pharm Technol Res.* 4:4.

Chemat F, Boutekedjiret C (2016). Extraction-steam distillation. *Reference module in Chemistry Molecular Sciences and Chemical Engineering*, 1–11.

Chen G, Lu Y, Krantz WB, Wang R, Fane, AG (2014). Optimization of operating conditions for a continuous membrane distillation crystallization process with zero salty water discharge. *J Membr Sci.* 450:1–11.

Chen J, Miao M, Campanella O, Jiang B, Jin Z (2016). Biological macromolecule delivery system for improving functional performance of hydrophobic nutraceuticals. *Curr Opin Food Sci.* 9:56–61.

Cheung H, Wang S, Ng T, Zhang Y, Lao L, Zhang Z, Tong Y, Chung F, Sze S (2017). Comparison of chemical profiles and effectiveness between Erxian decoction and mixtures of decoctions of its individual herbs: A novel approach for identification of the standard chemicals. *Chin Med.* 12:1.

Ciccolini V, Pellegrino E, Coccina A, Fiaschi AI, Cerretani D, Sgherri C, Quartacci MF, Ercoli L (2017). Biofortification with iron and zinc improves nutritional and nnutraceutical properties of common wheat flour and bread. *J Agric Food Chem.*

Claeys W, Verraes C, Cardoen S, De Block J, Huyghebaert A, Raes K, Dewettinck K, Herman L (2014). Consumption of raw or heated milk from different species: An evaluation of the nutritional and potential health benefits. *Food Control.* 42:188–201.

Da Costa JP (2017). A current look at nutraceuticals–key concepts and future prospects. *Trends Food Sci Technol.*

Damre PG (2015). Comparative evaluation of antimicrobial activity of herbal vs chemical root canal irrigants against E. faecalis-an in vitro study. *Int J Adv Res.* 3:1563–1572.

Dastagir G, Rizvi MA (2016). Review - Glycyrrhiza glabra L. (Liquorice). *Pak J Pharm Sci.* 29:1727–1733.

De Melo M, Silvestre A, Silva C (2014). Supercritical fluid extraction of vegetable matrices: Applications, trends and future perspectives of a convincing green technology. *J Supercrit Fluids.* 92:115–176.

Del Ben M, Polimeni L, Baratta F, Pastori D, Angelico F (2017). The role of nutraceuticals for the treatment of non–alcoholic fatty liver disease. *Br J Clin Pharmacol.* 83:88–95.

Eaglstein WH (2014). What are dietary supplements and nutraceuticals? *The FDA for Doctors.* Springer.

ElSohly HN, Croom JR EM, El-Feraly FS, El-Sherei MM (1990). A large-scale extraction technique of artemisinin from Artemisia annua. *J Nat Prod.* 53:1560–1564.

Enzing C, Ploeg M, Barbosa M, Sijtsma l (2014). Microalgae-based products for the food and feed sector: An outlook for Europe. *IPTS Institute for Prospective Technological Studies, JRC, Seville.*

Farzaneh V, Carvalho IS (2015). A review of the health benefit potentials of herbal plant infusions and their mechanism of actions. *Ind Crops Prod.* 65:247–258.

Foley SA, Szegezdi E, Mulloy B, Samali A, Tuohy MG (2011). An unfractionated fucoidan from Ascophyllum nodosum: Extraction, characterization, and apoptotic effects in vitro. *J Nat Prod.* 74:1851–1861.

Forsdike K, Pirotta M (2017). St John's wort for depression: Scoping review about perceptions and use by general practitioners in clinical practice. *J Pharm Pharmacol.*
Ganesan P, Noda K, Manabe Y, Ohkubo T, Tanaka Y, Maoka T, Sugawara T, Hirata T (2011). Siphonaxanthin, a marine carotenoid from green algae, effectively induces apoptosis in human leukemia (HL-60) cells. *Biochim Biophys Acta (BBA)-General Subjects.* 1810:497–503.
George M, Topaz M (2013). A systematic review of complementary and alternative medicine for asthma self-management. *Nurs Clin North Am.* 48:53–149.
Girdhar S, Pandita D, Girdhar A, Lather V (2017). Safety, quality and regulatory aspects of nutraceuticals. *Appl Clin Res Clin Trials Reg Affairs.* 4:36–42.
Glatstein M, Danino D, Wolyniez I, Scolnik D (2014). Seizures caused by ingestion of Atropa belladonna in a homeopathic medicine in a previously well infant: Case report and review of the literature. *Am J Ther.* 21:e196–e198.
Gomes L, Paschoalin V, Del Aguila E (2017). Chitosan nanoparticles: Production, physicochemical characteristics and nutraceutical applications. *Rev. Virtual Quim.* 9(1):387–409.
Gonzalez-Castejon M, Visioli F, Rodriguez-Casado A (2012). Diverse biological activities of dandelion. *Nutr Rev.* 70:534–547.
Gunawan S, Ismadji S, Ju Y-H (2008). Design and operation of a modified silica gel column chromatography. *J Chin Inst Chem Eng.* 39:625–633.
Hadley S, Petry JJ (2003). Valerian. *Am Fam Phy.* 67:1755–1758.
Halidou Doudou M, Degbey H, Daouda H, Leveque A, Donnen P, Hennart P, Dramaix-Wilmet M (2008). The effect of spiruline during nutritional rehabilitation: Systematic review. *Rev Epidemiol Sante Publique.* 56:425–431.
Hashemi SA, Madani SA, Abediankenari S (2015). The review on properties of aloe vera in healing of cutaneous wounds. *Biomed Res Int.* 2015:714216.
He M, Chen B, Hu B (2014). Recent developments in stir bar sorptive extraction. *Anal Bioanal Chem.* 406:2001–2026.
Hegde V, Kesaria DP (2013). Comparative evaluation of antimicrobial activity of neem, propolis, turmeric, liquorice and sodium hypochlorite as root canal irrigants against E. faecalis and C. albicans–an in vitro study. *Endodontology.* 25:38–45.
Ho D, Jagdeo J, Waldorf HA (2016). Is there a role for Arnica and Bromelain in prevention of post-procedure ecchymosis or edema? A systematic review of the literature. *Dermatol Surg.* 42:445–463.
Illuri R, Bethapudi B, Anandakumar S, Murugan S, Joseph JA, Mundkinajeddu D, Agarwal A, Chandrasekaran CV (2015). Anti-inflammatory activity of polysaccharide fraction of Curcuma longa extract (NR-INF-02). *Antiinflamm Antiallergy Agents Med Chem.* 14:53–62.
Indian Pharmacopoeia. (2007). Government of India. *Ministry of Health and Family Welfare.* 2:1020–1021.
Indian Pharmacopoeia. (2014). Govt. of India, *Ministry of Health and Family Welfare, Delhi, Ghaziabad,* 2014. Vol. II, 2066.
Jafari SM (2017). *Nanoencapsulation technologies for the food and nutraceutical industries.* London; San Diego, CA: Academic Press.
Joana Gil-Chávez G, Villa JA, Fernando Ayala–Zavala J, Basilio Heredia J, Sepulveda D, Yahia EM, González–Aguilar GA (2013). Technologies for extraction and production of bioactive compounds to be used as nutraceuticals and food ingredients: An overview. *Compr Rev Food Sci Food Saf.* 12:5–23.
Júnior SQ, Carneiro VHA, Fontenelle TPC, De Sousa Chaves L, Mesquita JX, De Brito TV, Prudêncio RS, De Oliveira JS, Medeiros J-VR, Aragão KS (2015). Antioxidant and anti-inflammatory activities of methanol extract and its fractions from the brown seaweed Spatoglossum schroederi. *J Appl Phycol.* 27:2367–2376.
Kako M, Miura T, Nishiyama Y, Ichimaru M, Moriyasu M, Kato A (1996). Hypoglycemic effect of the rhizomes of Polygala senega in normal and diabetic mice and its main component, the triterpenoid glycoside senegin-II. *Planta Med.* 62:440–443.

Kamal-Eldin A, Budilarto E (2015). Tocopherols and tocotrienols as antioxidants for food preservation. *Handb Antioxid Food Preserv.* 141–159.

Kapure PL, Makade KP, Sanap MD, Gandhi SJ, Ahirrao R, Pawar S (2015). Various extraction method and standardization parameter of amla and durva. *Pharma Sci Monit.* 6.

Karsch-Volk M, Barrett B, Kiefer D, Bauer R, Ardjomand-Woelkart, K, Linde, K (2014). Echinacea for preventing and treating the common cold. *Cochrane Database Syst Rev.* CD000530.

Kataoka H, Sakaki Y, Komatsu K, Shimada Y, Goto S (2017). Melting process of the peritectic mixture of lidocaine and ibuprofen interpreted by site percolation theory model. *J Pharm Sci.* 106(10):3016–3021.

Katiyar R (2017). Modeling and simulation of Mentha arvensis L. essential oil extraction by water-steam distillation process. *International Research Journal of Engineering and Technology.* 4(6):2793–2798.

Kehie M, Kumaria S, Tandon P, Ramchiary N (2015). Biotechnological advances on in vitro capsaicinoids biosynthesis in capsicum: A review. *Phytochem Rev.* 14:189–201.

Khora SS (2013). Marine fish-derived bioactive peptides and proteins for human therapeutics. *Int J Pharm Pharm Sci.* 5:31–37.

Kim S-K (2013). *Marine nutraceuticals: Prospects and perspectives.* New York: CRC Press.

Kim S-K, Himaya S (2013). Edible marine invertebrates. *Marine Nutraceuticals: Prospects and Perspectives.* 243. New York: CRC Press.

Kim S-K, Wijesekara I (2013). Marine-derived nutraceuticals. *Marine Nutraceuticals: Prospects and Perspectives,* 1. New York: CRC Press.

Koskela A, Reinisalo M, Petrovski G, Sinha D, Olmiere C, Karjalainen R, Kaarniranta K (2016). Nutraceutical with resveratrol and omega-3 fatty acids induces autophagy in ARPE-19 cells. *Nutrients.* 8:284.

Krivoruchko A, Nielsen J (2015). Production of natural products through metabolic engineering of Saccharomyces cerevisiae. *Curr Opin Biotechnol.* 35:7–15.

Kumar A, Zandi P (2014). Plant nutraceuticals for cardiovascular diseases with special emphasis to the medicinal herb fenugreek (Trigonella foenum-graecum L.). *Am J Soc Issues Humanit.* 4.

Kumar K, Kumar S (2015). Role of nutraceuticals in health and disease prevention: A review. *South Asian J Food Technol Environ.* 1:116–121.

Kumar Tekade R, Gs Maheshwari R, A Sharma P, Tekade M, Singh Chauhan A (2015). siRNA therapy, challenges and underlying perspectives of dendrimer as delivery vector. *Curr Pharm Design.* 21:4614–4636.

Lalu L, Tambe V, Pradhan D, Nayak K, Bagchi S, Maheshwari R, Kalia K, Tekade RK (2017). Novel nanosystems for the treatment of ocular inflammation: Current paradigms and future research directions. *J Control Release.* 268:19–39.

Lane T, Wassef NL, Poole S, Mistry Y, Lachmann H, Gillmore J, Hawkins PN, Pepys MB (2013). Infusion of pharmaceutical grade natural human C reactive protein is not pro inflammatory in healthy adult human volunteers. *Circ Res.*, CIRCRESAHA. 113.302770.

Lang T, Heasman M (2015). *Food Wars: The Global Battle for Mouths, Minds and Markets,* Routledge Earthscan.

Lee CH, Kim JH (2014). A review on the medicinal potentials of ginseng and ginsenosides on cardiovascular diseases. *J Ginseng Res.* 38:161–166.

Li-Chan EC (2015). Bioactive peptides and protein hydrolysates: Research trends and challenges for application as nutraceuticals and functional food ingredients. *Curr Opin Food Sci.* 1:28–37.

Li S, Shah NP (2017). Sulphonated modification of polysaccharides from Pleurotus eryngii and Streptococcus thermophilus ASCC 1275 and antioxidant activities investigation using CCD and Caco-2 cell line models. *Food Chem.* 225:246–257.

Liu L, Guan N, Li J, Shin H-D, Du G, Chen J (2017). Development of GRAS strains for nutraceutical production using systems and synthetic biology approaches: Advances and prospects. *Crit Rev Biotechnol.* 37:139–150.

Liu X, Bi J, Xiao, H, Mcclements DJ (2016). Enhancement of nutraceutical bioavailability using excipient nanoemulsions: Role of lipid digestion products on bioaccessibility of carotenoids and phenolics from mangoes. *J Food Sci.* 81.

Maccoss MJ, Noble WS, Käll L (2016). Fast and accurate protein false discovery rates on large-scale proteomics data sets with percolator 3.0. *J Am Soc Mass Spectrom.* 27:1719–1727.

Maeda H (2015). Nutraceutical effects of fucoxanthin for obesity and diabetes therapy: A review. *J Oleo Sci.* 64:125–132.

Maheshwari R, Tekade M, A Sharma P, Kumar Tekade R (2015a). Nanocarriers assisted siRNA gene therapy for the management of cardiovascular disorders. *Curr Pharm Des.* 21:4427–4440.

Maheshwari RG, Tekade RK, Sharma PA, Darwhekar G, Tyagi A, Patel RP, Jain DK (2012). Ethosomes and ultradeformable liposomes for transdermal delivery of clotrimazole: A comparative assessment. *Saudi Pharm J.* 20:161–170.

Maheshwari RG, Thakur S, Singhal S, Patel RP, Tekade M, Tekade RK (2015b). Chitosan encrusted nonionic surfactant based vesicular formulation for topical administration of ofloxacin. *Sci Adv Mater.* 7:1163–1176.

Mann B, Athira S, Sharma R, Bajaj R (2017). Bioactive peptides in yogurt. *Yogurt in Health and Disease Prevention*. 411. New York: Academic Press.

Martins N, Barros L, Santos-Buelga C, Silva S, Henriques M, Ferreira IC (2015). Decoction, infusion and hydroalcoholic extract of cultivated thyme: Antioxidant and antibacterial activities, and phenolic characterisation. *Food Chem.* 167:131–137.

Mchugh M, Krukonis V (2013). *Supercritical Fluid Extraction: Principles and Practice*, New York: Elsevier.

Mcneil B, Archer D, Giavasis I, Harvey L (2013). *Microbial Production of Food Ingredients, Enzymes and Nutraceuticals*, New York: Elsevier.

Mehmood MH, Aziz N, Ghayur MN Gilani AH (2011). Pharmacological basis for the medicinal use of psyllium husk (Ispaghula) in constipation and diarrhea. *Dig Dis Sci.* 56:1460–1471.

Mendiola JA, Herrero M, Castro-Puyana M, Ibáñez E (2013). Supercritical fluid extraction. *Nat Prod Extr Principles Appl.* 21:196.

Milledge JJ, Nielsen BV, Bailey D (2016). High-value products from macroalgae: The potential uses of the invasive brown seaweed, Sargassum muticum. *Rev Environ Sci Bio Technol.* 15:67–88.

Modolo G, Asp H, Vijgen H, Malmbeck R, Magnusson D, Sorel C (2008). Demonstration of a TODGA–based continuous counter–current extraction process for the partitioning of actinides from a simulated PUREX raffinate, Part II: Centrifugal contactor runs. *Solvent Extr and Ion Exch.* 26:62–76.

Monaco R, Quinlan R (2014). Novel natural product discovery from marine sponges and their obligate symbiotic organisms. *Biorxiv.* 005454.

Moon S, Lee MS, Jung S, Kang B, Kim SY, Park S, Son HY, Kim CT, Jo YH, Kim IH, Kim YS, Kim Y (2017). High hydrostatic pressure extract of ginger exerts antistress effects in immobilization-stressed rats. *J Med Food.*

Mullin JM, Valenzano MC, Diguilio K, Teter M, Mercado J, To J, Mixson B, Ferraro B, Manley I, Baker V (2014). Su1880 effects on CaCO-2 gastrointestinal epithelial barrier properties of the nutraceuticals zinc, berberine, butyrate, indole and quercetin. *Gastroenterology.* 146:S-492.

Myszka D, Trzaskowski W (2017). The importance of applying the percolation theory to the analysis of the structure of polycrystalline materials. *J Manuf Technol.* 42:7–14.

Nanjo C, Fujimoto T, Matsushita MM, Awaga K (2014). Ambipolar transport in phase-separated thin films of p-and n-type vanadylporphyrazines with two-dimensional percolation. *J Phys Chem C.* 118:14142–14149.

Neelakantan N, Narayanan M, De Souza RJ, Van Dam RM (2014). Effect of fenugreek (Trigonella foenum-graecum L.) intake on glycemia: A meta-analysis of clinical trials. *Nutr J.* 13:7.

Ngo D-H, Kim S-K (2013). Sulfated polysaccharides as bioactive agents from marine algae. *Int J Biol Macromol.* 62:70–75.

Nogueira JMF (2015). Stir-bar sorptive extraction: 15 years making sample preparation more environment-friendly. *TrAC Trends Anal Chem.* 71:214–223.

Nomura T, Kutchan TM (2010). Three new O-methyltransferases are sufficient for all O-methylation reactions of ipecac alkaloid biosynthesis in root culture of Psychotria ipecacuanha. *J Biol Chem.* 285:7722–7738.

Oladoja NA (2016). Appropriate technology for domestic wastewater management in under-resourced regions of the world. *Appl Water Sci.* 1–16.

Palacios T, Vitetta L, Coulson S, Madigan CD, Denyer GS, Caterson ID (2017). The effect of a novel probiotic on metabolic biomarkers in adults with prediabetes and recently diagnosed type 2 diabetes mellitus: Study protocol for a randomized controlled trial. *Trials.* 18:7.

Pallela R, Park I (2014). Nutraceutical and pharmacological implications of marine carbohydrates. *Adv Food Nutr Res.* 73:183–195.

Pandey A, Tripathi S (2014). Concept of standardization, extraction and pre phytochemical screening strategies for herbal drug. *J Pharmacogn Phytochem.* 2(5):115–119.

Plaza M, Turner C (2015). Pressurized hot water extraction of bioactives. *TrAC Trends Anal Chem.* 71:39–54.

Prakash D, Sharma G (2014). *Phytochemicals of Nutraceutical Importance* CABI.

Prameela K, Venkatesh K, Immandi SB, Kasturi APK, Krishna CR, Mohan CM (2017). Next generation nutraceutical from shrimp waste: The convergence of applications with extraction methods. *Food Chem.* 237:121–132.

Prasad S, Gupta SC, Tyagi AK (2017). Reactive oxygen species (ROS) and cancer: Role of antioxidative nutraceuticals. *Cancer Lett.* 387:95–105.

Raab A, Nelson J, Stiboller M, Franzmann P, Gajdosechova Z, and Feldmann J (2016). Essential and non-essential trace elements in nutraceuticals. *Planta Med.* 82:OA39.

Ramluckan K, Moodley KG, Bux F (2014). An evaluation of the efficacy of using selected solvents for the extraction of lipids from algal biomass by the soxhlet extraction method. *Fuel.* 116:103–108.

Rao RK, Arnold LK (1956). Alcoholic extraction of vegetable oils. II. Solubilities of corn, linseed, and tung oils in aqueous ethanol. *J Am Oil Chem Soc.* 33:82–84.

Reda E, Saleh I, El Gendy AN, Talaat Z, Hegazy M-E, Haggag E (2017). Chemical constituents of Euphorbia sanctae-catharinae Fayed essential oil: A comparative study of hydro-distillation and microwave-assisted extraction. *J Adv Pharm Res.* 1:155–159.

Ríos-Hoyo A, Romo-Araiza A, Meneses-Mayo M, Gutiérrez-Salmeán G (2017). Prehispanic functional foods and nutraceuticals in the treatment of dyslipidemia associated to cardiovascular disease: A mini-review. *Int J Vitam Nutr Res.* 1:1–14.

Roccato M (2016). Entering the Swedish nutraceutical market: A case study.

Ronis M, Pedersen KB, Watt J (2017). Adverse effects of nutraceuticals and dietary supplements. *Ann Rev Pharmacol Toxicol.* 58.

Roy S, Yasmin S, Ghosh S, Bhattacharya S, Banerjee D (2016). Anti-infective metabolites of a newly isolated Bacillus thuringiensis KL1 associated with kalmegh (Andrographis paniculata Nees.), a traditional medicinal herb. *Microbiol Insights.* 9:1–7.

Ruocco N, Costantini S, Guariniello S, Costantini M (2016). Polysaccharides from the marine environment with pharmacological, cosmeceutical and nutraceutical potential. *Molecules.* 21:551.

Salem F, Abduljalil K, Kamiyama Y, Rostami-Hodjegan A (2016). Considering age variation when coining drugs as high versus low hepatic extraction ratio. *Drug Metab Dispos.* 44:1099–1102.

Santos R, Dias S, Pinteus S, Silva J, Alves C, Tecelao C, Pedrosa R, Pombo A (2016). Sea cucumber Holothuria forskali, a new resource for aquaculture? Reproductive biology and nutraceutical approach. *Aquac Res.* 47:2307–2323.

Sarris J (2017). Clinical use of nutraceuticals in the adjunctive treatment of depression in mood disorders. *Australas Psychiatry.* 1039856216689533.

Sarris J, Laporte E, Schweitzer I (2011). Kava: A comprehensive review of efficacy, safety, and psychopharmacology. *Aust N Z J Psychiatry.* 45:27–35.

Schimpl FC, Da Silva JF, Goncalves JF, Mazzafera P (2013). Guarana: Revisiting a highly caffeinated plant from the Amazon. *J Ethnopharmacol.* 150:14–31.

Schmidt F, Koch BP, Witt M, Hinrichs K-U (2014). Extending the analytical window for water-soluble organic matter in sediments by aqueous Soxhlet extraction. *Geochim Cosmochim Acta.* 141:83–96.

Segueni N, Zellagui A, Moussaoui F, Lahouel M, Rhouati S (2016). Flavonoids from Algerian propolis. *Arab J Chem.* 9:S425–S428.

Semwal RB, Semwal DK, Vermaak I, Viljoen A (2015). A comprehensive scientific overview of Garcinia cambogia. *Fitoterapia.* 102:134–48.

Sgarbossa A, Giacomazza D, Di Carlo M (2015). Ferulic acid: A hope for Alzheimer's disease therapy from plants. *Nutrients.* 7:5764–5782.

Shao A (2014). Global market entry regulations for nutraceuticals, functional foods, dietary/food/health supplements. *Nutraceutical and Functional Foods Regulations in the United States and Around the World.* pp. 41–45, Academic Press.

Shao A (2017). Chapter 15-Global market entry regulations for nutraceuticals, functional foods, dietary/food/health supplements A2-Bagchi, Debasis'. In: Nair S (ed). *Developing New Functional Food and Nutraceutical Products.* 279–290.

Sharif K, Rahman M, Azmir J, Mohamed A, Jahurul M, Sahena F, Zaidul I (2014). Experimental design of supercritical fluid extraction–a review. *J Food Eng.* 124:105–116.

Sharma PA, Maheshwari R, Tekade M, Kumar Tekade R (2015). Nanomaterial based approaches for the diagnosis and therapy of cardiovascular diseases. *Curr Pharma Des.* 21:4465–4478.

Shenoy SR, Kameshwari MN, Swaminathan S, Gupta MN (2006). Major antifungal activity from the bulbs of Indian squill Urginea indica is a chitinase. *Biotechnol Prog.* 22:631–637.

Shinomol GK, Muralidhara, Bharath MM (2011). Exploring the role of "Brahmi" (Bocopa monnieri and Centella asiatica) in brain function and therapy. *Recent Pat Endocr Metab Immune Drug Discov.* 5:33–49.

Shirvani MA, Motahari-Tabari N, Alipour A (2017). Use of ginger versus stretching exercises for the treatment of primary dysmenorrhea: A randomized controlled trial. *J Integr Med.* 15:295–301.

Singh A, Chandra R, Ali MN (2015). Rate of extraction and phytochemical screening of selected medicinal herbs for herbal yoghurt. *Pharma Innov.* 3(4):15–17.

Šližytė R, Mozuraitytė R, Martínez-Alvarez O, Falch E, Fouchereau-Peron M, Rustad T (2009). Functional, bioactive and antioxidative properties of hydrolysates obtained from cod (Gadus morhua) backbones. *Process Biochem.* 44:668–677.

Soni N, Soni N, Pandey H, Maheshwari R, Kesharwani P, Tekade RK (2016). Augmented delivery of gemcitabine in lung cancer cells exploring mannose anchored solid lipid nanoparticles. *J Colloid Interface Sci.* 481:107–116.

Soni N, Tekade M, Kesharwani P, Bhattacharya P, Maheshwari R, Dua K, Hansbro P, Tekade R (2017). Recent Advances in oncological submissions of dendrimer. *Curr Pharm Des.* 23(21):3084–3098.

Souza-Silva ÉA, Jiang R, Rodríguez-Lafuente A, Gionfriddo E, Pawliszyn J (2015). A critical review of the state of the art of solid-phase microextraction of complex matrices I. Environmental analysis. *TrAC Trends Anal Chem.* 71:224–235.

Spietelun A, Marcinkowski Ł, de la Guardia M, Namieśnik J (2014). Green aspects, developments and perspectives of liquid phase microextraction techniques. *Talanta.* 119:34–45.

Stanisz E, Werner J, Zgoła-Grześkowiak A (2014). Liquid-phase microextraction techniques based on ionic liquids for preconcentration and determination of metals. *TrAC Trends Anal Chem.* 61:54–66.

Suárez-Martínez SE, Ferriz-Martínez RA, Campos-Vega R, Elton-Puente JE, de la Torre Carbot K, García-Gasca T (2016). Bean seeds: Leading nutraceutical source for human health. *CyTA-J Food.* 14:131–137.

Subramanian R, Subbramaniyan P, Ameen JN, Raj V (2016). Double bypasses soxhlet apparatus for extraction of piperine from Piper nigrum. *Arab J Chem.* 9:S537–S540.

Suleria HA, Butt MS, Anjum FM, Saeed F, Khalid N (2015a). Onion: Nature protection against physiological threats. *Crit Rev Food Sci Nutr.* 55:50–66.

Suleria HAR, Osborne S, Masci P, Gobe G (2015b). Marine-based nutraceuticals: An innovative trend in the food and supplement industries. *Mar Drugs.* 13:6336–6351.

Sweazea KL, Johnston CS, Knurick J, Bliss CD (2017). Plant-based nutraceutical increases plasma catalase activity in healthy participants: A small double-blind, randomized, placebo-controlled, proof of concept trial. *J Diet Suppl.* 14:200–213.

Szewczyk K, Zidorn C (2014). Ethnobotany, phytochemistry, and bioactivity of the genus Turnera (Passifloraceae) with a focus on damiana—Turnera diffusa. *J Ethnopharmacol.* 152:424–43.

Tabarsa M, Karnjanapratum S, Cho M, Kim J-K, You S (2013). Molecular characteristics and biological activities of anionic macromolecules from Codium fragile. *Int J Biol Macromol.* 59:1–12.

Tekade RK, Maheshwari R & Jain NK (2018). 9. Toxicity of nanostructured biomaterials A2— Narayan, Roger. *Nanobiomaterials.* Woodhead Publishing, USA.

Tsai S-P, Hickey R, Du J, Xu J, Schumacher J (2016). Processes for the acidic, anaerobic conversion of hydrogen and carbon oxides to oxygenated organic compound. Google Patents.

Ulbricht C, Conquer J, Costa D, Hamilton W, Higdon ER, Isaac R, Rusie E, Rychlik I, Serrano JM, Tanguay-Colucci S, Theeman M, Varghese M (2011). An evidence-based systematic review of senna (Cassia senna) by the Natural Standard Research Collaboration. *J Diet Suppl.* 8:189–238.

Wang J, Guleria S, Koffas MA, Yan Y (2016). Microbial production of value-added nutraceuticals. *Curr Opin Biotechnol.* 37:97–104.

Wang X, He Y, Lin L, Zeng F, Luan T (2014). Application of fully automatic hollow fiber liquid phase microextraction to assess the distribution of organophosphate esters in the Pearl River Estuaries. *Sci Total Environ.* 470:263–269.

Wenzel T, Stillhart C, Kleinebudde P, Szepes A (2017). Influence of drug load on dissolution behavior of tablets containing a poorly water-soluble drug: Estimation of the percolation threshold. *Drug Dev Ind Pharm.* 1–11.

Wildman RE (2016). *Handbook of Nutraceuticals and Functional Foods,* Florida: CRC Press.

Williams RJ, Mohanakumar KP, Beart PM (2015). *Neuro-Nutraceuticals: The Path to Brain Health via Nourishment Is Not so Distant.* Rio de Janeiro: Elsevier.

Williams SN, Nestle M (2015). *'Big Food': Taking a Critical Perspective on a Global Public Health Problem.* Abingdon: Taylor & Francis.

Wu H, Guo J-B, Du L-M, Tian H, Hao CX, Wang Z-F, Wang J-Y (2013). A rapid shaking-based ionic liquid dispersive liquid phase microextraction for the simultaneous determination of six synthetic food colourants in soft drinks, sugar-and gelatin-based confectionery by high-performance liquid chromatography. *Food Chem.* 141: 182–186.

Yalçinçjray Ö, Anli R (2015). The impact of storage conditions on the phenolic content and antioxidant activity of pomegranate liquors produced by maceration method from Hicaz pomegranate. *GIDA-J Food.* 40:209–216.

Yuan S, Zou C, Yin H, Chen Z, Yang W (2015). Study on the separation of binary azeotropic mixtures by continuous extractive distillation. *Chem Eng Res Des.* 93:113–119.

Yusoff ZM, Muhammad Z, Ahmad ND, Rahiman MHF, Taib MN (2016). Hybrid fuzzy plus PID controller of hydro-diffusion steam distillation essential oil extraction system: Design and performance evaluation. In: *AIP Conference Proceedings,* 2016. AIP Publishing, 050007.

Zhang Y, Liu C, Li J, Qi Y, Li Y, Li S (2015). Development of "ultrasound-assisted dynamic extraction" and its combination with CCC and CPC for simultaneous extraction and isolation of phytochemicals. *Ultrason Sonochem.* 26:111–118.

Zhang Y, Zhang L, Geng Y, Geng Y (2014). Hawthorn fruit attenuates atherosclerosis by improving the hypolipidemic and antioxidant activities in apolipoprotein e-deficient mice. *J Atheroscler Thromb.* 21:119–28.

Zheng Y, Hu H, Maya C (2016). Potential interaction between dietary fiber and iron in extruded food products. *Nutrients.* 8:779–795.

5

Taste-Masking Techniques in Nutraceutical and Functional Food Industry

Shankar D. Katekhaye, Bhagyashree Kamble, Ashika Advankar, Neelam Athwale, Abhishek Kulkarni, and Ashwini Ghagare

Contents

5.1 Introduction 124
5.2 Sensory evaluation strategies for taste masking 126
    5.2.1 Insent taste-sensing system 127
    5.2.2 α-ASTREE electronic tongue 127
    5.2.3 Carbon nanotubule field-effect transistor bioelectronic sensors 128
    5.2.4 Biomimetic sensors 128
5.3 Conventional techniques used in taste masking 129
    5.3.1 Sweeteners and flavors 129
        5.3.1.1 Sweeteners 129
        5.3.1.2 Flavorants 130
    5.3.2 Salt 130
    5.3.3 Bitter blockers 131
    5.3.4 Prodrugs 132
    5.3.5 Film coating 132
    5.3.6 Microencapsulation 133
    5.3.7 Ion-exchange resins 134
    5.3.8 Complexation 135
    5.3.9 Supercritical fluids 135
5.4 Recent technologies 136
    5.4.1 Nanohybrid technology 136
    5.4.2 Hot-melt extrusion 136
    5.4.3 Solvent-free cold extrusion 137
    5.4.4 Off-taste masking agents 137
    5.4.5 Lipid nanoparticles 137
    5.4.6 Porous microspheres 137
    5.4.7 Multi-particulate rupture 138

5.5　Patented techniques 138
　　5.5.1　Microcaps technology® 138
　　5.5.2　Opadry® 139
　　5.5.3　FlavoRite® 139
　　5.5.4　OXPzero® 139
　　5.5.5　Actimask® 139
　　5.5.6　Formulcoat® 140
　　5.5.7　Camouflage® 140
　　5.5.8　KLEPTOSE®Linecaps 140
　　5.5.9　Micromask® 140
5.6　Conclusion 141
References 141

5.1 Introduction

The pharmaceutical, functional food, and nutraceutical industries have been tremendously instrumental in the last two decades in drug development, economic growth, and health promotion. The food that we eat is our primary source of nutrition. Sometimes, due to various factors such as lifestyle changes, eating habits, and stress, we are unable to get all the nutrition our bodies require. People today have become increasingly health conscious. In an effort to achieve balanced nutrition, they consume dietary supplements, enriched products, and nutraceuticals. This has led to the meteoric rise of the nutraceutical and functional food industries. Consumers are also weary of expensive medical treatments (Das et al., 2012). Nutraceuticals and functional foods are attractive alternatives.

The term nutraceutical is a composite of two words, nutrition, which means alimentative food or food component, and pharmaceutical, which means drug like (Santini et al., 2017). Stephen DeFelice coined the term in 1979. A nutraceutical is any nonpoisonous extract from food that has been proved scientifically effective for the treatment and/or prevention of disease. Nutraceuticals are marketed in many forms, including pills, capsules, and powders (Prabu et al., 2012). Nutraceuticals are found in products from the food industry, herbal and dietary supplements, the pharmaceutical industry, and the newly clustered pharmaceutical/agribusiness/nutrition industry (Das et al., 2012).

There are two types of nutraceuticals. Potential nutraceuticals show inherent activity, but clinical study remains to be done. Established nutraceuticals have been clinically proven to act as nutrition sources (Pandey et al., 2010).

Similar to nutraceuticals, the concept of functional food arose in 1984, as a convenient and inexpensive alternative to chronic health problems. It was a product of collaborative food and science research. The U.S. Institute of Food Technology defines functional food as "Substances that provide essential nutrients often beyond quantities necessary for normal maintenance, growth and development, and/or other biologically active components that impart health benefits or desirable physiological effects" (Martirosyan and Singh, 2015). The functional food component may be a macronutrient, a micronutrient, or a non-nutritive food component (Roberfroid, 2000).

Nutraceuticals and functional foods pose challenges. Unpalatable taste is a hindering factor for consumers. Patient compliance and the ability to attract consumers are part of the mission of these industries. Product acceptability depends primarily on appearance, which includes color, shape, and flavor. The increase in the use of new techniques and methods to help improve these attributes has led to increased research. There has been a significant increase in the use of edible dyes and sweetening and flavoring excipients to make products more appealing and to improve palatability. Virtually every product on the market has used these substances to attract consumers.

Flavor plays a vital role. Consumers tend to avoid anything with a disagreeable taste. Taste is an organoleptic property that can be used to identify the flavor (Bhattacharjee et al., 2016). It is a phenomenon that occurs because of transfer of signals from taste receptors in the taste buds that are present on the tongue (Karaman, 2013). The four basic taste sensations are sweet, salty, sour, and bitter. A fifth taste sensation, umami, is a savory sensation that was identified recently by the Japanese. Other tastes attributed to foods are astringent taste, experienced with tea infusion, wine, and orange juice (Valentová et al., 2002); pungent taste, which is observed in paprika (Collera-Zúñiga et al., 2005), isothiocyanates, diallyl sulphides, and capsaicin (Tepper et al., 2003); and the fishy flavor of fish oil. An individual's taste perception depends on physiological factors, psychological processes, and environmental and cultural influences (Gaudette and Pickering, 2013). Substances that are sweet are considered pleasant, substances that are sour or salty are acceptable to many, and substances with bitter or metallic taste are unpalatable (Szejtli and Szente, 2005).

The effectiveness of any drug depends on proper dosage at the proper interval. Patient comfort leads to better compliance with optimal drug therapy. Compliance is a dominant challenge in both children and elderly patients, but good taste is more significant for children than the elderly (Shahiwala, 2011). Bitter taste is acceptable or even desirable in coffee, tea, bitter lemon, and cocoa products (Ley et al., 2005). Many active pharmaceutical ingredients (API) on the market have a repugnant taste (Walsh et al., 2014). This remains a challenge for the functional food and nutraceutical industries. Table 5.1 lists examples of off-tasting ingredients in functional foods. Their chemical structures are shown in Figure 5.1 (Cadwallader, 2015).

Improper dosage consumption due to unpalatability may lead to failure of treatment and may even worsen the condition. There is a need for a universal inhibitor of bitter components in all drugs but no such solution has yet been reported. Often bitter components may be responsible for therapeutic action or lack thereof. Masking this unpleasant taste is thus a significant goal.

Table 5.1 Some Functional Food Ingredients and Their Off-Tastes

Ingredient	Taste
Limonin (Citrus)	Bitter
Tannins in Cranberry	Astringent
Vitamin B$_1$ (Thiamine)	Meaty taste
Minerals	Metallic taste
Egg protein	Sulphurous
Rice protein	Rancid, smoky
Catechins	Bitter, astringent

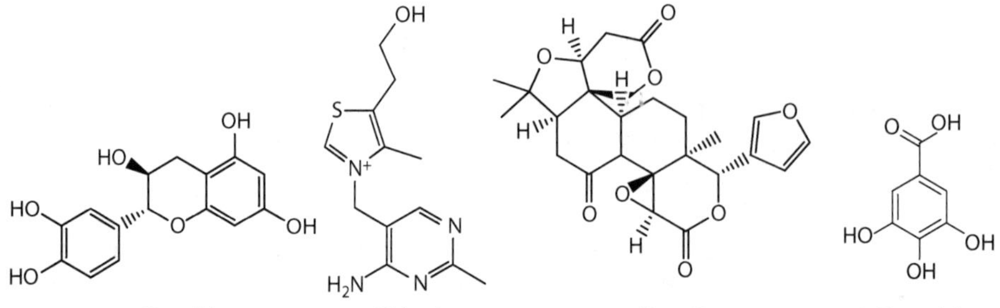

Figure 5.1 Chemical structures of some functional foods.

Taste masking of bitter tasting components is also economically driven. Improved palatability increases product profit margin (Shahiwala, 2011). Taste-masking agents conceal the authentic taste of a compound by various techniques. They are vital elements in oral drug delivery systems because the taste of the drug can be sensed only in these systems. These agents are thoroughly studied as excipients for use in the nutraceutical and functional food industries.

5.2 Sensory evaluation strategies for taste masking

It is necessary to evaluate the efficiency of the agents or techniques involved in masking the taste in nutraceuticals and functional foods. Conventional and recent novel techniques are used for chemical detection and sensory evaluation. Sensory evaluation that involves panelists or random subjects to interpret testing is sensitive to an individual's psychological state at that given moment and overall personal preference, which leads to conflict in most results. Chemical detection methods such as high-performance liquid chromatography (HPLC) can be used to obtain quantitative but not qualitative data (Lu et al., 2017). Hence, technologies that can mimic the human tongue for detection of taste are needed. Research activity outcomes have led to the development of newer techniques to mask the taste. We describe some of these techniques in the following sections.

5.2.1 Insent taste-sensing system

The Insent taste-sensing system was developed in 2007 by a scientist at Kuyushu University in Japan. It is marketed by Intelligent Sensor Technology Inc., Astugi, Japan. It was the first taste sensor that consisted of lipid/polymer membrane. It is a potentiometric multichannel taste sensor equipped with up to eight lipid membrane sensors that characterize different tastes. It is based on the principle of change of membrane potential caused by adsorption (Haraguchi et al., 2016; Woertz et al., 2011). In a study, researchers tested the ability of an electronic tongue with taste-masking properties on diclofenac using the Insent TS-5000Z system. The results obtained confirmed that it was a valuable tool for analyzing and predicting the taste of components in the early development stage (Guhmann et al., 2012).

5.2.2 α-ASTREE electronic tongue

The α-ASTREE electronic tongue is a quantitative taste sensor. It is a potentiometric based system with a seven-sensor probe, a reference electrode, and an autosampler (see Figure 5.2). It has the ability to evaluate the overall taste of a

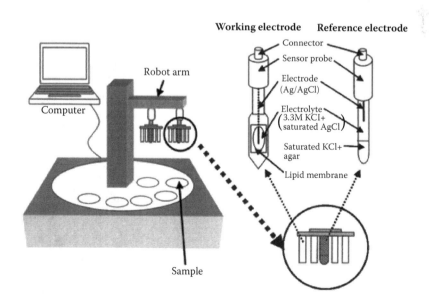

Figure 5.2 *α-ASTREE electronic tongue.*

product by utilizing output values from different electrodes (Haraguchi et al., 2016; Woertz et al., 2011). European scientists conducted a study to develop a quinine dosage form that was taste masked and evaluated it with the ASTREE Electronics Bitterness prediction module. They determined the concentration of polymer needed to mask the bitter taste of quinine (Kayumba et al., 2007).

5.2.3 Carbon nanotubule field-effect transistor bioelectronic sensors

By linking human sensory (olfactory/taste) receptors to multi-channel carbon nanotubule field-effect transistor (CNT-FET) bioelectronic sensors with channel splitters researchers developed a portable and multiplexed sensor to detect the target molecule simultaneously in mixtures. This device is highly sensitive and can be used as a tool for quality assessment of various foods and other substances (Son et al., 2017).

5.2.4 Biomimetic sensors

Biomimetic sensors contain potentiometric, voltammetric, and impedance spectrum sensors that modify with the biomimetic material. They analyze taste sensation and its transduction mechanism and obtain information on taste sensing.

The main disadvantage of biomimetic sensors is that they cannot replicate the biological features of the human tongue. Hence they are unable to identify a single component in a complicated mixture (Lu et al., 2016).

Figure 5.3 shows technologies used in taste masking and their classifications (Douroumis, 2007; Karaman, 2013; Kaushik and Dureja, 2014).

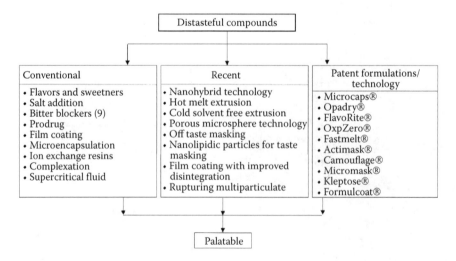

Figure 5.3 *Technologies used in taste masking.*

5.3 Conventional techniques used in taste masking

5.3.1 Sweeteners and flavors

Sweeteners and flavors impart pleasant taste to bitter-tasting compounds. They have been widely used for decades as additives and they remain the most widely used techniques today. Depending on the additive, sweeteners and flavors change the perception of taste or suppress it. This approach is very versatile. Its main disadvantage is that it is not very compelling against very bitter drugs (Walsh et al., 2014) and water-soluble drugs (Kaushik and Dureja, 2014). In industry, sweeteners and flavors are used in combination with other taste-masking techniques (Karaman, 2013).

5.3.1.1 Sweeteners

Sweet taste is acceptable to everyone, particularly the pediatric population. Sugar is the most common sweetening agent used in everyday life. It is used to alter the bitter taste of tea and coffee and the sour taste of lemon. Sweeteners are the primary ingredients incorporated in functional foods to mask unpalatable components.

Sweeteners can be classified as nutritive and non-nutritive. Nutritive sweeteners provide energy together with sweet taste. Non-nutritive sweeteners do not supply energy (American Dietetic Association, 2004). Examples of nutritive sweeteners include sugars, maple syrup, honey, and agave. They are known to show lower glycemic potential than refined sugars. They have been used without issue for centuries. Examples of non-nutritive sweeteners are stevia, saccharin, sucralose, sugar alcohols, and aspartame (Shankar et al., 2013). They contribute more to sweet taste and less to caloric value. Non-nutritive sweeteners can also be classified as natural and artificial sweeteners (Table 5.2). Natural sweeteners include sucrose, glucose, fructose, sorbitol, mannitol, glycerol, honey, and liquorice. Artificial sweeteners include saccharin, saccharin sodium, and aspartame (Tripathi et al., 2011).

Non-nutritive sweeteners are also classified by their chemical structure. Sucrose and glucose are sugars; sorbitol, mannitol, and xylitol are sugar alcohols; aspartame is a peptide; and monellin is a protein (Wu et al., 2016). The main disadvantage of these sweeteners as masking agents is that they can lead to obesity and other health-related problems (Edwards et al., 2016).

Lenik et al. (2016) conducted experiments that suggested that the taste-masking efficiency of sweeteners was altered by its structure. For the bitter drug diclofenac, sweeteners such as cyclodextrin, acesulfame potassium, and sodium saccharin showed better sweetening properties while lactose was deemed inappropriate (Lenik et al., 2016).

Table 5.2 Sweeteners and Their Relative Sweetness

Sweetening Agent	Relative Sweetness*	Significance	Reference
Aspartame	185–200	Unstable in solution	(Cardello et al., 1999; Kalaskar and Singh, 2014)
Saccharin	288	Unpleasant aftertaste	(Cardello et al., 1999; Kalaskar and Singh, 2014)
Stevia	152	—	(Cardello et al., 1999)
Glycyrrhizin	50	Expensive	(Kalaskar and Singh, 2014)
Sucralose	600	Synergistic sweetening effect	(Kalaskar and Singh, 2014)

* The relative sweetness is measured in relation to sucrose, which is taken as 1.

5.3.1.2 Flavorants

Different flavors are used in the functional food and nutraceutical industries. These flavors harmonize the constituent present and mentioned on the label as well as the source. They may be natural or artificial, depending on the need (Walsh et al., 2014). Direct use of flavoring agents results in problems relating to their consistency, stability, and aroma content. Hence methods such as microencapsulation and entrapment of flavors have been used in recent years (Gupta et al., 2016). These flavors attract customers and promote economic stability of the consumer base. Table 5.3 lists classifications of some flavors (Tripathi et al., 2011).

Table 5.4 summarizes flavoring agents used to mask specific drug tastes (Abraham and Mathew, 2014).

5.3.2 Salt

The addition of salt increases the salty perception and decreases the bitterness. Researchers have demonstrated that in bitter-sweet mixtures salt enhanced the sweetness, hence salts can be used to enhance taste. Researchers also found that the bitterness suppressed by sodium salts differed with the type of salt. For example, sodium salt suppressed the bitterness of KCl, urea, and

Table 5.3 Classification of Flavors

Flavor Type	Examples
Natural Flavors	Juices (raspberry)
	Extracts (liquorice)
	Spirits (lemon and orange)
	Syrups (blackcurrant)
	Tinctures (ginger)
	Aromatic waters (anise and cinnamon)
	Aromatic oils (peppermint and lemon)
Synthetic Flavors	Aqueous solutions (mint flavored, banana flavored)
	Alcoholic solutions (benzyl alcohol as solvent for flavoring agents)
	Flavored powders

Table 5.4 Flavoring Agents Used to Mask Drug Tastes

Drug Taste	Flavoring Agent Used
Sweet	Honey, vanilla, chocolate, mint, raspberry, bubble gum, wild cherry, fruit and berry
Bitter	Grapefruit, passion fruit, peach, orange, lemon, lime
Acidic sour	Lemon, lime, orange, cherry, grapefruit, liquorice
Alkaline	Mint, chocolate, cream, vanilla
Salty	Butterscotch, maple apricot, melon

amiloride, but it was less effective in suppressing the bitterness of quinine-HCl and caffeine (Keast et al., 2001).

Several mechanisms may be responsible for this suppression. Sodium may influence G-protein coupled receptors. It modulates various ion channels or pumps involved in the taste transduction pathway. Sodium may stabilize the cellular membrane and limit access to lipophilic bitter compounds to receptor sites in the membrane. It also inhibits direct access to those bitter compounds through the membrane into intracellular pathways (Keast et al., 2001). This method also has some limitations. High sodium intake can cause health issues such as high blood pressure. Zinc salt may impart astringent taste and may or may not reduce bitterness (Sun-Waterhouse and Wadhwa, 2013).

5.3.3 Bitter blockers

Bitter blockers are compounds that reduce bitterness. There is no compound yet on the market that can directly inhibit the bitterness of a compound (Gaudette and Pickering, 2013). Bitter blockers work by interfering with taste transduction signals that travel from mouth to brain. They are more effective than sweeteners and flavors. They are effective even at low concentrations (Walsh et al., 2014). With the absence of potent bitter blockers, there has been a significant rise in the development of new techniques such as microencapsulation, complexation, film coating, disintegration, and the prodrug approach. Table 5.5 lists examples of bitter blockers and their mechanisms (Gaudette and Pickering, 2013).

Table 5.5 Bitter Blockers, Target Molecules, and Modes of Action

Bitter Blocker	Target Molecule	Mode of Action
Cyclodextrin	Naringin, limonin	Molecular encapsulation
Flavanones	Caffeine, naringin, 6-methoxyflavanone	Hydrophobic interaction
Zinc salts	Caffeine, tetralone, quinine-HCl	Interaction with amino acids in taste receptor cells (TRC)
Magnesium	Quinine-HCl	–

Bitter blockers have shortcomings. There are no reports of universal bitter blockers. They have low regulatory acceptability. No proper mechanism of action is known (Walsh et al., 2014).

5.3.4 Prodrugs

The prodrugs approach is based on intramolecular processes that make use of density functional theory (DFT) methods and experimental and calculated reaction rates without the use of enzymes. Changing the competency of the drug to interact with taste receptors may reduce or cancel out bitterness. Bitter-tasting molecules require a polar group and hydrophobic activity. Reducing the number of these groups may decrease bitterness. Nonpolar bitter prodrugs are administered, they are metabolized into drugs, and exert biological action. For example, the addition of pyridinium moiety to an amino acid chain of a bitter compound reduces bitterness. Changing the conformation of a bitter-tasting drug can also affect its ability to bind to bitter receptors, e.g., L-tryptophan is bitter while D-tryptophan is sweet (Karaman, 2013). Table 5.6 lists examples of drugs and the modifications made to mask their tastes (Kalaskar and Singh, 2014; Sajal et al., 2008).

5.3.5 Film coating

Film coating is a widely used technique that masks the bitter taste of any component. It works by decreasing the contact of bitter compounds or by reducing the speed of interaction with taste receptors, thereby halting the perception of taste. A boundary around bitter-tasting ions or molecules is created by increasing viscosity of the polymer used and stearic hindrance (Sun-Waterhouse and Wadhwa, 2013).

Natural and synthetic polymers are used for coating, but natural polymers of plant, animal, or microbial origin are preferred for forming a barrier around food or food-related material. Natural and basic polymers used for these purposes are proteins and carbohydrates (Coupland and Hayes, 2014). In film coatings, colorants can be added to the coating material to improve appearance and provide a glossy finish. Table 5.7 lists polymer types and examples (Joshi and Petereit, 2013).

Table 5.6 Prodrug Modifications That Mask Taste

Drug	Modification Made
Chloramphenicol	Palmitate or phosphate ester
Clindamycin	Alkyl ester
Erythromycin	Alkyl ester
Lincomycin	Phosphate ester
Tetracycline	Benzoate salts
Triamcinolone	Diacetate ester

Table 5.7 Examples of Polymer Types

Polymer Type	Examples
Water soluble	Hydroxyethyl cellulose (HEC), hydroxypropyl methoxy cellulose (HPMC), sodium carboxymethyl cellulose (Na-CMC), polyvinyl alcohol, polyvinyl pyrrolidone (PVP)
Cationic	Amino dimethyl methacrylate copolymer (EUDRAJIT®), amino diethyl-methacrylate copolymer (Kollicoat®)
Anionic	Sodium alginate, shellac, carboxyl methyl cellulose (CMC), cellulose acetate phthalate (CAP), cellulose acetate butyrate (Barra et al. 2007)
Insoluble	Ethyl cellulose, cellulose acetate, polyvinyl acetate, ammonio methacrylate types A and B

5.3.6 Microencapsulation

Unlike film coating, microencapsulation involves coating small bitter particles with a thin layer of film (Kaushik and Dureja, 2014). It creates a border around the bitter-tasting molecule to decrease initial perception of the unpalatable taste. It is a widely used process in commercial products. It requires proper coupling of flavor with suitable coating material. Proper technique is also important.

Microencapsulation can be classified into two types. The first, microspheres, are prepared mechanically by physical encapsulation technologies, including spray drying, fluidized bed spray coating, spray cooling, melt extrusion, liposome entrapment, and molecular inclusion complexation. The second, microcapsules, are prepared chemically by coacervation or ion cross-linking processes.

The main advantage of microencapsulation is that it masks bad taste. It can be used for sensitive bioactive materials. It prevents degradation of the bitter component. It can also be used to immobilize cells or enzymes in food processing applications. This method provides stability during and after processing (Nedovic et al., 2011).

In 1982, Sobel et al. patented the suspension of microencapsulated ampicillin acid with addition of salt. Ampicillin shows antibacterial activity. Its water-soluble acid salt had unpalatable taste that made it unacceptable for oral use, hence microencapsulation was used. The component was coated with a mixture of ethyl cellulose and the resulting formulation showed palatable taste and good stability (Beatty, 1982).

In 2010, Wakil et al. masked the fishy flavor of fish oil containing eicosapentaenoic acid and increased its bioavailability by encapsulating it within plant spore exines as microcapsules (Wakil et al., 2010).

As with other techniques, microencapsulation has limitations (Kaushik and Dureja, 2014; Sobel et al., 2014). Complexity of flavor system can make it a difficult task. Flavor systems are typically volatile so stability is also an issue. Maintaining phase equilibrium, which is essential for stability, is difficult.

Simple sphere   Multiwalled sphere   Multi-core sphere

Figure 5.4 *Sphere types for microencapsulated particles.*

Retention and diffusion of flavor depend on molecular size, hence high-molecular-weight flavoring agents cannot be used because they show less diffusivity. Other issues include solubility of flavor in the coating material, low encapsulation efficiency, and environmental concerns due to use of organic solvents.

Figure 5.4 shows various sphere types for microencapsulated particles.

5.3.7 Ion-exchange resins

Ion-exchange resins (IERs) are insoluble, ionic compounds with two parts, a polymer matrix and a functional protein. The functional protein is an ion-active group that is exchanged. The protein may have a positively charged group known as a cation exchanger or a negatively charged group known as an anion exchanger. They can also be classified as strong or weak. They are used in the pharmaceutical formulation to provide stability and to mask the taste. They eliminate the bitter taste and do not delay the onset of action. Bitter compounds are used as functional proteins that are exchanged (Figure 5.5). They are bonded to the oppositely charged resin substrate, forming insoluble intermediates by the weak ionic interaction that does not dissociate under salivary pH and is not perceived on the tongue (Guo et al., 2009; Sohi et al., 2004). Table 5.8 lists types of IERs, their commercial preparations, and pH ranges (Sohi et al., 2004; Suhagiya et al., 2010).

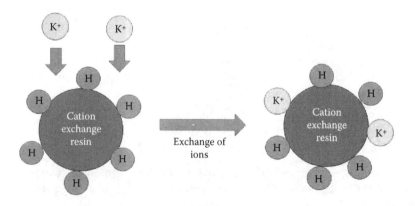

Figure 5.5 *Ion exchange in cationic resin.*

Table 5.8 Ion-Exchange Resins, Their Commercial Preparations, and pH Ranges

Ion-Exchange Resin	Commercial Preparations	pH Range and Use
Strong acid cation exchange resin	Amberlite IR 120, Dowex 50, Tulsion T-344	Entire pH range, used to mask basic drugs
Weak acid cation-exchange resin	Amberlite IRC 50, Indion 204, Tulsion T-335	> 6.0
Strong base anion exchange resin	Amberlite IR 400, Dowex 1, Indion 454	All pH
Weak base anion-exchange resin	Amberlite IR 4B, Dowex 2	< 7.0

IERs are mostly used to mask the unpalatable taste of the main components in chewing gum and chewing tablet taste (Suhagiya et al., 2010). In 1971, scientists associated with Abbott Laboratories filed a patent for chewable tablets including coated particles of pseudoephedrine-weak cation exchange resin. They masked the unpalatable taste of pseudoephedrine using polymethacrylic ion-exchange resin (Amberlite™ CG-50) (Saul and Paul, 1971). This concept can be used for nutraceuticals and functional food preparations. Some of the other resins used are Indion-464, Tulsion-335, C11HMR, Purolite, and Kyron-T-114 (Suhagiya et al., 2010).

This technique has limitations. It sometimes shows delay onset of action. It is also ineffective for drugs with low pH (Kaushik and Dureja, 2014).

5.3.8 Complexation

In complexation, a compound fits into the cavity of the complexing agent, which is also the host molecule, and forms a stable complex. The complexing agent masks the bitter taste by reducing the oral solubility or decreasing the number of the bitter compounds exposed to the tongue (Sohi et al., 2004). Cyclodextrin (CD) is an important complexing agent that is widely used in industry. CDs are cyclic oligosaccharides that have an external hydrophilic surface and an internal hydrophobic cavity. Compounds of proper size and polarity form inclusion complexes in the hydrophobic cavity by noncovalent interaction forces. Complexation depends on size, temperature, shape, hydrophobicity, and the structural form of the bitter compound. β-CD is widely used in industry due to its availability and large cavity (Lee et al., 2010; Sun-Waterhouse and Wadhwa, 2013).

Limitations of complexing agents include their dependence on physiochemical properties of the host compound. They are suitable only for low-dose drugs and they sometimes create a sour taste that must be masked by sweeteners (Kaushik and Dureja, 2014; Sohi et al., 2004). Other complexing agents that are used as taste-masking agents are pectin, sodium alginate and carrageenan (Sun-Waterhouse and Wadhwa, 2013).

5.3.9 Supercritical fluids

Supercritical fluids (SCFs) constitute a one-step process that masks taste with the formation of microparticles. The ability of SCFs to separate a component from a mixture and its control over physiochemical property are helpful in masking the taste. SCFs also reduce particle size and residual content consumption. They are used in the preparation of bitter compound-CD complexes. A medium is used to coat the compound with the polymer. During preparation, the compound and polymer are dissolved in an organic solvent and sprayed in a high-pressure chamber filled with SCF. Precipitated particles are removed with the help of precipitator.

SCFs are part of a one-step, non-toxic and nonhazardous process (Lee et al., 2010). They do not cause any change in chemical composition and can easily be removed from the products. They can be used to coat heat-labile compounds. Drawbacks are cost and limited polymer/drug solubility (Douroumis, 2007).

Many taste-masking options are available. Many agents also have been banned by different regulatory agencies such as FDI/CODEX/BIS, which list drugs considered unsafe for human consumption. One such example is cyclamate, which is banned by the U.S. FDA as a carcinogen. Regulatory agencies permit the use of safe additives in nutraceuticals and functional foods.

5.4 Recent technologies
5.4.1 Nanohybrid technology

The term nanohybrid is used to explain the integration of organic and inorganic components at the nano level. Nanohybrids are also called polymer hybrids (Chujo, 2007). In 2012, Lee et al. published a study in which they masked the bitter taste of sildenafil using nanohybrid technology. Bentonite was the inorganic clay material used. Its main component is montmorillonite (MMT). Bentonite has a negative charge due to the isomorphous cationic exchange of Al^{3+} which is present on MMT with Mg^{2+}. Sildenafil is a cationic drug that has an ionic interaction with the anionic MMT, forming a nanohybrid. The nanohybrid is also coated with polyvinylacetaldiethylaminoacetate (AEA) to increase the rate of drug release in the gut and decrease the release rate in the buccal cavity, thus masking the taste of that drug (Lee et al., 2012).

5.4.2 Hot-melt extrusion

Hot-melt extrusion (HME) is a novel technique that has been used in the last decade (Safdari et al., 2011). This approach increases dissolution of poorly soluble drugs without using solvent. The process includes mixing and melting a bitter component, a polymer, and other additives in a melt extruder and giving required shape to the mass after drying. The melt can also be extruded into fine fibers that can later be powdered. The powder formed has a drug–polymer interaction that masks the taste. It is important that the polymer be stable and soluble in the melt. Another requirement is that the polymer should not solubilize in the mouth pH. Scientists isolated the taste of paracetamol by hot-melt extrusion and evaluated it in vivo and in vitro (Coupland and Hayes, 2014; Douroumis, 2007; Kaushik and Dureja, 2014; Maniruzzaman et al., 2012; Repka et al., 2008).

5.4.3 Solvent-free cold extrusion

Solvent-free cold extrusion follows the same principle as hot-melt extrusion but it does not require any heating or solvent and it uses different binders. Scientists studied and investigated preparation of immediate release pellets through a solvent-free cold extrusion/spheronization using non-toxic solid lipid binder and compared it with binders used in wet extrusion.

This method has many advantages. It is suitable for thermosensitive and hygroscopic material. The process does not involve any temperature control. It is applicable for drugs sensitive to degradation and pseudomorphism because no solvent or drying step is required (Krause et al., 2009).

5.4.4 Off-taste masking agents

As discussed previously, sweeteners are the most widely used taste masking approach. Sweeteners may be natural or synthetic. Their chief disadvantage, an unpleasant off aftertaste, has led to a decrease in the use of sweeteners and contributed to the increase in other technologies. The off-taste masking agents are also known as flavor improvers (Ungureanu and Van Ommeren, 2012). U.S. patents were granted to a scientist for masking the taste of nonnutritive sweeteners using long chain fatty acids and tea extracts. (Johnson and Lee, 2007; Roy et al., 2007).

5.4.5 Lipid nanoparticles

Coating prevents a bitter compound from coming into contact with the receptors present in the tongue, thereby preventing the perception of bitter taste. Another approach is nanosizing the bitter compound and preparing nanoparticles. They can be solid lipid nanoparticles (SLNs) or polymeric nanoparticles. The advantage of SLNs is that the matrix is made from a lipid compound that decreases toxicity. Scientists presented with a challenge to formulate an oral formulation for quinine sulphate, which is extremely bitter, loaded quinine sulphate in SLN to mask the bitter taste and achieve dose precision (Dandagi et al., 2014).

5.4.6 Porous microspheres

Porous microspheres contain pores that may be present internally in the core or externally on the surface. The bitter tasting compound can be dissolved or dispersed in the microsphere core. In contrast to the traditional microsphere, porous microspheres show contrast drug absorption and drug release kinetics

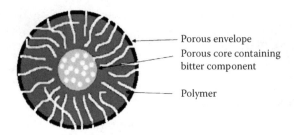

Figure 5.6 *Porous microsphere.*

property. They are formulated by initially preparing different solutions of polymer and bitter component and then mixing them together. The bitter component polymer ratio ranges from 0.05:1 to 3:1. Figure 5.6 shows a schematic diagram of the porous microsphere.

Porous microspheres have many advantages: During the process, there is no loss of active component or change in the ionic form of the component. They do not impart taste in the aqueous process. They can be used to yield products with extended release properties. The porous microsphere is inert. Standard equipment such as SLN granulating and drying apparatus can be used (Cai et al., 2013).

5.4.7 Multi-particulate rupture

The multi-particulate dosage form is a type of oral dosage form that contains numerous small units with some unique qualities. They are spherical in structure with a diameter of 0.05–2.00 mm (Dey et al., 2008). When taken orally they increase absorption and decrease side effects as they disperse freely in the gastrointestinal tract. This dosage decreases the rate of drug release in the mouth but ruptures itself only when it leaves the mouth. This property can be used for taste masking of bitter drugs. Some of the agents are ethyl cellulose and polymethacrylate.

To provide taste masking the material used should be robust enough to remain intact in the mouth and it should be capable of rupturing and releasing the drug once it leaves mouth (Appel et al., 2012).

The time of rupture depends on the type of coating and a swelling agent (Kaushik and Dureja, 2014).

5.5 Patented techniques

5.5.1 Microcaps technology®

Microcaps Technology is a technology patented by Aptalis Pharmaceutical Technologies (Kaushik and Dureja, 2014). This is a pharmaceutical process that uses microcapsules in which taste of the bitter component is masked. It also refers to the formulation containing microcapsules in effervescent formulations.

This technology provides taste-masked microcapsules for oral administration and H_2 antagonists like Ranitidine or a pharmaceutically accepted acid addition salt in taste-masked form. It also provides a method of manufacturing of oral pharmaceutical formulation containing microcapsules in which the component can be taste masked as well as a method that uses phase separation coacervation (Friend et al., 2000; Vummaneni and Nagpal, 2012).

5.5.2 Opadry®

Opadry is Colorcon's one-step original film-coating system that utilizes polymer, plasticizer, and pigment in dry concentrate. It has been a base for several other patents such as Opadry II, Opadry QX, Opadryamb II, Opadryfx, and Opadry 200. These systems differ in inlet air temperature and spray rate used in the process. They show good barrier performance and high coating productivity that can be used for taste masking of the bitter component. It has been used to mask the taste of marketed preparations of vitamins, ibuprofen, and caffeine. (Colorcon, n.d.; Kaushik and Dureja, 2014; Porter and Woznicki, 1987, 1988).

5.5.3 FlavoRite®

The FlavoRite technology was developed by ForTe-bv. It is established on the use of a complex of flavor masking agent and viscosity enhancing agent in the proprietary process. It aims to provide taste masking of bitter components by administering in solid and liquid dosage form for both phases of onset and after-taste. It is designed with pharmaceutical grade Generally Regarded as Safe (GRAS) excipients. It is a fitting technology for custom-made product development for product improvement of existing products. It can be economically used in any conventional cGMP pharmaceutical manufacturing facility (ForTe-bv, n.d.).

5.5.4 OXPzero®

OXPzero is a novel technology used for anionic components with a bitter or bad taste. It is an innovative salt system. It is one-step crystallization process in which a bitter component is enveloped in a layered structure that protects it. OXPzero is insoluble in the mouth but it readily dissolves in the stomach. The starting materials used in this technique are cheap and readily available which decreases manufacturing cost (Kaushik and Dureja, 2014).

5.5.5 Actimask®

Actimask is a technology developed by SPI Pharma (Kaushik and Dureja, 2014). They have taste masked acetaminophen and ibuprofen which are fitting choices for the formulation of the chewable or orally dispersible formulation. It is coated with aqueous coating material. The uniform hydrogel aqueous coating material provides a smooth surface that gives a formulation good mouthfeel and makes it easy to swallow (SPI Pharma, n.d.a,b).

5.5.6 Formulcoat®

Formulcoat is an exclusive technology developed by the supercritical fluid development centre at Pierre Fabre Médicament, France (Kaushik and Dureja, 2014). It is based on the use of supercritical CO_2. It works on the principle of green chemistry, i.e., absence of solvents and use of mild operating conditions. It is a continuous coating procedure used to mask the bitter taste and also increase the bioavailability and solubility of poorly soluble components (Pierre Fabre, n.d.).

5.5.7 Camouflage®

The Camouflage technology was developed by Cambrex Corporation, USA. It uses polymers to mask taste. The polymers used are tasteless and insoluble, which changes the organoleptic properties of bitter components. This technique enhances drug products that are colorless, tasteless, sugar-free, and can be used in a variety of dosage forms (Kaushik and Dureja, 2014).

5.5.8 KLEPTOSE®Linecaps

KLEPTOSE®Linecaps were developed by Roquette Freres, France. They originate from pea starch (maltodextrin) and are rich in soluble and linear amylose. This amylose allows the formation of inclusion complex that is needed for taste masking and other applications. Like cyclodextrin, amylose has a hydrophilic outer surface and a hydrophobic inner cavity. The bitter component enters this cavity, forming bonds, and makes itself inaccessible to the taste buds (Kaushik and Dureja, 2014).

This technology is suitable for a wide range of dosage forms like over-the-counter (OTC) formulations, food supplements, and pediatric and nutraceutical formulations.

Roquette Freres developed a range of KLEPTOSE for the purpose of taste masking and promoting stability, which includes native beta cyclodextrins, hydroxypropyl beta cyclodextrins, and methyl beta cyclodextrin along with pea maltodextrins (Roquette Freres, n.d.).

5.5.9 Micromask®

Micromask® is a patented coating technology (Particles Dynamics International, LLC) that imparts better taste, odor masking, and mouthfeel characteristics. It also shows excellent flow and compression characteristics. It offers barrier protection that reduces the likelihood of degradation and increases stability (Kaushik and Dureja, 2014).

Applications include chewable tablets, oral dissolving tablets, and oral dissolving powders (PD Holdings, n.d.).

5.6 Conclusion

Nutraceuticals are increasing in popularity day by day. The advancement of technology in pharmaceutical sciences has helped to address many challenges faced by nutraceuticals pertaining to the taste. We have made efforts to update the available techniques and technologies addressing the aforesaid problem. However in the ever-changing environment of technological advancement, more techniques are expected in nutraceuticals.

References

Abraham J, Mathew F (2014). Taste masking of paediatric formulation: A review on technologies, recent trends and regulatory aspects. *Int J Pharm Pharm Sci.* 6:12–19.

American Dietetic Association (2004). Position of the American Dietetic Association: Use of nutritive and nonnutritive sweeteners. *J Am Diet Assoc.* 104(2):255.

Appel LE, Friesen DT, LaChapelle ED, Konagurthu S, Falk RF, Reo JP (2012). Taste-masked drugs in rupturing multiparticulates. Google Patents.

Barra A, Coroneo V, Dessi S, Cabras P, Angioni A (2007). Characterization of the volatile constituents in the essential oil of *Pistacia lentiscus* L. from different origins and its antifungal and antioxidant activity. *J Agric Food Chem.* 55(17):7093–7098.

Beatty ML (1982). Suspension of microencapsulated bacampicillin acid addition salt for oral, especially pediatric, administration. Google Patents.

Bhattacharjee S, Majumdar S, Guha N, Dutta G (2016). Taste masking technologies in oral pharmaceuticals: Recent developments and approaches. *World J Pharm Pharm Sci.* 5(8):1752–1764.

Cadwallader K (2015). Flavor challenges and solutions for high protein functional foods and beverages, http://www.globalfoodforums.com/wp-content/uploads/2015/05/2015-PTT-Flavor-Challenge-Keith-Cadwallader.pdf

Cai Y, Chen Y, Hong X, Liu Z, Yuan W (2013). Porous microsphere and its applications. *Int J Nanomed.* 8(1):1111–1120.

Cardello H, Da Silva M, Damasio M (1999). Measurement of the relative sweetness of stevia extract, aspartame and cyclamate/saccharin blend as compared to sucrose at different concentrations. *Plant Foods Hum Nutr.* 54(2):119–129.

Chujo Y (2007). Organic–inorganic nano-hybrid materials [Translated]. *Kona.* 25:255–260.

Collera-Zúñiga O, Jiménez FGA, Gordillo RM (2005). Comparative study of carotenoid composition in three Mexican varieties of *Capsicum annuum* L. *Food Chem.* 90(1):109–114.

Colorcon (n.d.). Opadry complete film coating system, http://www.colorcon.com/products-formulation/all-products/film-coatings/immediate-release/opadry, accessed March 27, 2017.

Coupland JN, Hayes JE (2014). Physical approaches to masking bitter taste: Lessons from food and pharmaceuticals. *Pharm Res.* 31(11):2921–2939.

Dandagi PM, Rath SP, Gadad AP, Mastiholimath VS (2014). Taste masked quinine sulphate loaded solid lipid nanoparticles for flexible pediatric dosing. *Indian J Pharm Educ Res.* 48(4):93–99.

Das L, Bhaumik E, Raychaudhuri U, Chakraborty R (2012). Role of nutraceuticals in human health. *J Food Sci Technol.* 49(2):173–183.

Dey N, Majumdar S, Rao M (2008). Multiparticulate drug delivery systems for controlled release. *Trop J Pharm Res.* 7(3):1067–1075.

Douroumis D (2007). Practical approaches of taste masking technologies in oral solid forms. *Expert Opin Drug Deliv.* 4(4):417–426.

Edwards CH, Rossi M, Corpe CP, Butterworth PJ, Ellis PR (2016). The role of sugars and sweeteners in food, diet and health: Alternatives for the future. *Trends Food Sci Technol.* 56:158–166.

ForTe-bv (n.d.). FlavoRite® taste masking technology, http://www.forte-bv.nl/index.php?id=21, accessed March 28, 2017.

Friend DR, Ng S, Sarabia RE, Weber TP, Geoffroy J-M (2000). Taste-masked microcapsule compositions and methods of manufacture. Google Patents.

Gaudette NJ, Pickering GJ (2013). Modifying bitterness in functional food systems. *Crit Rev Food Sci Nutr.* 53(5):464–481.

Guhmann M, Preis M, Gerber F, Pöllinger N, Breitkreutz J, Weitschies W (2012). Development of oral taste masked diclofenac formulations using a taste sensing system. *Int J Pharm.* 438(1):81–90.

Guo X, Chang RK, Hussain MA (2009). Ion-exchange resins as drug delivery carriers. *J Pharm Sci.* 98(11), 3886–3902.

Gupta S, Khan S, Muzafar M, Kushwaha M, Yadav AK, Gupta AP (2016). Encapsulation: Entrapping essential oil/flavors/aromas in food. *Encapsulations.* 2:229.

Haraguchi T, Yoshida M, Kojima H, Uchida T (2016). Usefulness and limitations of taste sensors in the evaluation of palatability and taste-masking in oral dosage forms. *Asian J Pharm Sci.* 11(4):479–485.

Johnson W, Lee T (2007). Long chain fatty acids for reducing off-taste of non-nutritive sweeteners. Google Patents.

Joshi S, Petereit H-U (2013). Film coatings for taste masking and moisture protection. *Int J Pharm.* 457(2), 395–406.

Kalaskar R, Singh R, (2014). Taste masking: A novel technique for oral drug delivery system. *Asian J Pharm Res Dev.* 2(3):1–14.

Karaman R (2013). Prodrugs for masking bitter taste of antibacterial drugs—A computational approach. *J Mol Model.* 19(6):2399–2412.

Kaushik D, Dureja H (2014). Recent patents and patented technology platforms for pharmaceutical taste masking. *Recent Pat Drug Deliv Formul.* 8(1):37–45.

Kayumba P, Huyghebaert N, Cordella C, Ntawukuliryayo J, Vervaet C, Remon JP (2007). Quinine sulphate pellets for flexible pediatric drug dosing: Formulation development and evaluation of taste-masking efficiency using the electronic tongue. *Eur J Pharm Biopharm.* 66(3):460–465.

Keast RS, Breslin PA, Beauchamp G.K (2001). Suppression of bitterness using sodium salts. *Chimia Int J Chem.* 55(5):441–447.

Krause J, Thommes M, Breitkreutz J (2009). Immediate release pellets with lipid binders obtained by solvent-free cold extrusion. *Eur J Pharm Biopharm.* 71(1):138–144.

Lee C-W, Kim S-J, Youn Y-S, Widjojokusumo E, Lee Y-H, Kim J, Lee Y-W, Tjandrawinata RR (2010). Preparation of bitter taste masked cetirizine dihydrochloride/β-cyclodextrin inclusion complex by supercritical antisolvent (SAS) process. *J Supercrit Fluids.* 55(1):348–357.

Lee J-H, Choi G, Oh Y-J, Park JW, Choy YB, Park MC, Yoon YJ, Lee HJ, Chang HC, Choy J-H (2012). A nanohybrid system for taste masking of sildenafil. *Int J Nanomed.* 7:1635–1649.

Lenik J, Wesoły M, Ciosek P, Wróblewski W (2016). Evaluation of taste masking effect of diclofenac using sweeteners and cyclodextrin by a potentiometric electronic tongue. *J Electroanal Chem.* 780, 153–159.

Ley JP, Krammer G, Reinders G, Gatfield IL, Bertram H-J (2005). Evaluation of bitter masking flavanones from Herba Santa (*Eriodictyon californicum* (H. & A.) Torr., Hydrophyllaceae). *J Agric Food Chem.* 53(15):6061–6066.

Lu L, Hu X, Zhu Z (2017). Biomimetic sensors and biosensors for qualitative and quantitative analyses of five basic tastes. *TrAC Trends Anal Chem.* 87:58–70.

Maniruzzaman M, Boateng JS, Bonnefille M, Aranyos A, Mitchell JC, Douroumis D (2012). Taste masking of paracetamol by hot-melt extrusion: An in vitro and in vivo evaluation. *Eur J Pharma Biopharma.* 80(2):433–442.

Martirosyan DM, Singh J (2015). A new definition of functional food by FFC: What makes a new definition unique? *Funct Foods Health Dis.* 5(6):209–223.

Nedovic V, Kalusevic A, Manojlovic V, Levic S, Bugarski B (2011). An overview of encapsulation technologies for food applications. *Procedia Food Sci.* 1:1806–1815.

Pandey M, Verma RK, Saraf SA (2010). Nutraceuticals: New era of medicine and health. *Asian J Pharma Clin Res.* 3(1):11–15.

PD Holdings (n.d.). MicroMask, http://pdhllc.com/products/micromask, accessed March 28, 2017.

Pierre Fabre (n.d.). Supercritical fluid services, http://www.pierre-fabre.com/en/supercritical-fluids-services, accessed March 27, 2017.

Porter SC, Woznicki EJ (1987). Maltodextrin coating. Google Patents.

Porter SC, Woznicki EJ (1988). Maltodextrin coating. Google Patents.

Prabu SL, Suriyaprakash TNK, Kumar CD, Kumar SS (2012). Nutraceuticals and their medicinal importance. *Int J Health Allied Sci.* 1(2):47.

Repka MA, Majumdar S, Kumar Battu S, Srirangam R, Upadhye SB (2008). Applications of hot-melt extrusion for drug delivery. *Expert Opin Drug Deliv.* 5(12):1357–1376.

Roberfroid MB (2000). Concepts and strategy of functional food science: The European perspective. *Am J Clin Nutr.* 71(6):1660s–1664s.

Roquette Freres (n.d.). Solubility and stability improvement and taste masking, https://www.roquette.com/pharma/oral-dosage/solubility-stability-taste-masking, accessed March 23, 2017.

Roy G, Johnson W, Lee T (2007). Tea extracts for reducing off-taste of non-nutritive sweeteners. Google Patents.

Safdari DF, Salehian T, Ganji F, Beygi M (2011). The effect of chewing sugar free gum after elective cesarean-delivery on return of bowel function in primiparous women. *QOM Univ Med Sci J.* 4(4):16–20.

Sajal JK, Uday SR, Surendra V (2008). Taste masking in pharmaceuticals: An update. *J Pharm Res.* 1(2):126–130.

Santini A, Tenore GC, Novellino E (2017). Nutraceuticals: A paradigm of proactive medicine. *Eur J Pharm Sci.* 96:53–61.

Saul B, Paul SD (1971). Chewable tablets including coated particles of pseudoephedrine-weak cation exchange resin. Google Patents.

Shahiwala A, (2011). Formulation approaches in enhancement of patient compliance to oral drug therapy. *Expert Opin Drug Deliv.* 8(11):1521–1529.

Shankar P, Ahuja S, Sriram K (2013). Non-nutritive sweeteners: Review and update. *Nutrition.* 29(11):1293–1299.

Sobel R, Gundlach M, Su C-P (2014). Novel concepts and challenges of flavor microencapsulation and taste modification. In: Gaonkar AG, Vasisht N, Khare AT, Sobel R (eds). *Microencapsulation in the Food Industry: A Practical Implementation Guide.* Amsterdam: Academic Press; 421–442.

Sohi H, Sultana Y, Khar RK (2004). Taste masking technologies in oral pharmaceuticals: Recent developments and approaches. *Drug Dev Ind Pharm.* 30(5):429–448.

Son M, Kim D, Ko HJ, Hong S, Park TH (2017). A portable and multiplexed bioelectronic sensor using human olfactory and taste receptors. *Biosens Bioelectron.* 87:901–907.

SPI Pharma, n.d.a. Actimask—Acetaminophen, https://www.spipharma.com/products/actimask-acetaminophen, accessed July 27, 2017.

SPI Pharma (n.d.b.). Actimask—Ibuprofen, https://www.spipharma.com/products/actimask-ibuprofen, accessed March 27, 2017.

Suhagiya V, Goyani A, Gupta R (2010). Taste masking by ion exchange resin and its new applications: A review. *Int J Pharm Sci Res.* 1(4):22–37.

Sun-Waterhouse D, Wadhwa SS (2013). Industry-relevant approaches for minimising the bitterness of bioactive compounds in functional foods: A review. *Food Bioproc Tech.* 6(3):607–627.

Szejtli J, Szente L (2005). Elimination of bitter, disgusting tastes of drugs and foods by cyclodextrins. *Eur J Pharm Biopharm.* 61(3):115–125.

Tepper BJ, Keller KL, Ullrich NV (2003). Genetic variation in taste and preferences for bitter and pungent foods: Implications for chronic disease risk. In: Hofmann TL, Ho CT, Pickenhagen W (eds). *Challenges in Taste Chemistry and Biology*, Vol. 867. American Chemical Society Symposium Series. Washington, DC: American Chemical Society; 60–74.

Tripathi A, Parmar D, Patel U, Patel G, Daslaniya D, Bhimani B (2011). Taste masking: A novel approach for bitter and obnoxious drugs. *J Pharma Sci Biosci Res.* 1(3):36–142.

Ungureanu IM, Van Ommeren E (2012). Off-taste masking. Google Patents.

Valentová H, Skrovánková S, Panovská Z, Pokorný J (2002). Time-intensity studies of astringent taste. *Food Chem.* 78(1):29–37.

Vummaneni V, Nagpal D (2012). Taste masking technologies: An overview and recent updates. *Int J Res Pharma Biomed Sci.* 3(2):510–525.

Wakil A, Mackenzie G, Diego-Taboada A, Bell JG, Atkin SL (2010). Enhanced bioavailability of eicosapentaenoic acid from fish oil after encapsulation within plant spore exines as microcapsules. *Lipids.* 45(7):645–649.

Walsh J, Cram A, Woertz K, Breitkreutz J, Winzenburg G, Turner R, Tuleu C, Initiative EF (2014). Playing hide and seek with poorly tasting paediatric medicines: Do not forget the excipients. *Adv Drug Deliv Rev.* 73:14–33.

Woertz K, Tissen C, Kleinebudde P, Breitkreutz J (2011). Taste sensing systems (electronic tongues) for pharmaceutical applications. *Int J Pharm.* 417(1):256–271.

Wu X, Onitake H, Haraguchi T, Tahara Y, Yatabe R, Yoshida M, Uchida T, Ikezaki H, Toko K (2016). Quantitative prediction of bitterness masking effect of high-potency sweeteners using taste sensor. *Sens Actuat B: Chem.* 235:11–17.

6

The Effect of Bitter Components on Sensory Perception of Food and Technology Improvement for Consumer Acceptance

Geeta M. Patel and Yashawant Pathak

Contents

6.1 Introduction 146
6.2 Types of taste 147
    6.2.1 Sweet 147
    6.2.2 Sour 147
    6.2.3 Salty 147
    6.2.4 Bitter 147
    6.2.5 Umami 148
6.3 Importance of taste 149
6.4 Aspects of taste perception 149
6.5 Bitter perception 150
    6.5.1 A common myth: Bitter is in back, sweet is in front 150
    6.5.2 Mechanisms of bitter taste perception 151
    6.5.3 Bitter taste perception 152
    6.5.4 Bitter taste receptors as potential therapeutic targets 153
6.6 Genetics of sweet and bitter perception 153
    6.6.1 Naturally occurring alleles 153
    6.6.2 Cross-species comparisons 154
    6.6.3 Family and twin studies 154
6.7 Taste perception and behavior 155
    6.7.1 Challenges in the population 155
    6.7.2 Bitter taste and food rejection 156
6.8 Consumer acceptance of functional foods 156
    6.8.1 Socio-demographic determinants 156
    6.8.2 Cognitive and attitudinal determinants 157

6.9   Challenges in consumer acceptance and successful marketing of products 157
    6.9.1   Role of sensory science in decision making 158
    6.9.2   Sensory evaluation and quality of food 159
6.10 Taste improvement technology... 159
    6.10.1   Taste inhibition.. 160
    6.10.2   Suppression of mixture.. 160
    6.10.3   Masking by encapsulation... 161
    6.10.4   Using food technology to create new flavors........................ 162
    6.10.5   Role of umami in food science 162
6.11 Conclusion 162
References 163

6.1 Introduction

Food and food components are essential to life. Food habits differ by geographic region. The traditional methods for processing and preservation of food in these areas also vary. The scientific study of food is important. The functional food sector is the fastest-growing sector in the food industry. Extensive critical and operational efforts by leading food, pharmaceutical, and biotechnology industries began in the 1990s (Childs and Poreezes 1997).

Food science is the application of physical, biological, and behavioral sciences to the processing, preserving, and marketing of foods. Food science highlights technology for processing but also considers the nutritional value and good taste of food.

Nutrients or other chemical compounds of the foods activate specialized receptor cells in the oral cavity when the sense of taste is stimulated. The risk of making poor food selection is not only cause wasting of energy but it also incur metabolic harm. Taste serves two functions. It enables us to evaluate foods for toxicity and nutrients to decide what to eat and it prepares the body for metabolism of foods (Breslin 2013).

The tongue and soft palate are responsible for perception of taste. Specific chemicals in foods or drinks interact with taste receptors located there. These receptors are found on specialized epithelial cells and have neuron-like properties such as release of neurotransmitters, depolarization, and the ability to form synapses to afferent neurons (Zhao et al. 2005). These taste receptors are Types I, II, and III. Types I and II stain cells separately during their histological preparation. Type I and II cells are called dark cells and light cells, respectively. Type III cells are intermediate in appearance between light and dark cells. Taste receptor cells contain neuropeptides that may modify the activity of neighboring cells (Finger et al. 2005).

Functional food molecules stimulate the taste buds in epithelia of the oral cavity and pharynx that perceive taste (Pedersen et al. 2002). Moreover, the primal sense of "acceptable" or "unacceptable" food for what is ingested is driven by taste. Taste stimuli are typically released when food is chewed, disintegrates, and dissolves into saliva and is pre-digested by oral enzymes (Mattes 2011). Smell and strong sensations combine with taste to form flavor, which allows us to analyze and recognize food as known or new. If the food is novel, we can use this sensory indication to learn about the physiological outcomes of the ingested food. A number of taste qualities of other nutrients have been suggested, including specific taste perceptions from starch, water, calcium, maltodextrins, and fatty acids. However, there is presently little agreement on how humans perceive these chemicals and accordingly on whether we would report our oral experiences with them as unique tastes.

6.2 Types of taste

Five primary types of taste have been identified based on the message transported from the tongue to the brain. Taste can be classified as sweet, sour, salty, bitter, and umami. Many recipes and dishes are made up of a mixture of different tastes. Taste buds differentiate tastes by detecting interaction with different molecules or ions (Yarema et al. 2006).

6.2.1 Sweet

What we perceive as sweetness is generally due to the presence of sugar and its derivatives such as lactose or fructose. Activation of the sensory cells that respond to sweetness can be caused by other types of molecules or substances such as amino acids and alcohol in fruit juices or alcoholic drinks. Figure 6.1 summarizes primary actions and common sources of various tastes.

6.2.2 Sour

Sour taste is mainly found in citrus fruits such as lemon and limes and in sour milk products such as yogurt, cheese, and sour cream. Various fermented substances also have sour taste. The sensation of sour taste is caused by H^+ split off by an acid dissolved in a watery solution. Sour taste aids circulation, waste elimination, and digestion. It also nourishes the vital tissues.

6.2.3 Salty

Salty taste is obtained from food that contains table salt. The chemical basis of this taste is salt crystals, which consist of sodium and chloride. Other minerals such as potassium or magnesium salts may also account for a salty sensation.

6.2.4 Bitter

Various substances are responsible for bitter taste. Bitter taste is composed of air and ether. It is light, cooling, and dry in nature. The main sources of

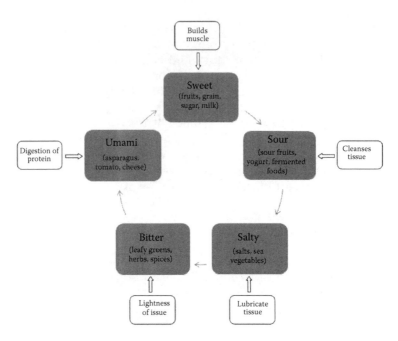

Figure 6.1 *Primary actions and common sources of various tastes. (Modified from Yarema, T., Rhoda, D., and Brannigan, J. 2014. Eat—Taste—Heal: An Ayurvedic Guidebook and Cookbook for Modern Living. New York: Five Elements Press.)*

this taste are leafy green vegetables, herbs and spices, coffee, tea, and certain fruits, specifically grapefruit, olives, and bitter melon. Bitter taste helps stimulate the appetite and also brings out the flavor of other tastes. It is a potent detoxifying agent that is antibiotic, anti-parasitic, and antiseptic.

Research has shown that TAS2Rs (taste receptors, type 2, also known as T2Rs) such as TAS2R38 are responsible for the human ability to taste bitter substances (Maehashi et al. 2008). These taste receptors are identified not only by their ability to taste certain bitter ligands but also by the morphology of the receptor itself, which is surface bound and monomeric (Lindemann 2001).

6.2.5 Umami

Umami is often described as a "savoury" or "meaty" taste that is differ from sweet, salty, bitter, and sour. The term was coined in 1908 by a Kikunae Ikeda, a chemist at Tokyo University, when he began to identify the active principle in kombu and identified the principle in the same year. Ikeda had noticed this taste in vegetables such as asparagus and tomatoes and some cheeses and meat. He analyzed dashi, the Japanese stock that is the basis of Japanese cooking, and eventually identified the amino acid glutamate that is in seaweed as the source of this taste sensation. Glutamates in crystal form are known as MSG, the flavor enhancer that is so prevalent in Japanese cooking.

In the late 1900s, umami was internationally recognized as the fifth basic taste based on psychophysical, electrophysiological, and biochemical studies. Three umami receptors (T1R1 + T1R3, mGluR4, and mGluR1) were identified (Kurihara 2015). Since umami has its own receptors rather than arising out of a combination of the traditionally recognized taste receptors, scientists now consider umami to be a distinct taste (Torii et al. 2013).

Glutamic acid occurs naturally in some foods, such as mushrooms and anchovies. It develops in foods after cooking (such as seared beef), aging (such as Parmesan cheese), or fermentation (soy sauce).

6.3 Importance of taste

Foods not only nourish the body but also the soul. Taste and smell sensations prepare our bodies for digestion. The taste and smell sensations trigger the salivary glands and produce digestive juices. Taste also increases the desire for healthy foods. Taste is important in choice of food. Foods that taste good and are rich in fat, sugar, and salt are highly palatable and consumable. Taste is also crucial for our health because it provides information about our food. For example, the flavor and smell of food sometimes indicate whether food, such as fish and meat, are fresh. A single bite of food may detect an off-smell and an off-taste. We get signals from the brain that food is spoiled. People who cannot taste or smell food often lose weight because they have little or no desire to eat.

Human beings are omnivores. The variety of foods and variation in nutrient content and the risk and hazards of accidental toxin ingestion are complicated by different feeding approaches. An omnivore consumes a variety of foods but has fewer nutritional decisions and faces fewer hazards from toxins compared, for example, to pandas, which mainly eat a specialized diet of bamboo. As a result, their gustatory systems have diminished sensations. Giant pandas have lost the amino acid taste receptor gene *TAS1R1* (Li et al. 2005). In contrast, carnivorous mammals have retained the amino acid taste receptor but have lost many others. For example, all cats have lost their canonical sweet taste receptor gene, *TAS1R2* (Jiang et al. 2012). Carnivorous mammals such as sea lions have even more taste receptor pseudo genes and appear to have lost a large number of taste receptors. Many species appear to have lost some or all of their taste receptors because they do not need the specific nutrient detectors.

6.4 Aspects of taste perception

The perception of taste comes from the everyday experiences we have with functional foods and their compounds. Taste is a multi-characteristic sensation. Taste perceptions have the characteristics of quality and strength. Taste can indicate

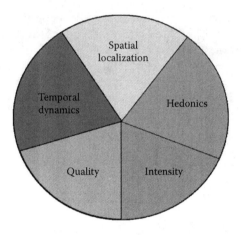

Figure 6.2 Various aspects of taste. (Modified from Breslin, PAS 2013. An evolutionary perspective on food review and human taste. Curr. Biol. 23(9), R409–R418.)

location and timing (Lugaz et al. 2005; Shikata et al. 2000; Delwiche et al. 2000; Stevens 1996), such as when bitter tastes remain too long in the back of the throat (Figure 6.2). If stimuli responsible for multiple taste qualities are combined into a cocktail containing sucrose, monosodium glutamate, sodium chloride, citric acid, and quinine sulfate, someone who drinks this cocktail would experience it as simultaneously sweet, savory, salty, sour, and bitter (Breslin and Beauchamp 1995).

The taste system is thus able to examine the individual components of a complex mixture of foods, consistent with the idea that it analyzes foods for nutritional ingredients. However, this does not prevent the components of a taste mixture interacting with one another to alter cognizance. Certain combinations of stimuli interact in the taste buds and receptor cells. Some such as salts and toxins show inhibition. Many combinations of strong stimuli can interact cognitively to suppress or enhance one another (Breslin et al. 1997). For example, gorillas have been found to tolerate more bitter plant tannins if a high sugar content is also present. There are also numerous chances of taste–taste interactions (Remis and Kerr 2002; Breslin 1996).

6.5 Bitter perception

6.5.1 A common myth: Bitter is in back, sweet is in front

There has long been a misunderstanding that the tongue has specific areas for each flavor or aroma where sweet, bitter, or sour taste can be sensed. This misconception is based on an incorrect and improper reading of diagrams of the tongue. These zones of the tongue are still found in many textbooks and literature.

In reality, all five tastes—sweet, sour, salty, bitter, and savory—can be sensed by all parts or zones of the tongue. The side parts of the tongue are more sensitive than the middle part of tongue for all tastes, with the exception of bitter taste. The back of the tongue is more sensitive to bitter taste perception. This means we can spit out poisonous or rotten foods or food products before they enter the throat and are swallowed (Breslin and Huang 2006).

6.5.2 Mechanisms of bitter taste perception

Consumption and ingestion of food start with the tongue. The tongue behaves as a "gatekeeper." It differentiates between good and harmful substances and helps guide us in our food choices.

The tongue is a very complex organ containing thousands of taste buds (Figure 6.3). These taste buds are small structures that mostly reside on papillae (or raised bumps) on the upper surface of the tongue and on the palate (Iwatsu et al. 2012). Taste buds are the sensory organs within these epithelia. A taste bud is a tiny rosette-shaped cluster of approximately 80–100 receptor cells. The chemicals from the ingested foods are analyzed and detected by transmembrane receptors.

Scientists classify these cells into four subsets, types I through IV. Type I are the most abundant taste cells in taste buds. They act as support cells, mediating biological processes following intense taste stimulation. Type I cells can also detect salt taste. Type II cells are the most extensively studied taste cells. They have specific receptor proteins on their surfaces that allow each

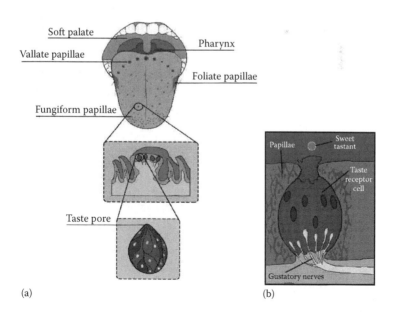

Figure 6.3 *(a) Taste papillae and taste buds of the human tongue. (b) Harbors of taste buds. (Modified from Breslin, PAS 2013. An evolutionary perspective on food review and human taste. Curr. Biol. 23(9), R409–R418.)*

cell to sense sweet, bitter, or umami taste (Young et al. 2009; Janssen and Depoortere 2013). Type III cells detect sour taste. The function of type IV cells is not fully understood. Release of neurotransmitters, a kind of chemical, by specific receptor triggering is responsible for the activation of the specific region of the brain where taste is identified, recognized, and processed (Zhang et al. 2003).

The binding of substances with a bitter taste, e.g., quinine, phenylthiocarbamide (PTC), also takes place on G-protein-coupled receptors that are coupled to gustducin. The signaling cascade is the same as for sweet. Humans have genes encoding 25 different bitter receptors (T2Rs). Each taste cell responsive to bitter expresses a number (4–11) of these genes. This is in sharp contrast to the olfactory system where a single odor-detecting cell expresses only a single type of odor receptor. Nevertheless, a single taste cell seems to respond to certain bitter-tasting molecules in preference to others. The sensation of taste, like all sensations, resides in the brain. Transgenic mice that express T2Rs in cells that normally express T1Rs (sweet) respond to bitter substances as though they were sweet. They also express a receptor for a tasteless substance in cells that normally express T2Rs (bitter) and that are repelled by the tasteless compound. It is the activation of hard-wired neurons that determines the sensation of taste, not the molecules or the receptors themselves.

6.5.3 Bitter taste perception

Researchers have found that distinct groups of type II taste cells contain receptors that distinguish between the sweet taste of sugar and the bitter taste of coffee. These receptors come from a family of proteins known as G-protein-coupled receptors (GPCRs)—T1R2, T1R3, and T2R—that is distantly related to metabotropic glutamate, pheromone, extracellular-calcium sensing, and γ-aminobutyric-acid type B receptors (Hoon 1999; Bachmanov 2001; Nelson 2002; Li 2002). GPCRs are responsible for sensing a wide arrangement of substances located in the immediate surrounding area of cells (Kitagawa et al. 2001; Max 2001; Montmayeur et al. 2001; Sainz et al. 2001; Zhao 2003).

The activation of a GPCR by a particular substance triggers a cascade of signals within the cell that results in diverse cellular responses, as is the case during taste perception. T1R2 and T1R3 receptors specifically identify a spectrum of sweet tastants with a wide range and variety of chemical structures, including synthetic sweeteners, sugars, and sweet-tasting proteins. T2R receptors recognize bitter compounds. Sweet substances activate sweet taste receptors such as T1R2 and T1R3 by signaling proteins situated within the cell such as α-gustducin, PLC-β2, IP3R, and TRPM5 (Nakamura et al. 2012). Interestingly, scientists also investigated whether the same collection of signaling proteins is required for bitter taste perception. The removal of any of these receptors results in a decrease or complete loss of sensitivity for sweet or bitter tastes. Because these signaling proteins, along with the receptors themselves, are thought to be found exclusively in taste cells, scientists have explained them as "protein expression markers," which are able to differentiate taste cells from other types of cells in the body. However, in the last few years, it has been observed that these "protein expression markers" are present in organs of the body other than the tongue (http://sitn.hms.harvard.edu/flash/2013/the-bittersweet-truth-of-sweet-and-bitter-taste-receptors/).

6.5.4 Bitter taste receptors as potential therapeutic targets

Bitter taste receptors located in the stomach will reject harmful toxic substances in food. Recent investigations have found that the gut stimulates the production of hormones actively involved in stimulation of appetite and it is due to activation of bitter taste receptors. One study on mice indicated that, when bitter tastants were administered by insertion of a tube through the stomach, bitter taste receptors influenced the release of ghrelin, which is an appetite-stimulating hormone that resulted in short-term food intake (Zhang et al. 2003). This was immediately followed by a prolonged decrease in food ingestion, correlating with an observed delay in emptying of the stomach, leading to a sensation of satisfaction. The relationship between the ingestion of bitter compounds and a feeling of fullness indicates and suggests new way for researchers and scientists to design and developed treatments for obesity such as "bitter pills."

Bitter taste receptors were also targeted for treating asthma. Bitter taster receptors are also unexpectedly found on airway smooth muscle. The activation of these bitter taste receptors (TAS2Rs) causes smooth muscle relaxation by a different mechanism than that from β_2-adrenegic receptors. This might be responsible for the emergence of a novel class of bronchodilator. Extraoral bitter taste receptors also have been investigated for off-target drug effect mediators. Recently, the structural basis for bitter taste receptor activation and its role in metabolic diseases such as diabetes were reported. Bitter taste receptors appear to be promising therapeutic targets for metabolic diseases.

6.6 Genetics of sweet and bitter perception

Generally, bitter taste perception is considered bad and undesirable, but it shares several features with sweet taste perception. Bitter taste components that humans perceive as bad have diverse structure. Both bitter and sweet compounds bind to gene GPCRs. For the sweet and umami receptors the genes are limited in number to only a few because it is a small family. For the bitter component receptors the family is large, consisting of 30–50 genes.

6.6.1 Naturally occurring alleles

There is a difference in bitter taste perception in mice and humans. Natural alleles of bitter receptors exist and are heritable in both mice and humans. They are also specific to some bitter compounds but not for others (Whitney 1986; Snyder 1931; Lush 1981; Lush 1984; Lush and Holland 1988). Amino acid substitutions were discovered in one mouse bitter taste receptor that predicted sensitivity to a specific compound that could be considered confirmation that naturally occurring alleles are alive (Chandrasekhar et al. 2000). These genetic variants were less responsive in cell-based assays found in bitter-insensitive mouse strains. This study reveals that alleles of a taste receptor can alter cellular and behavioral responses to bitter compounds.

One study indicated that phenylthiocarbamide and propylthiouracil are the chemicals responsible for individual differences in the ability to taste different bitter components of the *TAS2R38* receptor gene. Various later investigations supported this observation and indicated that most people have a wide variation in taste distribution and that its threshold varies from person to person in different age groups (Guo and Reed 2001).

6.6.2 Cross-species comparisons

Various experiments indicate that different species have the ability to detect different bitter compounds. This is mainly due to differences in the range of bitter receptor genes (Glendinning 1994; Shi et al. 2003). For example, mice are not sensitive to the bitterness of phenylthiocarbamide but humans have a bitter sensitivity and a taster receptor gene for it. Animals have receptors with sensitivity for harmful and dangerous toxins in their specific environments.

Bitterness may provide information that a substance is poisonous. Bitterness warns the consumer who takes a bite of food to "go slow," particularly if something tastes unpleasant. For example, rodents that became sick due to experimental testing effects take very small bites of food. In certain cases, bitter perception in one species may provide the ability to eat plants that may be poisonous to other species. For example, some species of lemur eat bamboo, which contains cyanide, which is harmful and fatal to human beings and other species of lemur (Glander et al. 1989). The ability of lemurs to ingest cyanide is due to their bitter taste perception. Poison detection and avoidance can be studied by comparing various species. It may help us understand the role of bitter taste perception.

6.6.3 Family and twin studies

Biologically related people such as parents and their children, twins, or siblings were examined for similarity in acceptance in the perception of taste qualities and were genetically determined for the investigation of differences in the taste perception (Reed 2006). Similarity between blood relatives was observed as a measure of similar characteristics. This is due to their similar genetic variation, which is hereditary. Most populations have some genetic variation in the DNA that causes slight or greater differences in protein function. For example, twins could have the same characteristics such as eye and face color due to genetic similarity, and we assume that this may be due to their identical characteristics. If other characteristics are considered, such as hobbies and habits, then genetically similar twins may differ as do other randomly chosen people. On this basis, we can say that music or reading preference is not hereditary and is not related to genetic differences. This can be useful in taste research because the similarity in taste can be compared between twins and genetically identical populations. Twins generally live together in childhood and early life and have identical exposure to breast milk and foods at home. Differences in taste perception can be assumed to be part of genetic differences. For this reason, study of twins is important and mainly used to measure and compute heritability for taste as a research tool.

For a genetically informative study, different family members such as parents, their children, and their grandchildren can also be used as other pairs for investigating shared genetic variation.

6.7 Taste perception and behavior

6.7.1 Challenges in the population

Health and nutrition can be affected by the differing responses of human beings to sweet and bitter taste components. Taste perception may account for health outcomes in the population due to genetic differences among people. In certain cases food rejection may be due to higher concentration of bitter chemicals in food. It may be considered behavioral interpretation for rejection and avoidance of toxic and harmful chemicals available in foods, mainly rancid fat, hydrolyzed protein, and plant alkaloids (Basson et al. 2005). Certain bitter components in food are beneficial and advisable for health issues, but other bitter components should be avoided because they are harmful and fatal for humans (Maga 1990). Examples of such compounds include triterpenes (citrus fruits), phenols (found in tea, citrus fruits, wine, and soy), and organosulfur compounds (broccoli and cabbage).

Consumption of fruits and vegetables can be challenging for some. The bitter taste of food is a major reason to avoid it and that lowers its consumption. For example, "healthy and nutritious" vegetables such as broccoli and cabbage are sometimes not appealing to children due to their bitter taste (Drewnowski and Gomez-Carneros 2000). The overall bitter taste of diets for several conditions, including obesity, heart disease, and cancer, may be affected by differences in bitter taste perception (Liu 2004).

People who can or cannot taste phenylthiocarbamide are found in virtually all human populations worldwide. This might be taken as evidence that this polymorphism has been subject to balancing selection or a selection for heterozygous individuals in the population. People are unlikely to eat foods that are harmful to health.

6.7.2 Bitter taste and food rejection

Unusual bitterness has a harmful effect on health. Certain groups of chemicals in food that have an unpleasant bitter taste may present dietary dangers. Some bitter-tasting compounds may result from microbial fermentation (Maga 1990). Bitterness of certain plant foods such as potatoes, beans, cabbage, sprouts, cucumbers, pumpkins, and spinach has been reported (Rouseff 1990). Efforts were made in the past to cultivate fewer good plant foods out of safety concerns, not for reasons of taste.

The biology of the perception of bitter taste is not yet well understood. It is a long-term challenge to explain how compounds can have the same bitter taste even though they are not structurally similar. The different chemical structures of various compounds such as amino acids, peptides, saccharin, ureas, thioureas, phenylthiocarbamide, phenols, and polyphenols are not structurally similar but share a bitter taste. Different or multiple bitter-taste receptors may be due to differences in chemical structures of these reported compounds. Linkage to G proteins may be another probability for the same mechanism in taste perception of some sweet and bitter tastes. The taste of any compound can be changed from slightly bitter to sweet by some modification of the chemical structure of the compound.

A new family of gustducin-linked bitter taste receptors consisting of 40–80 receptors has been suggested (Adler et al. 2000). One supportive study indicated that humans are responsive to bitter-taste receptors such as hT^2R-4 for denatonium, while mice are responsive to the MT2R-5 receptor for cycloheximide (Chandrashekar et al. 2000).

6.8 Consumer acceptance of functional foods

Socio-demographic characteristics and cognitive and attitudinal factors are some of the multiplicity and diversity factors that influence consumer acceptance of functional foods. Invariably marketing studies reveal that consumer taste perception is the important factor in the selection of various foods consumed routinely for good health and nutritional value. Some cancer researchers have suggested that certain consumers select broccoli sprouts for their nutritional value even though they contain a high amount of glucosinolate, a potential bitter compound (Fahey et al. 1997).

6.8.1 Socio-demographic determinants

A U.S. study conducted in 1992–1996 showed that the primary consumers of functional foods were mainly highly educated females 30–35 years old from a higher income class. A more recent quantitative study by the International Food Information Council (IFIC) provides data about consumption of healthy and nutritious food by females age 55 and older (Table 6.1) (Verbeke 2005).

Table 6.1 Socio-Demographic Parameters for Female Functional Food Consumers in the United States and Europe

Age	Education/Income
United States	
35–55	Highly educated/higher income
35–60	Higher education/higher income
45–74	Graduate
55 and older	Graduate
Europe	
55 and older	Low education
55 and older	Higher education/higher income

Source: István Siroá, Emese Kápolna, Beáta Kápolna, Andrea Lugasi (2008). Functional food. Product development, marketing and consumer acceptance—A review, *Appetite*, 51, 456–467.

Gender is an issue that has been examined in the context of consumer acceptance of functional foods. A 1997 study by Childs and Poreezes shows females are the largest group of users and purchasers of functional foods (Childs and Poreezes 1997). Presence of young children in the household is another important socio-demographic factor. Socio-demographic factors influence the choice of foods that are higher risk of spoilage as well as quality awareness. Reviews that consider socio-demographic factors that influence consumption of functional foods suggest that determinants such as age, gender, education, and presence of young children and family members with health issues influence the choice and selection of functional foods.

6.8.2 Cognitive and attitudinal determinants

Choice of functional foods is mainly explained by different parameters such as knowledge, attitude, and beliefs of the consumer (Verbeke 2005). In 1999 the IFIC determined that consumer knowledge and beliefs were the primary motivating factors that affect food purchase and consumption. The IFIC also reported that lack of knowledge was the main reason for not consuming functional foods.

Consumer "belief" deals mainly with "health benefit belief" and "perceived role of functional foods for health." Consumer acceptance and intention to buy are also influenced by the price of functional foods.

6.9 Challenges in consumer acceptance and successful marketing of products

Product developers use various tools such as chemical tests, microbiological procedures, and physical equipment to determine the elasticity, hardness, viscosity, and color intensity. Product quality may be assigned to food products on the basis of appearance. Traditional grading methods are used to assign quality. Trained experts are assigned to grade product quality on the basis of various quality determinants such as appearance, flavor, and texture of the products. Grading methods can determine presence or absence of defects in the sample. These traditional methods have limitations. They cannot determine product acceptance by the consumer. The assessment of product quality is subjective and it is difficult to assign a quantitative score (Claassen and Lawless 1992).

In certain cases a product identified by its flavor profile as having no defects and another product with different sensory characteristics may get identical quality gradings. Sensory evaluation is an important tool in such cases. Users of products generally experience them entirely with their senses of taste, flavor, and smell and not with different equipment or chemical tests involved in their production. Today, many companies believe that use of tools such as chemical, physical and microbiological tests are not sufficient for producing quality products and, in certain cases, may lead to product failure.

6.9.1 Role of sensory science in decision making

Decisions to purchase consumable items and functional foods are determined by consumer attitudes, which can be correlated with their cognitive, conative, affective, and economic considerations. Better product development that will fulfill consumer satisfaction and be accepted by them may be determined by joint analysis which helps in marketing, research, and development. Today companies use both sensory science and methodology in marketing and R&D that offers improvement in innovative practices responsible for market success. Consumer taste in marketing and development of products in the food industry is crucial. Business performance is not affected directly by market orientation R&D as suggested by some of the economics literature (Grunert et al. 1997). Consumer orientation is the main fundamental concept and method for the evaluation of the product in R&D but not in the marketing approach.

Intrinsic traits of the physical product such as odor, taste, size, and appearance are evaluated by instrumental, physiochemical, and sensory analysis. This analysis is the primary method of incorporation with marketing where the priority is given first to consumer perceptions of sensory quality rather to than that of real taste evaluations for most food products. Consumer thinking about the taste of a product rather than about what they actually taste is more important for any marketer. Typically concentration, perceived concentration, coded sensation, and overt response in the sensory evaluation process are reviewed. In the laboratory products are coded in an environment far from the real marketplace. Quality and control quality assurance processes depend on established methods within the industry. Various services for product development, its improvement, and marketing research also contribute to quality establishment.

6.9.2 Sensory evaluation and quality of food

Sensory evaluation is the scientific branch of knowledge that produces, measures, analyzes, and interprets different responses to products, which are perceived through sight, smell, touch, taste, and hearing. Quality and nutritional value are less important than palatability in the selection of food products.

Food quality is important for successful marketing of food. What is arguably more important is that the quality of taste, aroma, aftertaste, tactual properties, and appearance are acceptable to the consumer so that they hunger for more. If we find that consumers like the product best in turn, then we can accept the food quality score. The quality score is based on the degree of desirable attributes and absence of undesirable characteristics, which are mostly determined by the consumer's sensory organs. Sensory evaluation is a good method for determining product quality.

6.10 Taste improvement technology

Food science and technology is a relatively new field. It has matured in its national and international development with respect to its technical aspects. Today, food science is dealing with physical, chemical and biological sciences because they directly affect the processing and preservation of food. Currently, food researchers are involved in reducing, blocking, and masking the effect of bitter components present in foods to make them more palatable and acceptable. Hayes says that the bitter taste perception is "highly complicated," consisting of 25 bitter receptor genes. Food technology considers engineering, scientific, and technical problems required for transferring raw materials and other ingredients into safe, nutritious, and qualitative food products (Desrosier 1997).

Today the food industry faces the very costly common problem of off-taste. For several decades, salt, and spices have been used to help mask undesirable taste, mainly bitter and sour. New taste-masking approaches are being investigated to replace these substances and the use of functional ingredients in nutraceuticals has also increased. Undesirable taste is essentially present in food naturally, for example, bitter tastes from vegetables and strong sourness in certain foods. Other off-tastes result from microbial metabolites, enzymatic degradation, heat treatment, and oxidation of lipids or the addition of ingredients such as vitamins, minerals, or antioxidants, which can be associated with remaining off-notes.

Methods to reduce off taste in foods either remove the off-taste source through fermentation or hydrolysis or mask the off taste or bitter taste by adding one or more ingredient or complex formation. A number of natural bioactive compounds available in food remove bitterness and also have positive effects on health such as flavonoids, polyphenols, peptides, minerals, and terpenes. Taste improvement can be accomplished by removing responsible chemicals or rendering competing chemicals less bitter. Addition of flavor or flavor enhancement is also an important factor that increases consumer acceptance (Galindo-Cuspinera, 2011). For effective taste masking researchers must first understand the types of interaction that occur in food compounds as well as interactions that occur in the mouth. Taste masking can be achieved either by taste inhibition via peripheral interactions in the mouth or by suppression of mixture through central cognitive interactions at the brain level. Various taste-masking methods use these two approaches as well as technology involved with source elimination of undesired taste such as encapsulation or target elimination of the bitter compound.

6.10.1 Taste inhibition

Individual taste and its quality are detected mainly through a mechanism called transduction for different receptors. A number of possible interactions may occur once we put two or more compounds in the mouth. The taste of one compound can be increased or reduced by possible interaction of compounds or interference of one compound with others to bind with taste receptors through transduction mechanisms. Many such efforts have been made. Researchers have concentrated on blocking specific taste receptors, especially bitter ones, to reduce off taste. It is too difficult to isolate single receptors among the reported 25 bitter taste receptors for blocking of bitter taste (Schiffman 1999; Galindo-Cuspinera and Breslin 2006).

Recent research has concentrated on the development of various receptor-based assays for specific compounds or chemicals that can be tasted directly on target receptors and accessed for activation or reduction of taste signals. A drawback of this method is that it helps determine activity in compounds that are not consumed by humans due to safety concerns. Different solutions have been investigated by NIZO (Netherlands Institute for Dairy Research), including various methods of food-grade fractionation for screening of natural taste blockers by trained sensory panels. This natural source can be found in fruits, vegetables, and fermentation products. Such natural sources are the promising compounds that mask bitter-taste compounds.

6.10.2 Suppression of mixture

Different responses are produced by single tastants than are produced by those same tastants present in mixtures. A target compound can easily be masked by a strong-tasting taste masker. For example, the bitter taste of medicine or a pharmaceutically active compound can be masked by using sugar or different types of resinous materials by complexation.

If we consider this specific case, reduction or suppression of bitter taste is done at a higher level by combining taste and smell to deliver the compound, which does not necessarily require sugar interaction with taste receptors or other components. The brain is the main site for the suppression of mixture, but it is not limited to taste–taste interactions only. Flavor may strengthen the taste of a compound so it is also possible that it may reduce taste perception. For example, the flavor of angelica oil has been shown to decrease sweet taste (Stevenson 1999).

The mixture suppression technique has recently become important for reducing or suppressing the bitter taste of chemicals. The mixture suppression method can improve or eradicate the off taste of any compound on a trial-and-error basis. The NIZO company developed Olfactoscan®, a screening tool for the step-by-step investigation of flavor–flavor and flavor–taste interactions. This technique delivers flavor or taste pulses in a continuous manner that combines with stream of flavor compounds at a controlled rate for panelists to evaluate. Several off-taste compounds have been prepared by NIZO using this technique, including off-flavor potato, orange juice, and several milk products (Burseg and de Jong 2009).

6.10.3 Masking by encapsulation

Encapsulation or microencapsulation is a method or process by which small particle droplets can be surrounded or entrapped by a coating agent. This method is used for delivery of many pharmaceutical compounds to improve their taste or release them in a controlled manner. It is also used to entrap enzymes, biologically active molecules, and living cells in foods. The protective material used for coating or encapsulation must be food grade, biodegradable, and nontoxic. It should also be able to form a barrier between the internal phase and its surroundings. Polysaccharides are the major materials used for encapsulation of foods in the food industry. Proteins and lipids are also sometimes used for encapsulation.

Spray drying is widely used in the food industry. It produces products in a flexible, continuous, and economical way. Most materials are encapsulated by spray drying. Other methods used include spray chilling, freeze drying, melt extrusion, and melt injection. There are several reason why spray drying is so widely used. It may provide barriers between sensitive bioactive materials and the environment. It allows differentiation between taste and flavor, masks taste or flavor, and improves bioavailability and stability of food products.

Encapsulation of active compounds provides good stability during processing and in the final products. Encapsulation of volatile activity such as aroma causes less evaporation and degradation. Encapsulation is also one of the preferred methods used to mask polyphenols with bitter and astringent taste.

It prevents reaction of oxygen and water with the components present in food products. This method is also used for enzyme immobilization, fermentation, and metabolite production processes in the food industry (Wandrey et al. 2009; Gibbs et al. 1999).

Encapsulation is the most efficient method for masking the bitter taste of compounds. It is more efficient than taste inhibition or the mixture suppression method for masking or removing bitter compounds. Various technologies for microencapsulation include spray drying, coacervation phase separation, and spray congealing. Coacervation methods provide a platform to a biopolymer complex to form a layer around the bitter chemical and its bitter taste. R&D programs coordinated at NIZO masked bitter taste by microencapsulation at the effective layer. For example, microencapsulating by coacervation masked the bitter taste of naringin, a bitter chemical found in oranges.

6.10.4 Using food technology to create new flavors

Research and development teams at flavor crafting companies carefully design artificial and natural flavors. Often an R&D team begins by deciding which flavor to create, from blueberry pancake to strawberry to salted caramel. Then they move on to extracting and analyzing the chemicals that produce that flavor in the real thing. For example, they isolate the dominant chemicals that create the initial "strawberry" taste and then strengthen the taste with other chemicals to produce a more distinct and realistic flavor, adding aromas and other components to the mixture. (http://www.eurmscfood.nl/the-role-of-food-technology-in-improving-flavour/).

6.10.5 Role of umami in food science

The taste of old recipes can be enhanced by improving or developing new flavors. Finding new flavors and increasing consumer acceptance are challenges for food scientists, particularly for low-fat, low-sodium, and low-sugar products. This may feel like a tall order because some of the best flavor-boosters other than natural and artificial flavors are sugar, fat, and salt. Foods such as mushrooms and steak contain chemicals responsible for the distinct umami taste. New research has found that umami is a unique taste that may also act as a powerful flavor booster. As a result, some food scientists have started using umami to preserve the taste of low-sodium products (http://www.eurmscfood.nl/the-role-of-food-technology-in-improving-flavour/).

6.11 Conclusion

Our gustatory system ensures our survival. The taste, flavor, and palatability of food help us identify and select nutritious and safe foods. The main sensory properties of food are flavor; temperature, which affects the sensation of heat and cold; and appearance and mouthfeel, which affect the sense of touch. The demands of taste, cost, and convenience are the factors that drive consumer food choice. The identification of human bitter-taste receptors is bound to have an industry-wide effect given that the debittering of foods remains a very common problem.

Trained laboratory personnel differentiate taste by testing and scoring it. A special testing area should be used for the sensory evaluation of foods, and judges should be separated from one another to ensure independent judgment of taste. Several methods and technologies are used to improve taste and flavor of compounds as part of consumer acceptance. Taste-masking techniques are applied to reduce or eradicate the bitter or unpleasant taste of bitter components to improve consumer acceptance. Food researchers are involving in investigating various means to block or suppress the bitter taste of food components to make them more palatable for consumers.

References

Adler E, Hoon MA, Mueller KL, Chandrashekar J, Ryba NJP, Zuker CS (2000). A novel family of mammalian taste receptors. *Cell.* 100:693–702.

Bachmanov AA (2001). Positional cloning of the mouse saccharin preference (Sac) locus. *Chem Senses.* 26:925–933.

Basson MD, Bartoshuk LM, Dichello SZ, Panzini L, Weiffenbach JM, Duffy VB (2005). Association between 6npropylthiouracil (PROP) bitterness and colonic neoplasms. *Dig Dis Sci.* 50(3):483–489.

Breslin PA, Beauchamp GK (1995). Suppression of bitterness by sodium: variation among bitter taste stimuli. *Chem Senses.* 20:609–623.

Breslin PA, Beauchamp GK (1997). Salt enhances flavor by suppressing bitterness. *Nature.* 387:563.

Breslin PA, Huang L (2006). Human taste: peripheral anatomy, taste transduction, and coding. *Adv Otorhinolaryngol.* 63:152–190.

Breslin PAS (1996). Interactions among salty, sour and bitter compounds. *Trends Food Sci Tech.* 7:390–399.

Breslin PAS (2013). An evolutionary perspective on food review and human taste. *Curr Biol J Sci Dir.* 23(9):R409–R418.

Burseg K, de Jong C (2009). Application of the Olfactoscan method to study the ability of saturated aldehydes in masking the odor of methional. *J Agric Food Chem.* 57(19):9086–9090.

Chandrashekar J, Mueller KL, Hoon MA, Adler E, Feng L, Guo W (2000). T2Rs function as bitter taste receptors. *Cell.* 100(6):703–11.

Childs NM, Poreezes GH (1997). Foods that help prevent diseases: consumer attitudes and public policy implications. *J Consum Mark.* 14(6):433–447.

Claassen MR, Lawless, HT (1992). A comparison of descriptive terminology systems for the sensory analysis of flavor defects in milk. *J Food Sci.* 57:596–621.

Delwiche JF, Lera MF, Breslin PA (2000). Selective removal of a target stimulus localized by taste in humans. *Chem Senses.* 25:181–187.

Desrosier NW (1997). Elements of food technology. AvI publishing company, I.N.C, west port, Connecticut.

Drewnowski A, Gomez-Carneros C (2000). Bitter taste, phytonutrients, and the consumer: A review. *Am J Clin Nutr.* 72:1424–1435.

Fahey JW, Zhang Y, Talalay P (1997). Broccoli sprouts: An exceptionally rich source of inducers of enzymes that protect against chemical carcinogens. *Proc Natl Acad Sci U S A.* 94:10367–10372.

Finger TE, Danilova V, Barrows J, Bartel DL, Vigers AJ, Stone L (2005). ATP signaling is crucial for communication from taste buds to gustatory nerves. *Science.* 310:1495–1499.

Galindo-Cuspinera, V (2011). *Taste Masking: Trends and Technologies*, available at https://www.preparedfoods.com/articles/109539-taste-masking-trends-and-technologies.

Galindo-Cuspinera V, Breslin PA (2006). The liaison of sweet and savory. *Chem Senses.* 31(3):221–225.

Gibbs S, Kermasha I, Alli CN (1999). Mulligan, encapsulation in the food industry: A review *Int J Food Sci Nutr.* 50:213–224.

Glander KE, Wright PC, Seigler DS, Randrianasolo V, Randrianasolo B (1989). Consumption of cyanogenic bambooby a newly discovered species of bamboo lemur. *Am J Primatol.* 19:119–124.

Glendinning JI (1994). Is the bitter rejection response always adaptive? *Physiol Behav.* 56(6):1217–1227.

Grunert KG, Brunsø K, Bisp S (1997). Food-related lifestyle: development of a cross-culturally valid instrument for market surveillance. In L. R. Kahle, & L. Chiagouris (Eds.), *Values, Lifestyles, and Psychographics* (pp. 337–354). Mahwah, NJ: Lawrence Erlbaum.

Guo SW, Reed DR (2001). The genetics of phenylthiocarbamide. *Percep Ann Hum Biol.* 28(2):111–142.

Hoon MA (1999). Putative mammalian taste receptors: a class of taste-specific GPCRs with distinct topographic selectivity. *Cell.* 96:541–551.

István Siroá, Emese Kápolna, Beáta Kápolna, Andrea Lugasi (2008). Functional food. Product development, marketing and consumer acceptance—A review. *Appetite.* 51:456–467.

Iwatsu K, Ichikawa R, Uematsu A, Kitamura A, Uneyama H, Torri K (2012). Detecting sweet and umami tastes in the gastrointestinal tract. *Acta Physiol.* 204:169–177.

Janssen S, Depoortere (2013). Nutrient sensing in the gut: new roads to therapeutics? *Trends Endocrinol Metabol.* 24:92–100.

Jiang P, Josue J, Li X, Glaser D, Li W, Brand JG, Margolskee RF, Reed DR, Beauchamp GK (2012). Major taste loss in carnivorous mammals. *Proc Natl Acad Sci.* 109:4956–4961.

Kitagawa M, Kusakabe Y, Miura H, Ninomiya Y, Hino A (2001). Molecular genetics identification of a candidate receptor gene for sweet taste. *Biochem Biophys Res Comm.* 283:236–242.

Kurihara K (2015). Umami the fifth basic taste: History of studies on receptor mechanisms and role as a food flavor. *BioMed Res Int.* 1–10.

Li X, Li W, Wang H, Cao J, Maehashi K, Huang L, Bachmanov AA, Reed DR, Legrand-Defret in V, Beauchamp GK (2005). Pseudogenization of a sweet-receptor gene accounts for cats' indifference toward sugar. *PLoS Genet.* 1:27–35.

Li X (2002). Human receptors for sweet and umami taste. *Proc Natl Acad Sci.* 99:4692–4696.

Lindemann B (2001). Receptors and transduction in taste. *Nature.* 413(6852):219–25.

Liu RH (2004). Potential synergy of phytochemicals in cancer prevention: mechanism of action. *J Nutr.* 134:3479S–85S.

Lugaz O, Pillias AM, Boireau-Ducept N, Faurion A (2005). Time-intensity evaluation of acid taste in subjects with saliva high flow and low flow rates for acids of various chemical properties. *Chem Senses.* 30:89–103.

Lush IE (1984). The genetics of tasting in mice: III. Quinine *Genet Res.* 44(2):151.

Lush IE (1981). The genetics of tasting in mice: I. Sucrose octaacetate, *Genet Res.* 38(1):9.

Lush IE, Holland G (1988). The genetics of tasting in mice: V. Glycine and cycloheximie *Genet Res.* 52(3):207.

Maehashi K, Matano M, Wang H, Vo LA, Yamamoto Y, Huang L (2008). Bitter peptides activate hTAS2Rs, the human bitter receptors. *Biochem Biophys Res Commun.* 365(4):851–5.

Maga JA (1990). Compound structure versus bitter taste. In: Rousseff RL, ed. *Bitterness in foods and beverages; developments in food science 25.* Amsterdam: Elsevier, 35–48.

Mattes RD (2011). Accumulating evidence supports a taste component for free fatty acids in humans. *Physiol Behav.* 104:624–631.

Max M (2001). Tas1r3, encoding a new candidate taste receptor, is allelic to the sweet responsiveness locus Sac. *Nat Genet.* 28:58–63.

Montmayeur JP, Liberles SD, Matsunami H, Buck LBA (2001). Candidate taste receptor gene near a sweet taste locus. *Nat Neurosci.* 4:492–498.

Nakamura Y, Goto TK, Tokumori K, Yoshiura T, Kobayashi K, Honda H, Ninomiya Y, Yoshiura K (2012). The temporal change in the cortical activations due to salty and sweet tastes in humans: fMRI and time-intensity sensory evaluation. *Neuroreport.* 23:400–404.

Nelson G (2002). Mammalian sweet taste receptors. *Cell.* 106:381–390.

Pedersen AM, Bardow A, Jensen SB, Nauntofte B (2002). Saliva and gastrointestinal functions of taste, mastication, swallowing and digestion. *Oral Dis.* 8:117–129.

Reed DR (2006). Diverse tastes: Genetics of sweet and bitter perception. *Physiol Behav.* 30;88(3):215–226.

Remis MJ, Kerr ME (2002). Taste Responses to Fructose and Tannic Acid among Gorillas (Gorilla gorilla gorilla). *Int J Primatol.* 23:251–261.

Rouseff RL (1990). Bitterness in food products: An overview. In: Rouoseff RL, ed. *Bitterness in foods and beverages. Developments in food science. Vol 25.* Amsterdam: Elsevier, 1–14.

Sainz E, Korley JN, Battey JF, Sullivan SL (2001). Identification of a novel member of the T1R family of putative taste receptors. *J Neurochem.* 77:896–903.

Schiffman SS (1999). Selective inhibition of sweetness by the sodium salt of +/-2-(4-Methoxyphenoxy) propanoic acid. *Chem Senses.* 24:439–447.

Shi P, Zhang J, Yang H, Zhang YP (2003). Adaptive diversification of bitter taste receptor genes in mammalian evolution. *Mol Biol Evol.* 20(5):805–814.

Shikata H, McMahon DB, Breslin PA (2000). Psychophysics of taste lateralization on anterior tongue. *Percept Psychophys.* 62:684–694.

Snyder LH (1931). Inherited taste deficiency. *Science.* 74:151.

Stevens JC (1996). Detection of tastes in mixture with other tastes: Issues of masking and aging. *Chem Senses.* 21:211–221.

Stevenson RJ (1999). Confusing tastes and smells: how odours can influence the perception of sweet and sour tastes. *Chem Senses.* 24(6):627–635.

Torii K, Uneyama H, Nakamura E (2013). Physiological roles of dietary glutamate signaling via gut–brain axis due to efficient digestion and absorption. *J Gastroenterol.* 48(4):442–451.

Verbeke W (2005). Consumer acceptance of functional foods: Socio-demographic, cognitive and attitudinal determinants. *Food Qual Prefer.* 16:45–57.

Wandrey C, Bartkowiak A, Harding SE (2009). Materials for Encapsulation In: Zuidam NJ, Nedovic VA (Eds.) *Encapsulation Technologies for Food Active Ingredients and Food Processing.* Springer: Dordrecht, The Netherlands, 31–100.

Whitney G, Harder DB (1986). Singlelocus control of sucrose octaacetate tasting among mice. *Behav Genet.* 16(5):559–74.

Yarema T, Rhoda D, Brannigan, J. (2006). Ayurvedic concepts in a nutshell, The 6 tastes: Our guide map to optimal nutrition, New York, Five Elements Press. Pp-18. Available at http://www.eattasteheal.com/eth_6tastes.htm

Young RL, Sutherland K, Pezos N, Brierley SM, Horowitz M, Rayner CK, Blackshaw LA (2009). Expression of taste molecules in the upper gastrointestinal tract in humans with and without type 2 diabetes. *Gut.* 58:337–346.

Zhang Y, Hoon MA, Chandrashekar J, Nueller KL, Cook B, Wu D, Zuker CS, Ryba MJP (2003). Coding of sweet, bitter, and umami tastes: different receptor cells sharing similar signaling pathways. *Cell.* 112:293–301.

Zhao FL, Shen T, Kaya N, Lu SG, Cao Y, Herness S (2005). Expression, physiological action and co expression patterns of neuropeptides Y in rat taste-bud cells. *Proc Natl Acad Sci.* 102(31):11–100.

Zhao GQ (2003). The receptors for mammalian sweet and umami taste. *Cell.* 115:255–266.

7

Sensory Qualities and Nutraceutical Applications of Flavors of Terpenoid Origin

Ana Clara Aprotosoaie, Irina-Iuliana Costache, and Anca Miron

Contents

7.1 Introduction 167
7.2 Sensory qualities of terpenes 169
7.3 Nutraceutical applications of terpene flavors 174
    7.3.1 Citral 180
        7.3.1.1 Biological properties of citral 180
    7.3.2 Geraniol 184
        7.3.2.1 Biological properties of geraniol 184
    7.3.3 Menthol 187
        7.3.3.1 Biological properties of menthol 189
7.4 Conclusions 194
References 194

7.1 Introduction

Terpenes constitute the largest and most structurally diverse class of natural chemicals with more than 55,000 molecules reported to date (Caputi and Aprea 2011; Brahmkshatriya and Brahmkshatriya 2013). They are commonly found in higher plants, but they have also been isolated from other biological kingdoms (animal and microbial species) (Caputi and Aprea 2011). Terpenes have important biological and ecochemical functions in plants. Some terpenes are involved in cellular function and maintenance (photosynthetic pigments, electron carriers, mediators of polysaccharide assembly), but most are secondary vegetal metabolites that play diverse functional roles. They are components of plant defense responses to biotic and abiotic stresses and mediators of the plant–environment interactions, infochemicals (attractans or repellents), or phytohormones. Many of these are due to the fact that the volatile terpenes are responsible for the characteristic scents of aromatic plants (Paduch et al. 2007; Caputi and Aprea 2011; Brahmkshatriya and Brahmkshatriya 2013).

The main types of terpenes are hemiterpenes (C_5H_8), monoterpenes ($C_{10}H_{16}$), sesquiterpenes ($C_{15}H_{24}$), diterpenes ($C_{20}H_{32}$), sesterterpenes ($C_{25}H_{40}$), triterpenes ($C_{30}H_{48}$), tetraterpenes (carotenoids, $C_{40}H_{64}$), and polyterpenes [(C_5H_8)n, rubber] (Ludwiczuk et al. 2017). More than 1,000 monoterpenes, 7,000 sesquiterpenes and 3,000 diterpenes are known to date (Brahmkshatriya and Brahmkshatriya 2013).

All terpenes derive biosynthetically from two isoprene units, isopentenyl diphosphate (IPP) and its isomer, dimethylallyl diphosphate (DMAPP), via two distinct compartmentalized pathways. The mevalonic acid (MVA) pathway operates in cytosol, the endoplasmic reticulum, and peroxisomes. The 2-C-methyl-D-erythritol-4-phosphate (MEP) route, also called the deoxyxylulose-5-phosphate (DXP), is active in plastids (Caputi and Aprea 2011; Singh and Sharma 2015). The repetitive condensation of IPP with DMAPP or a prenyl diphosphate catalyzed by prenyl transferases generates the main progenitors of terpenes, namely geranyl pyrophosphate (GPP) for monoterpenes, farnesyl pyrophosphate (FPP) for sesquiterpenes, geranylgeranyl pyrophosphate (GGPP) for diterpenes and tetraterpenes, and squalene for triterpenes (Breitmaier 2006; Caputi and Aprea 2011). Although head-to-tail linkage is the most common, head-to-head fusion also occurs (triterpenes, tetraterpenes). Head-to-middle addition is characteristic for irregular monoterpenes (Ludwiczuk et al. 2017). The primary terpene skeletons result from the derivatization of these above-mentioned precursors through the intervention of terpene synthases. The subsequent change of primary terpene backbones via different chemical reactions (reductions, oxidations, cyclizations, rearrangements, ring cleavages) leads to the broad array of terpene molecules (Koziol et al. 2014).

The plants that contain terpenes have long been used in traditional medicine. Numerous studies have highlighted the tremendous bioactivity of terpenes, such as antioxidant, antimicrobial, antiparasitic, anti-inflammatory, antiviral, anti-allergenic, antispasmodic, antihyperglycemic, antitumor, anxiolytic, cardioprotective, immunomodulatory, local anesthetic, and insecticide properties. In addition to their huge potential as medicines, terpenes have significant applications in the pharmaceutical, food, cosmetic, chemical and perfume industries (Caputi and Aprea 2011; Koziol et al. 2014). Monoterpenes and sesquiterpenes are the principal odoriferous terpenes; they essentially determine the aroma of herbs and spices and are some of the most important used flavouring agents and fragrances. They have the ability to enhance the sensory appeal of foods and beverages and to mask the unpleasant taste/odor of vitamins, minerals, antioxidants, and other active ingredients from pharmaceuticals or nutraceuticals (Caputi and Aprea 2011).

This chapter provides an overview of the potential applications of the main terpene flavors as nutraceuticals. We describe the sensory qualities of these flavors and discuss the biological properties of representative compounds in the context of recent studies.

7.2 Sensory qualities of terpenes

Mono- and sesquiterpenes are common constituents of essential oils. Monoterpene hydrocarbons are mainly found in essential oils obtained from leaves while oxygenated monoterpenes predominate in the essential oils obtained from flowers and fruits. Bark and root essential oils contain primarily sesquiterpene derivatives (Caputi and Aprea 2011). The aroma range of these compounds is very broad. There is a direct relationship between the flavor type and intensity and the chemical structure of compounds. Mono- and sesquiterpene hydrocarbons generally have a lower odorant quality. They have limited use in the flavor and fragrance industries. They are used as raw materials in the synthesis of some flavoring/fragrance agents, in room deodorizers or the like, or for the reconstitution of essential oils. Oxygenated terpenes, also called terpenoids, exert very pleasant fragrances. They are extremely useful flavoring materials in different industries, including the pharmaceutical industry. Terpenoids such as alcohols and their esters with lower acids (mainly acetates), aldehydes, and ketones are the most important odoriferous compounds (Surburg and Panten 2006). Figures 7.1 and 7.2 illustrate the relevant flavoring mono- and sesquiterpenes. Table 7.1 describes their sensory profiles.

Molecular structural characteristics and molecular shape are essential for the interaction with olfactive receptors and the generation of a particular odor. For some monoterpenols, the polar hydroxyl group at C-1 (citronellol, geraniol, nerol) or at C-3 (linalool) is the key structural feature that determines specific scent and odor potency. The replacement of the hydroxyl group with acetyl, as in the case of the corresponding esters, commonly decreases the odor potency and changes the sensory character of nerol from a citrus, lemon-like note to a clove-like, phenolic smell. The presence of a carbonyl function at C-8 in the structure of citronellol, geraniol, and nerol results in a musty note. The substitution of C-8 with a carboxy group changes the aroma of geraniol and nerol to a waxy, greasy, plasticine-like aroma and determines the loss of citronellol odor (Elsharif et al. 2015; Elsharif and Buettner 2016; Elsharif and Buettner 2017).

Molecular stereochemistry influences the olfactory features of terpenes as well as their physiological properties. Many of these structures are optically active, and the enantiomers of the same compound may be isolated from different sources. Thus, $S(+)$-carvone isolated from the essential oils of caraway (*Carum carvi*) and dill (*Anethum graveolens*) imparts a herbaceous bread-like odor and spicy minty caraway taste, while $R(-)$-carvone, the main component in spearmint essential oil (*Mentha spicata*), provides herbaceous minty odor and taste (Table 7.1). $R(+)$-limonene (citrus peel essential oils) has a sweet, clean citrus, orange-like fragrance, and $S(-)$-limonene (*Mentha* sp., conifers) exerts a less pleasant orange smell, with a touch of turpentine (Breitmaier 2006; Buchbauer and Ilic 2013). Sensory differences can occur between natural and synthetic derivatives. These differences can be used to control the adulteration of natural compounds with their synthetic analogues.

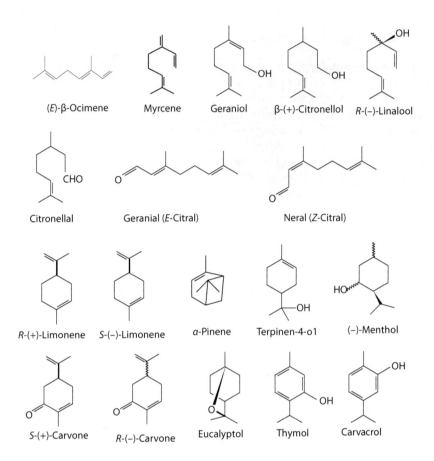

Figure 7.1 Chemical structures of some natural monoterpenes flavors.

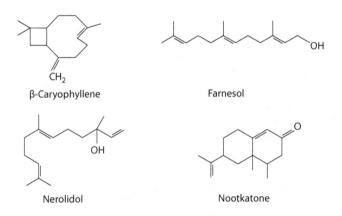

Figure 7.2 Chemical structures of some natural sesquiterpene flavors.

Table 7.1 Representative Odoriferous Monoterpenes and Sesquiterpenes

Chemical Class	Compound	Sensory Character
Monoterpene hydrocarbons	Myrcene	Odor: Herbaceous, woody, with a rosy celery and carrot nuance Taste: Woody, citrusy with a mango and slight leafy minty touch
	(E)β-ocimene	Odor, taste: Sweet herbaceous
	(Z)β-ocimene	Odor: Citrus tropical Taste: Green, tropical with floral, vegetable nuances
	R(+)-limonene	Odor: Sweet, citrus peely Taste: Sweet, citrus fruity
	S(−)-limonene	Odor, taste: Pine-like, terpene, herbal, peppery
	α-Pinene	Odor: Fresh, pine-like, camphoreous, woody Taste: Woody, camphoraceous with turpentine note
	β-Pinene	Odor: Fresh, pine-like, minty, spicy Taste: Fresh, woody, resinous with camphoreous, spicy tones
	α-Phellandrene	Odor: Herbal, citrus, green, woody, black pepper–like Taste: Fresh, citrusy, green
	β-Phellandrene	Odor: Minty Taste: Burning
Monoterpenols	Borneol	Odor: Woody-camphoraceous, dry minty Taste: Burning
	α(−)-Citronellol	Odor: Rosy, sweet, green with citrus notes Taste: Floral rose with fruity citrus nuances
	β(+)-Citronellol	Odor: Sweet, fresh rose-like Taste: Bitter
	Geraniol	Odor: Flowery, rose-like Taste: Sweet floral rose-like with citrus, fruity, waxy nuances
	R(−)-Linalool	Odor: Woody, lavender-like Taste: Floral, woody with a spicy, green tropical nuance
	S(+)-linalool	Odor: Sweet floral, herbaceous with petitgrain-like, citrus, fruity nuances Taste: Spicy citrus
	(−)-Menthol	Odor: Cooling, minty, refreshing Taste: Cooling, camphoraceous, clean, spicy
	Nerol	Odor: Fresh floral, sweet, green, citrusy with a spicy depth Taste: Bitter lemon, green, fruity
	α-Terpineol	Odor: Lilac, woody floral Taste: Citrus, woody, soapy
	Terpinen-4-ol	Odor: Nutmeg-like, mild spicy, woody-earthy, lilac-like Taste: Cooling, woody, spicy with a citrus undertone

(Continued)

Table 7.1 (Continued) Representative Odoriferous Monoterpenes and Sesquiterpenes

Chemical Class	Compound	Sensory Character
Monoterpene phenols	Carvacrol	Odor: Pungent, herbal, camphoreous, smoky Taste: Spicy, herbal, woody
	Thymol	Odor: Spicy herbal, thyme-like with camphoraceous nuance Taste: Medicinal, woody and spicy
Monoterpene aldehydes	Citronellal	Odor: Sweet, floral, rosy, citrus, green Taste: Floral, green, rosy, citrus
	Geranial	Odor: Citrus, lemon-like Taste: Citrusy
	Neral	Odor: Sweet citrus lemon peely Taste: Lemon-like
Monoterpene ketones	S(+)-Carvone	Odor: Herbaceous, bread-like, reminiscent of caraway seeds Taste: Spicy, minty, caraway
	R(−)-Carvone	Odor, taste: Herbaceous minty
	(−)-Menthone	Odor: Fresh-cooling, minty, herbal with a green anise nuance Taste: Fresh-cooling, minty with herbaceous nuance
	(+)-Menthone	Odor: Minty with a slightly musty nuance Taste: Bitter
	S(+)-Piperitone	Odor, taste: Minty, cooling
	R(−)-Piperitone	Odor, taste: Minty, pungent
	α-Thujone	Odor, taste: Cedar
	β-Thujone	Odor, taste: Cedar with spicy, woody nuances
Monoterpene-esters	(−)Bornyl acetate	Odor, taste: Sweet balsamic, woody, fresh, pine-like
	(+)Bornyl acetate	Odor, taste: Camphoreous, herbal, pine-like
	Geranyl acetate	Odor: Floral, rosy, waxy, green with a cooling nuance Taste: Waxy, green, floral, soapy with citrusy and winery nuances
	Neryl acetate	Odor: Floral, rosy, citrus, fruity with a tropical note Taste: Floral, rosy, fruity
	Linalyl acetate	Odor: Sweet green floral with a fresh, woody, citrus nuance Taste: Floral, waxy, woody, with herbal and spicy nuances
	(−)-Menthyl acetate	Odor: Fresh fruity, minty, rose Taste: Cooling, minty
Monoterpene oxides	Eucalyptol (1,8-cineole)	Odor: Fresh camphoraceous Taste: Pungent, cooling, spicy
Sesquiterpene hydrocarbons	Cadinene	Odor, taste: Woody, fresh, herbal
	α-Caryophyllene	Odor, taste: Woody
	β-Caryophyllene	Odor: Spicy, woody, camphoreous, citrus Taste: Spicy, peppery, woody, camphoreous
	Cedrene	Odor, taste: Woody, cedar, sweet

(Continued)

Table 7.1 (Continued) Representative Odoriferous Monoterpenes and Sesquiterpenes

Chemical Class	Compound	Sensory Character
	α-Farnesene	Odor: Woody, green, citrus, floral
		Taste: Fresh green, with celery and hay notes
	(E)β-Farnesene	Odor, taste: Woody, citrus, sweet
	(Z)β-Farnesene	Odor, taste: Green citrus
	Longifolene	Odor, taste: Woody, sweet, rose, coniferous
Sesquiterpene alcohols	Cedrol	Odor, taste: Woody, cedar, dry sweet
	Farnesol	Odor: Fresh, floral linden, sweet
		Taste: Floral, juicy, green
	Nerolidol	Odor: Floral, green, citrus-like with woody nuances
		Taste: Green, floral with woody, citrusy and melon nuances
	(−)-Patchoulol	Odor, taste: Woody, earthy, camphoreous, powdery
	(Z)-α-Santalol	Odor: Woody, sandalwood
		Taste: Spicy, woody
	(Z)-β-Santalol	Odor, taste: Woody
	Vetiverol	Odor, taste: Woody, sweet, balsamic
Sesquiterpene aldehydes	α-Sinensal	Odor, taste: Citrus orange
	β-Sinensal	Odor, taste: Orange, sweet, fresh, juicy
Sesquiterpene ketones	Nootkatone	Odor: Grapefruit, sweet peely with woody nuances
		Taste: Grapefruit, citrus, bitter

Source: Surburg, H., and J. Panten: *Common Fragrance and Flavor Materials. Preparation, Properties and Uses,* 5th ed. 2006. Copyright Wiley-VCH Verlag GmbH & Co. KGaA. Reproduced with permission; Good Scent Company: http://www.thegoodscentscompany.com/search2.html (accessed June 1, 2017).

(−)-Patchoulol is one of the main contributors to the precious woody scent of patchouli (*Pogostemon cablin*). It has an earthy, camphoreous odor with woody balsamic notes. Its unnatural isomer, (+)-patchoulol, has a weaker odor that does not resemble patchouli but rather β-santalol with a green undertone (Leffingwell 2006; NPCS Board of Consultants & Engineers 2007).

The breakdown products of higher isoprenoids (diterpenes, triterpenes, and tetraterpenes) are also outstanding odorant materials. Valuable fragrance compounds such as irones result from the degradation of the triterpene cycloiridals. The irones are natural violet-like odorants found in orris root essential oil (*Iris pallida*) and oak moss extracts. The chemical degradation of triterpene (+)-ambrein generates a complex mixture of compounds that determines the specific fragrance of ambergris, one of the most precious raw materials in perfumery. (−)-Ambergris oxide, one of the most important ambergris-related compounds, can be obtained from diterpene sclareol (*Salvia sclarea*, clary sage). Some labdanic diterpenoids, such as labdanolic acid (*Cistus labdanifer*, rock rose) and larixol (*Larix* sp., larch) or pentacyclic triterpene oleanolic acid, can also generate derivatives with amber odor (Serra 2015). C13-norisoprenoid compounds such as ionones and damascenones result from oxidative cleavage of carotenoids. They contribute essentially to the aroma of fruits and flowers during ripening and blossoming and are important sources of flavors for the food industry (Caputi and Aprea 2011; Serra 2015). α- and β-ionones have been found in carrots, raspberries, roasted almonds, black tea, blackberries, blackcurrants, many herbs, and fruits (RIFM Expert Panel 2007). These compounds have an intense floral odor with woody, balsamic and sweet, green-fruity nuances that strongly resemble violets (Sell 2006). The ionones are used as flavoring agents in foodstuffs and for fragrances in cosmetics, toiletries, household cleaners, detergents, and perfumes (RIFM Expert Panel 2007). In combination with anethole, menthol or liquorice, α-ionone can be used to flavor confectionery products. β-Ionone is the starting material for the synthesis of retinol (Caputi and Aprea 2011). Damascenones derive from allenic carotenoids (neoxanthin) (NPCS Board of Consultants & Engineers 2007). They have a woody, sweet, fruity sweet aroma with green floral nuances. β-Damascenone contributes essentially to the highly valued fragrance of Bulgarian rose essential oil. It also occurs in apricots, roses, grapes, kiwi, raspberries, blackberries, black tea, and red wine (Caputi and Aprea 2011). Damascenones are used as flavoring and fragrance agents. In perfumery, they confer naturalness and brilliance to the scented combinations (Surburg and Panten 2006).

7.3 Nutraceutical applications of terpene flavors

Of the terpenes, carotenoids and phytosterols are directly used as nutraceuticals. Both categories of phytochemicals are dietary constituents with nutritive and medicinal values. Carotenoids are responsible for the yellow, orange, and red colors of many human foods (vegetables, fruits, juices) and are involved in the photoprotection and regulation of cell membrane fluidity. In vitro and in vivo investigations, human epidemiological studies, and interventional trials revealed multiple beneficial health effects of carotenoids, including antioxidant, chemoprevention in cancer, cardioprotection, antiobesity, protection on bone components, prevention and risk reduction of cataract and age-related macular degeneration, and UV photoprotection (Rao and Rao 2007; Stahl and Sies 2012; Abdel-Aal et al. 2013). Phytosterols are commonly found in nuts, vegetable oils, seeds, and cereals. The evidence strongly supports the usefulness of phytosterols in the prevention of cardiovascular diseases and cancer due to their cholesterol-lowering properties and anticancer effects (Woyengo et al. 2009).

Although terpene flavors have a broad range of biological properties, they are non-nutritive dietary components; however, the flavors play a key role in nutrition. These compounds enhance the sensory appeal of foods and essentially contribute to the adequate meal preparation and appearance, and are also considered "soul food" (Ludwiczuk et al. 2017). In this regard, we can appreciate fragrant terpenes as nutraceuticals. The main flavors of terpene origin are presented in Table 7.2. We include some important phenylpropanoids used as flavoring agents because they are also important constituents of essential oils.

Table 7.2 Main Natural Flavoring Terpenes and Phenylpropanoids

Flavoring Compound		Sources	Uses	Biological Properties	References
Terpene	Borneol	*Cinnamomum camphora* (camphor tree), *Coriandrum sativum* (coriander), *Dryobalanops camphora* (Borneo camphor), *Lavandula angustifolia* (lavender), *Laurus nobilis* (laurel), *Ocimum basilicum* (basil), *Rosmarinus officinalis* (rosemary), *Zingiber officinale* (ginger)	Flavoring agent (foods, beverages); fragrance (cosmetics, household cleaners, detergents)	Anti-inflammatory, antinociceptive, antiviral, neuroprotective, vasorelaxant, type I allergy inhibitor	Paduch et al. 2007; Woyengo et al. 2009; Burdock 2010; Da Silva et al. 2013; De Cássia da Silveira e Sá et al. 2013
	Carvacrol	*Lavandula multifida* (fernleaf lavender), *Monarda didyma* (bee balm), *Origanum vulgare* (oregano), *Origanum dictamnus* (hop marjoram), *Origanum* sp., *Satureja* sp., *Thymbra* sp., *Thymus* sp.	Flavoring agent	Antibacterial, antifungal, anti-inflammatory, antioxidant, antitumor neuroprotective	Buchbauer and Ilic 2013; Aprotosoaie et al. 2017
	Carvone	S(+)-isomer: *Anethum graveolens* (dill), *Carum carvi* (caraway), *Citrus reticulata* (mandarin orange), *Chrysanthemum indicum* (Indian chrysantemum), *Cymbopogon martini* var. *sofia* (gingergrass), *Lindera umbellate* (kuromoji) R(−)-isomer: *Mentha spicata* (spearmint)	Flavoring agent (foods, beverages, cosmetics, personal care products); feed flavoring; zootechnical feed additive	Anti-inflammatory, antimicrobial, antitumor, immunomodulatory, skin penetration enhancer	Surburg and Panten 2006; de Carvalho and da Fonseca 2006; Paduch et al. 2007; De Cássia da Silveira e Sá et al. 2013; EFSA 2014

(Continued)

Sensory Qualities and Nutraceutical Applications of Flavors of Terpenoid Origin

Table 7.2 (Continued) Main Natural Flavoring Terpenes and Phenylpropanoids

Flavoring Compound	Sources	Uses	Biological Properties	References
Citral	*Backhousia citriodora* (lemon myrtle), *Calyptranthes paniculata*, *Cymbopogon* sp., *Dracocephalum moldavica* (Moldavian dragonhead), *Leptospermum liversidgei* (lemon-scented tea tree), *Lindera citriodora*, *Litsea cubeba* (mountain pepper), *Ocimum gratissimum* (clove basil)	Flavoring and fragrance agent (food industry, cosmetics, household cleaners, detergents); perfumery; pharmaceutical industry	Analgesic, antiadipogenic, anti-inflammatory, antimicrobial, antitumor, expectorant, sedative, spasmolytic	Surburg and Panten 2006; Burdock 2010; Modak and Mukhopadhaya 2011; Buchbauer and Ilic 2013
Citronellol	α(−)-Isomer: *Pelargonium graveolens* (rose geranium), *Rosa × damascena* β(+)-isomer: *Boronia citriodora* (lemon-scented boronia), *Eucalyptus citriodora* (lemon-scented eucalyptus), *Nigella sativa* (black cumin) Racemate: *Cymbopogon citratus* (lemon grass), *Cymbopogon winterianus* (citronella grass), *Lippia alba* (bushy matgrass)	Flavoring agent (foods, beverages); perfumery (floral compositions)	Analgesic, anticonvulsant, antidiabetic, anti-inflammatory, antimicrobial, antinociceptive, hypotensive, spasmolytic, vasorelaxant, wound healing	Opdyke 1979; Burdock 1996; Surburg and Panten 2006; De Cássia da Silveira e Sá et al. 2013; Brito et al. 2015; Ribeiro-Filho et al. 2016; Elsharif and Buettner 2017;
Eucalyptol (1,8-cineole)	*Eucalyptus globulus*, *Hyssopus cuspidatus*, *Lavandula angustifolia* (lavender), *Lavandula latifolia* (spike lavender), *Melaleuca alternifolia* (tea tree), *Mentha* sp., *Rosmarinus officinalis* (rosemary), *Salvia officinalis* (sage)	Flavoring agent (beverages, foods, confectionery products, cosmetics); pharmaceuticals	Anti-inflammatory, antioxidant, antitumor, expectorant, spasmolytic, skin penetration enhancer	Opdyke 1979; Burdock 1996; Surburg and Panten 2006; Caputi and Aprea 2011; Koziol et al. 2014; Seol and Kim 2016

(Continued)

Table 7.2 (Continued) Main Natural Flavoring Terpenes and Phenylpropanoids

Flavoring Compound	Sources	Uses	Biological Properties	References
Geraniol	*Aeollanthus myrianthus* (nindi), *Cymbopogon martinii* (palmarosa), *Cymbopogon winterianus* (citronella grass), *Dracocephalum moldavica* (Moldavian dragonhead), *Monarda fistulosa* (wild bergamot), *Pelargonium graveolense* (rose geranium), *Rosa × damascena* (rose), *Thymus pubescens*	Flavoring agent; fragrance; perfumery	Anti-inflammatory, anti-herpetic, antioxidant, antimicrobial, antitumor, vasodilator, skin penetration enhancer	Surburg and Panten 2006; Paduch et al. 2007; Chen and Viljoen 2010; Koziol et al. 2014
R(+)-Limonene	*Artemisia* sp., *Citrus* sp. (*Citrus × sinensis*), *C. limon*, *C. reticulate*, *C. paradisi*), *Lippia* sp.	Flavoring agent (foods, beverages); fragrance (cosmetics, household and industrial cleaning products); perfumery; pharmaceuticals	Antacid, anti-inflammatory, antioxidant, antimicrobial, antinociceptive, antitumor, chemopreventive in cancer, vasorelaxant, skin penetration enhancer, solvent of cholesterol-containing gallstones	Burdock 1996; Surburg and Panten 2006; Erasto and Viljoen 2008; Cardoso Lima et al. 2012; De Cássia da Silveira e Sá et al. 2013; d'Alessio et al. 2014
Linalool	*Aniba rosaeodora* (rosewood), *Cinnamomum camphora* (L.) Sieb var. *linaloolifera* (ho), *Coriandrum sativum* (coriander), *Lavandula angustifolia* (lavender), *Lavandula latifolia* (spike lavender), *Lippia alba* (bushy lippia), *Orthodon linaloolifera*	Flavoring agent (foods, beverages); fragrance (cosmetics, household detergents, furniture care products, waxes); perfumery; pharmaceutical industry	Analgesic, anticonvulsant, anti-inflammatory, antitumor, anxiolytic, cholesterol-lowering activity, local anesthetic, sedative, spasmolytic, skin penetration enhancer	Aprotosoaie et al. 2014

(Continued)

Table 7.2 (Continued) Main Natural Flavoring Terpenes and Phenylpropanoids

Flavoring Compound	Sources	Uses	Biological Properties	References
(−)-Menthol	*Mentha piperita* (peppermint), *Mentha arvensis* (Japanese mint), *Mentha* sp., *Micromeria fruticosa* (white micromeria), *Pycnanthemum virginianum* (common mountain mint), *Satureja parviflora*	Flavoring agent (foods, beverages, chewing gum, sweets, teas, cosmetics, toiletries, tobacco products), pharmaceuticals	Cooling effect, anti-inflammatory, analgesic, antimicrobial, antipruritic, antiviral, broncholytic, choleretic, cytotoxic, local anesthetic, secretolytic, spasmolytic, skin penetration enhancer	Burdock 1996; Surburg and Panten 2006; Kamatou et al. 2013; Koziol et al. 2014
Nerol	Bulgarian rose oil, *Cymbopogon* sp., *Humulus lupulus* (hope)	Flavoring and food additive; perfumery	Antifungal, anxiolytic, sedative, spasmolytic	Burdock 1996; Surburg and Panten 2006
Nerolidol	*Baccharis dracunculifolia*, *Brassavola nodosa*, *Cannabis sativa*, *Citrus aurantium* subsp. *amara*, *Cymbopogon* sp., *Jasminum* sp. (jasmine), *Melaleuca quinquenervia*, *Piper claussenianum*, *Zanthoxylum gardneri*	Flavoring and fragrance agent (cosmetics, soaps, household products, oral hygiene products); perfumery	Antibacterial, antifungal, antitumor, anti-ulcer, skin penetration enhancer	Buchbauer and Ilic 2013
Nootkatone	*Citrus paradise* (grapefruit), *Chrysopogon zizanoides* (vetiver), *Cupressus nootkatensis* (Nootka cypress)	Flavoring agent (beverages, soft drinks, tobacco); perfumery	Anti-inflammatory, antiobesity, antiplatelet, antitumor, neuroprotective	Surburg and Panten 2006; Chen et al. 2006; Labuda 2009; Caputi and Aprea 2011; Seo et al. 2011; Leonhardt and Berger 2015
Terpinen-4-ol	*Calamintha nepeta* (lesser calamint), *Coriandrum sativum* (coriander), *Coridothymus capitatus*, *Elletteria cardamomum* (cardamom), *Juniperus wallichiana* (black juniper), *Melaleuca alternifolia* (tea tree), *Origanum majorana* (marjoram), *Myristica fragrans* (nutmeg)	Flavoring agent (cosmetics, pharmaceuticals); perfumery	Analgesic, anti-inflammatory, antimicrobial, antitumor, antiprotozoal	Burdock 1996; Carson et al. 2006; Chizzola 2013

(*Continued*)

Table 7.2 (Continued) Main Natural Flavoring Terpenes and Phenylpropanoids

Flavoring Compound		Sources	Uses	Biological Properties	References
		Pandanus odoratissimus (screw pine), *Rosmarinus officinalis* (rosemary), *Satureja montana* (winter savory), *Xylopia parviflorum* (boss-bitterhout), *Zingiber corallinum* (coral ginger)			
	Thymol	*Lippia graveolens* (Mexican oregano), *Monarda didyma* (golden balm), *Origanum* sp., *Trachyspermum ammi* (ajowan), *Thymus* sp.	Flavoring agent (foods, beverages); pharmaceuticals; oral care products; perfumery	Analgesic, anticonvulsant, anti-inflammatory, antioxidant, antileishmanial, antimicrobial, antimutagenic, radioprotective, spasmolytic, skin penetration enhancer, wound healing	Surburg and Panten 2006; Paduch et al. 2007; Zachariah 2008; Arana-Sánchez et al. 2010; Riella et al. 2012; Nagoor et al. 2017
Phenylpropanoids	Anethole (*trans*)	*Clausena anisata* (maggot killer), *Foeniculum vulgare* (fennel), *Illicium verum* (star anise), *Osmorhiza longistylis* (amberoot), *Pimpinella anisum* (anise), *Syzygium anisatum* (anise myrtle)	Flavoring agent (food and confectionery products, alcoholic beverages; cosmetics, oral hygiene products); feed additive; perfumery; pharmaceuticals	Anti-inflammatory, antitumor, antidiabetic, antithrombotic, hypotensive, gastroprotective, local anesthetic, neuroprotective, wound healing	Aprotosoaie et al. 2016
	Eugenol	*Cinnamomum zeylanicum* (cinnamon), *Eugenia caryophyllata* (clove), *Illicium anisatum* (Japanese star anise), *Myristica fragrans* (nutmeg), *Ocimum basilicum* (basil), *Ocimum gratissimum* (clove basil), *O. sanctum* (holy basil)	Flavoring agent (foods, cosmetics, tobacco products); pharmaceuticals; perfumery	Analgesic, anticonvulsant, anti-inflammatory, antioxidant, antimicrobial, antiviral, local anesthetic, sedative	Surburg and Panten 2006; Mahapatra et al. 2009; Moreira-Lobo et al. 2010; Chizzola 2013

Next we discuss three of the most frequently used flavoring agents.

7.3.1 Citral

Citral (3,7-dimethyl-2,6-octadienal) is an acyclic α,β-unsaturated monoterpene aldehyde that occurs in nature as a mixture of two geometric isomers: geranial (*E*-citral, citral a) and neral (*Z*-citral, citral b), in a ratio of about 1:2 or 1:3. (*E*)-isomer is more stable than (*Z*)-isomer (Weerawatanakorn et al. 2015). The major natural sources of citral are the essential oils obtained from the following species: *Backhousia citriodora*, Myrtaceae (more than 80%) (Sultanbawa 2016), *Litsea cubeba*, Lauraceae (78–87%) (Si et al. 2012), and *Cymbopogon* sp., Poaceae (50–85%) (Rana et al. 2016).

In its pure form, citral is a yellowish liquid, miscible in organic solvents. Geranial has a strong lemon-like fragrance, while neral has a similar smell, but less intense and sweeter.

Citral is rapidly absorbed from the gastrointestinal tract. A large part of the dose applied cutaneously is lost due to high volatility, but what remains on the skin is absorbed quite well. It metabolizes rapidly and it is excreted as metabolites, predominantly through the kidneys. Under nonocclusive conditions, the loss of citral by evaporation competes with cutaneous absorption, resulting in less exposure than the applied dose. Citral is widely used as flavoring and fragrance agent in the food, cosmetics, and pharmaceutical industries. It adds citrus flavor to aroma composition. It is used in perfumery and to perfume detergents and household products. It is a raw material used in the synthesis of vitamin A, ionones, citronellol, linalool, nerolidol, geraniol, farnesol, and bisabolol, as well as in masking the smell of tobacco.

Current European Union regulations require that the presence of citral be mentioned on the product label if concentrations are above 10 ppm for cosmetics that remain on the skin and 100 ppm for the cosmetics that are applied, then washed off. In humans, NOEL (No Observed Effect Level) for dermal sensitization induction for citral is 1400 μg/cm (Surburg and Panten et al. 2006; Lalko and Api 2008).

7.3.1.1 Biological properties of citral

7.3.1.1.1 CNS effects Low doses of citral exert sedative and relaxant motor effects in mice, while slight anxiogenic properties have been demonstrated at high doses (200 mg/kg body weight, intraperitoneal) (Do Vale et al. 2002).

Citral appears to have different effects on spatial learning and reference memory in rats. This is dose dependent. Low doses of citral (0.1 mg/kg body weight) can improve spatial learning capacity in rats and increase spatial memory. High doses (1 mg/kg body weight) produce rather the opposite effect, namely repressing the spatial learning and space memory ability in rats. The repressive action is more pronounced in later learning stages.

The effects of citral on the spatial learning and memory do not appear to be a consequence of a direct activity, but instead result from other pathways that influence spatial memory. For example, citral can influence the synthesis of retinoic acid by regulating the activity of retinaldehyde dehydrogenases. It is well known that retinoic acid, a potent signaling molecule, not only controls embryonic development but is also involved in the regeneration of the adult nervous system. It plays an important role in the development and maintenance of spatial memory. Citral has a biphasic effect on the concentration of retinoic acid in the rat hippocampus. Low doses induce a remarkable increase in hippocampal concentration of retinoic acid, and high doses reduce it (Yang et al. 2009; Quintans-Júnior et al. 2011).

Citral decreases locomotor activity and potentiates the effects of imipramine in mice. After a chronic treatment (15 days), citral (100 mg/kg body weight, intraperitoneal) potentiates the hypnotic action of barbiturates in experimental animals (Do Vale et al. 2002).

7.3.1.1.2 Analgesic and anti-inflammatory activities

Citral has significant central and peripheral antinociceptive properties. It plays a more active role in inflammatory nociception than in neurogenic nociception.

In rodents, citral (50, 100, and 200 mg/kg body weight) produces a significant inhibition of acetic acid–induced writhing response and determines antinociceptive effects on both first and second phases of formalin-induced nociception. High doses of citral (100 and 200 mg/kg body weight) also reduce the edematous responses in carrageenan-induced paw edema in rats. In addition, the compound inhibits leukocyte migration in the peritonitis model. These effects may be related to the modulation of pro-inflammatory cytokine production/release and/or the intervention in the arachidonic acid cascade (Quintans-Júnior et al. 2011). Lee et al. (2008) showed that the anti-inflammatory effects of citral may be due to the inhibition of NO (nitric oxide) production via suppresion of NF-κB (nuclear factor kappa B) activation. Citral is also a suppressor of cyclooxygenase-2 (COX-2) and integrin $\beta 7$ expressions (Ortiz et al. 2010).

Cyclooxygenase is a key enzyme in PG biosynthesis. It contains two isoforms, COX-1 and COX-2. COX-1 is constitutive in most cells, whereas COX-2 is absent, but it is inducible by inflammatory stimuli. COX-2 expression is regulated differently in various cell types and it plays an important role in inflammation, tumorigenesis, development, and circulatory homeostasis (Katsukawa et al. 2010). Integrin $\beta 7$ is a glycoprotein that is involved in adhesion of leukocytes to endothelial cells and the migration of leukocytes during inflammatory responses.

Citral activates and modulates transient receptor potential (TRP) ion channels from sensory neurons that are mainly involved in chemical sensing. It acts as a partial agonist of TRPV1 (transient receptor potential cation channel subfamily V member 1), TRPV3, TRPM8 (transient receptor potential cation channel subfamily M member 8), and TRPA1 (transient receptor potential cation channel subfamily A member 1) ion channels. It is known that TRPV (transient receptor potential cation channel subfamily V) receptors function as transducers of thermal and chemical stimuli, being involved in nociception, and their antagonists can function as analgesic agents. At the same time, TRPM (transient receptor potential cation channel subfamily M) receptors are important in the perception of low temperatures; they regulate magnesium reabsorption in the kidneys and participate in the transduction of the sensation in taste cells or in cell adhesion processes. TRPA1 receptors act as mechanical stress sensors. They have a role in controlling cell growth and signaling. They are activated by allylisothiocyanate, cinnamic aldehyde, nicotine, acrolein, and hydrogen peroxide. TRPA1 antagonists are effective in blocking the painful response in inflammation. The most relevant effects are citral's long-lasting inhibition of TRPV1-3 and TRPM8 following activation. Prolonged sensory inhibition leads to a more effective action of topical treatment with citral in the pain types that implies sensory nerves and skin, such as allodynia or itch. The effects of citral on the ion channels may be explained by hydrophobic interactions with constituent proteins and/or with the phospholipid/protein interface (Stotz et al. 2008).

The combined use of citral and naproxen, a well-known synthetic nonsteroidal anti-inflammatory drug, produces an additive antinociceptive effect. It is likely that, in addition to inhibiting NO release, the interaction between citral and naproxen at a systemic level might involve the inhibitory effects on prostaglandin synthesis (mechanism of naproxen action), as well as on the cationic channels mentioned above and demonstrated for citral. It is important to note that citral does not cause detectable gastric damage within 6 h after administration, the time period after which gastric injuries occur when administering naproxen. The citral-naproxen combination also has fewer gastro-intestinal side effects than naproxen itself (Blaschke et al. 2006).

7.3.1.1.3 Antiadipogenic effects Citral administration (10, 15, 20 mg/kg body weight, 28 days) significantly reduces body weight gain and lipid accumulation, and improves insulin sensitivity and glucose tolerance in a diet-induced model of obesity in rats. A plausible mechanism of antiadipogenic action of citral consists of its ability to inhibit retinaldehyde metabolism via retinaldehyde dehydrogenase antagonism. Retinaldehyde plays a significant role in adipocyte differentiation and metabolism. It increases fat metabolism and secretion of adiponectin, a key hormone in the regulation of energy homeostasis and lipid metabolism; in addition, it exerts insulin-sensitizing effects (Modak and Mukhopadhaya 2011).

Citral has been found to act as a dual activator of peroxisome proliferator-activated receptors (PPAR) α and γ. Its anti-obesity and anti-inflammatory effects may also be related to this activity. These receptors are ligand-activated transcription factors that belong to the family of nuclear hormone receptors. They play key roles in lipid and carbohydrate metabolism, cell proliferation and differentiation, and inflammation. They are considered important molecular targets in the treatment of so-called lifestyle related diseases (Blaschke et al. 2006; Katsukawa et al. 2010).

It is possible for citral to regulate in vivo sensitivity to PPARs activation. Considerable amounts of natural ligands of these receptors, such as fatty acids, occur under circumstantial conditions, including diet, hence the conclusion that citral can be useful as a component of the daily diet but not as a pharmacological agent. This is because, compared to synthetic drugs, the action potential of citral on PPARs is much lower (Katsukawa et al. 2010).

7.3.1.1.4 Antimicrobial effects

Citral elicits inhibitory effects on Gram-positive and Gram-negative bacteria, but also on various strains of fungi. The antimicrobial activity is increased in alkaline pH (Onawunmi 1989). It exhibits inhibitory activity against pathogenic and food-spoilage bacteria such as *Escherichia coli, Salmonella typhimurium, Listeria monocytogenes, Staphylococcus aureus,* and *Cronobacter sakazakii* (Shi et al. 2016). At low concentrations (0.08–0.1%), citral acts bactericidally on *E. coli* strains. The compound is also active on rifampicin-resistant strains of *Salmonella typhimurium* and methicillin-resistant *Staphylococcus aureus* pathogens (Saddiq and Khayyat 2010). Some of the reported antibacterial mechanisms for citral are morphological alterations in the cell membrane and cell wall, reduction of intracellular ATP and in cytoplasmic pH, and hyperpolarization of the cell membrane (Shi et al. 2016). Citral has a significant in vitro antifungal activity against *Candida* spp. (de Bona da Silva et al. 2008). Other fungal strains that have been found sensitive to citral are *Aspergillus flavus, Candida, Microsporum gypseum, Trichophyton mentagrophytes, Penicillium italicum,* and *Rhizopus stolonifer.* Some authors have shown that fungicidal action is achieved by the generation of complexes between citral and electron donors from the fungal cells, a process leading to apoptosis, or cell death (Saddiq and Khayyat 2010).

7.3.1.1.5 Other effects

Citral is an inhibitor of CYP2B6 (cytochrome P450 family 2 subfamily B member 6) activity. This is an isoenzyme that belongs to the CYP450 (cytochrome P450) group. It is found in hepatic and some extrahepatic tissues and participates in the metabolism of some drugs, such as bupropione, cyclophosphamide, sibutramine, tamoxifen, and efavirenz. High concentrations of citral can thus alter the bioavailability of the drugs that are metabolized by CYP2B6. Since this isoenzyme may also contribute to the activation of some procarcinogens (aflatoxin B1), citral may also be useful as a chemopreventive agent. However, as in vitro inhibition of enzyme's activity cannot be translated in vivo, the investigation of the in vivo interaction between citral or other monoterpenoids and CYP2B6 substrates is necessary in order to determine its clinical significance (Seo et al. 2008).

7.3.2 Geraniol

Geraniol is an acyclic monoterpenol found in flowers and other vegetative organs, where it coexists with geranial and nonal aldehydes, geraniol's oxidation products. Red spider females (*Tetranychus urticae*) also emit geraniol. The main sources of geraniol are the essential oils isolated from *Monarda fistulosa*, Lamiaceae (more than 95%); *Aeollanthus myrianthus*, Lamiaceae (66.6%); *Cymbopogon martinii*, Poaceae (53.5%); *Rosa* sp., Rosaceae (44.4%); and *Cymbopogon nardus*, Poaceae (24.8%).

Geraniol is an oily, transparent or slightly yellow liquid, insoluble in water, but soluble in most organic solvents. It has an odor similar to that of a rose and the taste is described as sweet floral, citrus and fruity with woody and waxy nuances (Chen and Viljoen 2010). The compound is used as flavoring agent for beverages and confectionery products (candies, ice creams) (de Carvalho et al. 2014). In small amounts, geraniol accentuates citrus notes in flavor compositions. As a fragrance, it is used in all flowery-rose like combinations (Surburg and Panten 2006). It is present in 76% of the deodorants on the European market, in 41% of the household cleaning formulations, and in 33% of natural cosmetics. In the EU, the compound is included on the list of odorants that must be mentioned on cosmetic and detergent labels. Although geraniol is not electrophilic and should not exhibit a sensitizing capacity, some cases of allergic contact dermatitis were recorded. Geraniol oxidizes due to air exposure or CYP-mediated metabolic cutaneous activation, thus generating allergenic compounds. Of these compounds, geranial, nonal, and 6,7-epoxygeraniol were found to be moderate sensitizers, while 6,7-epoxygeranial is a strong sensitizer (Hagvall et al. 2012).

In recent years, there has been a growing trend to produce flavors from monoterpenes through biotechnological conversion using plant cell or tissue cultures. In this regard, it was proved that various microorganisms are able to metabolize geraniol into different derivatives. For example, geraniol forms (*R*)(+)-citronellol by microbiological reduction, under the action of *Saccharomyces cerevisiae*, and it generates (*S*)citronellol by stereoselective reduction. Under the action of *Penicillium digitatum*, geraniol is converted to citral. The worldwide production of geraniol exceeds 1,000 tons per year (Chen and Viljoen 2010).

7.3.2.1 Biological properties of geraniol

7.3.2.1.1 Antimicrobial activity Geraniol has both antibacterial and antifungal properties. Some authors appreciate that geraniol is more effective as an antibacterial than an antifungal agent. Geraniol was most active on Gram-negative bacteria such as *E. coli*, *Salmonella enterica*, and *S. typhimurium* and on Gram-positive bacteria such as *Listeria monocytogenes*. The major respiratory tract pathogens, including *Haemophilus influenzae*, *Streptococcus pneumoniae*, *S. pyogenes*, and *Staphylococcus aureus*, were also sensitive to geraniol (Chen and Viljoen 2010). Geraniol exhibits a strong antimycobacterial activity against *Mycobacterium tuberculosis* (minimum inhibitory concentration, MIC = 12.5 µg/mL).

The antibacterial activity of geraniol is due to its deleterious effects on structure and function of bacterial membrane and cell wall (Andrade-Ochoa et al. 2015). Geraniol can also restore the susceptibility of some Gram-negative bacteria such as *Enterobacter aerogenes, Pseudomonas aeruginosa*, and *Acinetobacter baumannii* to quinolones, β-lactams, and chloramphenicol, increasing the efficacy of these antibiotics by targeting the efflux pumps (Lorenzi et al. 2009).

Geraniol demonstrates good inhibitory activity against different fungal pathogens, including *Candida albicans* (MIC = 19.5 mM), *Cryptococcus neoformans* (MIC = 100 µL/L), *Colletorichum camelliae* (MIC = 440 µg/mL), and *Tricophyton rubrum* strains (MIC = 16–256 µg/mL) (Chen and Viljoen 2010; De Oliveira Pereira et al. 2015). The inclusion of geraniol in vaginal wash formulations (25 µg/mL) provides protection against vaginal candidiasis. Together with the monoterpene phenols thymol and carvacrol, geraniol displays antibiofilm properties. At concentrations of 0.06%, it causes more than 80% inhibition of *Candida albicans* biofilm development. The combinations of geraniol or essential oils containing geraniol with synthetic antifungals determine a more effective therapy of mycoses. Strong synergistic interaction has been demonstrated for the combination of *Pelargonium graveolense* essential oil or geraniol with ketoconazole against *Tricophyton* spp. dermatophytes (Shin and Lim 2004).

The antifungal activity of geraniol results from its ability to disrupt microbial membrane permeability and function and to inhibit ergosterol biosynthesis (Chen and Viljoen 2010; De Oliveira Pereira et al. 2015).

7.3.2.1.2 Antioxidant activity In vitro antioxidant assays showed the free radical scavenging properties of geraniol. The compound protects rat alveolar macrophages against oxidative stress via different mechanisms, including antilipoperoxidant potential, the improvement of the status of endogenous enzymatic (superoxid-dismutase) and non-enzymatic antioxidants (glutathione), and the inhibition of NO release (Chen and Viljoen 2010).

7.3.2.1.3 Antitumor activity The therapeutic, preventive, or chemosensitizing effects of geraniol have been demonstrated in vitro and in vivo on various human cancer models for breast, lung, colon, prostate, pancreatic, hepatic, skin, kidney, and oral cancers. Geraniol (400 µM) strongly inhibits the growth of human colon cancer cells (70% inhibition). The antiproliferative effects of geraniol are mainly related to its capability to decrease the rate of DNA synthesis and to produce the accumulation of the cells in the S phase of the cell cycle. At the same time, the compound causes a 50% decrease in ornithine decarboxylase activity, a key enzyme in the biosynthesis of polyamines, a process that is intensified in the development of cancerous events. Geraniol leads to a

40% reduction in the intracellular pole of putrescine. Geraniol also activates the intracellular catabolism of polyamines, as indicated by the enhancement of polyamine acylation. (Carnesecchi et al. 2001). Its antiproliferative effects were also demonstrated on hepatoma and melanoma cell growth. In mammalian cells, geraniol can interact with 3-hydroxy-3-methylglutarylcoenzyme A(HMG-CoA)-reductase, an enzyme that catalyzes the formation of mevalonate, a precursor required for cell proliferation. It suppresses HMG-CoA reductase synthesis by altering the translation process of enzyme transcripts (Peffley and Gayen 2003; Chen and Viljoen 2010). Other studies have shown that geraniol inhibits the proliferation, cell cycle progression, and the activity of cyclin-dependent kinase-2 in MCF-7 breast cancer cells independently of the effects on the HMG-CoA reductase activity.

The intake of geraniol (36 mmoL/kg of diet) decreases mammary tumor multiplicity induced by 7,12-dimethylbenz[α]antracen in rats (Yu et al. 1995). Also, chemopreventive effects of geraniol have been demonstrated during the initial phases of hepatocarcinogenesis in rats. Geraniol (25 mg/100 g body weight, 8 weeks treatment) inhibits cells proliferation and DNA damage, and increases apoptosis of hepatic pre-neoplastic lesions (Ong et al. 2006).

Geraniol (400 μM) sensitizes colon cancer cells to the cytotoxic action of 5-fluorouracil (5-FU). It enhances the cellular uptake of 5-FU by a membrane-permeabilizing effect and the blockade of the morphological and functional differentiation of cancer cells (Carnesecchi et al. 2002). At concentrations of 150 μM, geraniol reduces the expression of thymidylate synthase and thymidine kinase in colon cancer cells. These enzymes are important targets in chemotherapy and are involved in the cancer cell resistance to 5-FU. A reduction of their level is associated with the increase in 5-FU cytotoxicity (Mans et al. 1999). Geraniol also enhances the cytotoxic effects of docetaxel, doxorubicin, paclitaxel, etoposide, and cisplatin in PC-3 prostate cancer cells (Cho et al. 2016).

Although the molecular mechanisms of geraniol anticancer effects remain largely unknown, recent experimental data revealed that geraniol affects different signaling molecules or pathways mainly involved in the cell cycle and apoptosis, such as extracellular signal-regulated kinase (ERK1/2); protein kinase C (PKC); cyclins A, B, D; cyclin-dependent kinases CDK1, CDK2, CDK4; caspases 3, 8, and 9; and AMP-activated protein kinase (pAMPK) (Cho et al. 2016).

7.3.2.1.4 Anti-Inflammatory activity In vitro, geraniol exhibited a marked inhibition of neutrophil adhesion induced by TNF-α (tumor necrosis factor alpha) (inhibitory concentration 50%, $IC_{50} < 0.00625\%$), suggesting a potential ability of this compound to negatively modulate neutrophil function in inflammatory responses (Abe et al. 2013). In a model of 12-O-tetradecanoyl phorbol-13-acetate-induced inflammation in mouse skin, the topical treatment with geraniol (250 μg) inhibits COX-2 expression and the activation of p38

MAPK (mitogen-activated protein kinase), a cell signaling pathway with a crucial role in inflammation. Geraniol also decreases the production of proinflammatory cytokines such as TNF-α, IL-6 (interleukin 6), and IL-1β (interleukin 1β) (Khan et al. 2013).

7.3.2.1.5 Gastroprotective activity
Geraniol (7.5 mg/kg) protects against gastric and duodenal lesions induced by different ulcerogenic agents (ethanol, ischemia-reperfusion, cisteamine) in rats. The antiulcer effects are mediated by antioxidant properties, activation of mucosa-protective factors (endogenous prostaglandins, NO pathway, sulfhydryl compounds, and mucus production), and the stimulation of calcitonin gene-related peptide through the activation of the transient receptor potential vanilloid channel (de Carvalho et al. 2014).

7.3.2.1.6 Anthelmintic effects
A nematicidal action was demonstrated for geraniol on *Meloidogyne incognita, Caenorhabditis elegans* nematodes. It showed larvicidal activity against *Anisakis simplex* marine nematodes and *Contracaecum* roundworms (Chen and Viljoen 2010).

7.3.2.1.7 Enhancement of percutaneous absorption
Geraniol improves transdermal delivery of some drugs (sodium diclofenac, caffeine, 5-FU) by enhancing their penetration into stratum corneum. The compound also increases the permeation of antimicrobial silver sulphadiazine through burn eschars (Chen and Viljoen 2010).

7.3.3 Menthol

Menthol (2-isopropyl-5-methylcyclohexanol, p-menthan-3-ol) is a monocyclic monoterpenol and the major component of mint essential oils (*Mentha* sp., Lamiaceae). The most important commercial essential oils containing menthol are obtained from the following mint species: *Mentha arvensis* var. *piperascens* (Japanese mint) (77–89% menthol) (Singh et al. 2005), *M. arvensis* (cornmint) (60–80% menthol) (Reineccius 2005), and *Mentha × piperita* (peppermint) (30–55% menthol) (Bruneton 1995).

The menthol molecule presents four pairs of optical isomers: (+)- and (–)-menthol, (+)- and (–)-isomenthol, (+)- and (–)-neomenthol, (+)- and (–)-neoisomenthol. (–)-Menthol is the principal isomer found in nature and it has (1R, 2S, 5R) configuration. It is a solid substance, crystalline, transparent or white, slightly soluble in alcohol, ether, chloroform, and fatty oils. The odor and the taste are characteristically minty, cooling, and refreshing. The other isomers of menthol differ in their sensory profiles; their aromas do not present the cooling and refreshing character (Surburg and Panten 2006; Kamatou et al. 2013).

The biosynthesis of (–)-menthol is complex and requires eight enzymatically catalyzed steps. These steps primarily involve the formation and subsequent cyclization of the universal precursor of monoterpenes, geranyl diphosphate, to the (4S)-limonene. Following hydroxylation at C-3, several redox and isomerization reactions occur in a general scheme of allylic oxidation and reduction leading to the appearance of three chiral centers on the substituted cyclohexane ring and the formation of (–)-menthol (Croteau 1986).

(–)-Menthol and its racemate are the principal compounds of commercial interest. The major part of worldwide production of (–)-menthol is obtained from cornmint essential oil and dementholized cornmint essential oil. After the removal of (–)-menthol by crystallization, the dementholized cornmint oil still contains 40–50% free menthol. (–)-β-Pinene, (+)-citronellal, (–)-piperitone, (+)-3-carene, and m-cresol are precursors in the industrial synthesis of (–)-menthol (Surburg and Panten 2006; Kamatou et al. 2013). Racemic menthol is obtained by chemical synthesis starting from thymol or pulegone (Surburg and Panten 2006).

Annual worldwide consumption of menthol exceeds 7000 tonnes. Menthol is one of the most important flavoring agents. It is used in the pharmaceutical, food, alcoholic/nonalcoholic beverage and tobacco industries, and in confectionery, oral hygiene products (mouthwashes, toothpastes), cosmetics, household products, and detergents. In perfumery, it is used per se or in the form of menthylesters that emphasize floral notes (especially the rose ones), and in organic chemistry as chiral material in asymmetric syntheses.

Menthol is available in a variety of pharmaceuticals. It is recommended in the treatment of oropharyngeal irritation, respiratory disorders, muscle and joint pain, migraines, itching, and sunburn (the latter commonly associated with aloe).

According to FDA regulations, menthol is a safe and effective topical agent. Concentrations of up to 16% have been approved by the FDA for external use. In concentrations of 40% menthol causes skin erythema and burns. It may cause systemic allergic reactions and allergic contact dermatitis. In infants and young children preparations containing menthol or peppermint essential oil are not applied to the face, especially the nasal area, due to the possibility of bronchospasm leading to asthmatic episodes or even occurrence of respiratory insufficiency (Patel and Yosipovitch 2007; Kamatou et al. 2013).

The maximum skin exposure resulting from the use of (–)-menthol in the fine aromas is 0.58%, given that odoriferous ingredients account for up to 20% of the final product. In the case of racemic menthol, the maximum level is 0.052%. In cosmetic formulations, (–)-menthol is incorporated in concentrations up to 0.4%, with the maximum daily skin exposure level 0.0102 mg/kg body weight. In the case of racemic menthol, these values are 0.9% and 0.0229 mg/kg body weight, respectively (Bhatia et al. 2008).

As an additive in the food industry, the acceptable daily intake (ADI), according to FAO regulations, is 0–4 mg/kg body weight (Van der Schaft 2007).

The pharmacokinetic profile of menthol reveals a rapid absorption. Up to 100% of the ingested menthol dose seems to be absorbed. Metabolic studies have shown that oral doses of menthol are mainly metabolized in the liver and excreted through the kidneys and the bile. Metabolites are simple conjugates of glucuronic acid and oxidation products resulting from the action of CYP450. The absorbed menthol is predominantly eliminated in the form of glucuronides within 48 h (Patel and Yosipovitch 2007).

7.3.3.1 Biological properties of menthol

Mint species and mainly peppermint are among the most used plants in traditional medicine to treat or ameliorate different conditions (mainly gastrointestinal, respiratory, and dermatological disorders). A broad array of bioactivities have been demonstrated for menthol and for essential oils that contain menthol, including antimicrobial, analgesic, antipruritic, anti-inflammatory, anticancer, antiviral, chemopreventive, skin penetration enhancing effects.

7.3.3.1.1 Menthol and the olfactory system

The inhalation of low menthol concentrations induces a distinct cooling sensation that is mediated by the stimulation of the trigeminal nerve fibers belonging to the sensory system of the olfactory epithelium. The characteristic peppermint flavor contributes to the stimulation of olfactory nerves. This cooling sensation can be useful to alleviate nasal congestion. The pleasant aroma associated with menthol can be particularly beneficial in topical cosmetic and facial formulations, increasing patient acceptability. However, high concentrations of menthol may cause an irritant sensation (pungent or even a sensation of burning and pain) that induces nasal congestion. Interestingly, those suffering from anosmia can detect menthol due to the pungent sensation, which suggests that the integrity of the olfactory system is not essential for menthol detection. The ability to detect the intensity and pleasure associated with menthol diminishes with aging (Eccles 1994). The sensory impact of menthol at oral administration and in topical applications depends on both its concentration and duration of exposure.

In the oral cavity, menthol has a complex action due to the sensitization of both warm and cold receptors as well as the modulation of taste receptor activity. It enhances cold oral sensations but may also decrease or enhance warmth sensations. The stimulation of taste receptors with menthol prevents for some time (up to 10 min) their response to subsequent menthol exposure. The oral administration of menthol may also produce a sensation of nasal cooling and nasal airflow. Topical application of menthol in low concentrations causes a cooling effect whereas high concentrations of menthol produce irritation. The sensitivity of body surfaces to the cooling action of menthol differ. The following order was mentioned: eye > tongue > buccal cavity > ano-genital area > axilla > inside forearm–breast > thigh, back > hands, feet > palms, soles (Eccles 1994).

7.3.3.1.2 Modulation of thermoreceptors

When applied to the skin, low concentrations of menthol cause a cooling sensation similar to low temperature exposure. In this respect, (−)-menthol results in an effect 50 times stronger than its isomer, (+)-menthol. The ability of this monoterpenoid to produce such sensations is due to the stimulation of specific thermoreceptors. In 2002, two independent teams of researchers identified, through different methods, the menthol receptor, TRPM8, a cationic channel formed of 1104 amino acids. This type of receptor can be activated by menthol and cold thermal stimuli (8–28°C), menthol acting as an agonist, as the TRPV1 vanilloid receptors are activated by high temperatures (over 43°C) or capsaicin, the spicy compound in hot pepper. TRPM8 is a member of the family of TRP excitatory ion channels (see Section 7.3.1.1.2). These thermoreceptors exhibit a wide spectrum of thermal sensitization, from noxious high temperatures to very low temperatures, also harmful. They are found in the primary small-diameter sensory fibers. For some of them, a localization in keratinocytes was also demonstrated. Under certain circumstances, they may also be expressed in non-neural epithelial cells (Patel and Yosipovitch 2007). TRPM8 can be a major chemosensory receptor of the trigeminal nerve and it may be involved in the sensory impact of menthol.

Menthol can also stimulate the heat-activated receptors TRPV3, but at concentrations much higher than those required to activate TRPM8 (Patel and Yosipovitch 2007). Low micromolar concentrations of menthol activate TRPA1 channels (see Section 7.3.1.1.2) while at millimolar concentrations (1 mM), menthol inhibits these receptors (Kolassa 2013). The repeated exposure to menthol may desensitize the TRPM8 receptors. The lipid second messenger, phosphatidylinositol 4,5-bisphosphate, appears to play a central role in the activation and desensitization of TRPM8 receptor (Patel and Yosipovitch 2007).

Menthol lowers the pain detection threshold caused by very low temperatures and intensifies the painful response to noxious cold stimuli applied at concentrations higher than the detection threshold. In terms of thermal heat stimuli, menthol has no effect on the detection of heat or the threshold for detecting pain on high temperature heat stimuli, although contradictory data have also been reported. Discrepancies between studies can be explained by the differences in menthol concentration, duration of application, and methods used to record the sensation (Patel and Yosipovitch 2007).

7.3.3.1.3 Antipruritic activity Bromm et al. (1995) have shown that skin cooling at 2–4°C reduces the intensity of histamine-induced pruritus. The application of menthol in a concentration of 1% leads to a similar effect. In contrast, 10% menthol has no effect on the duration or intensity of histamine-induced pruritus. The optimal concentration for calming pruritus has not yet been established, although 1% is frequently used in clinical practice.

The mechanism by which menthol acts antipruriginously is still unknown. Some authors have suggested that the activation of A-Δ fibers centrally inhibits pruritus. The implication of an imbalance of the endogenous opioidergic system in pruritus pathophysiology is also possible. Different opioid receptors have opposite effects on pruritus. Both types of compounds, μ-opioid receptor agonists and κ-opioid receptor antagonists, can induce pruritus, while μ-receptor antagonists and κ-receptor agonists can reduce it. Menthol acts as a κ-opioid receptor agonist, so this mechanism could explain the antipruritic properties of the compound. There is, however, a category of patients suffering from chronic pruritus (atopic dermatitis, uremia, psoriasis) whose symptoms have diminished with cold showers. It is thus likely that the cooling sensation produced by menthol upon cutaneous application will also serve as a possible mechanism for reducing the itch sensation in these patients (Patel and Yosipovitch 2007).

7.3.3.1.4 Analgesic effects In concentrations of 1% or less, menthol depresses skin sensory receptors, while at concentrations between 1.25% and 16% it stimulates sensory receptors and thus it acts as a revulsive agent. At high concentrations (30% or more), menthol can induce the so-called cold pain, by activation and sensitization of cold-sensitive C nociceptors and activation of low-temperature-specific A-Δ fibers. It was shown that the hyperalgesic effects of menthol can result not only from the sensitization of C nociceptors but also depend on the balance between the activation of A-Δ and C-nociceptor fibers. Similar to capsaicin, menthol can also cause analgesia by desensitizing nociceptive C fibers. Other authors have postulated that the analgesic effects of menthol can be explained by TRPM8 activation and/or TRPA1 inhibition. The latter of these receptors is activated by a variety of harmful stimuli, including low temperatures, pungent natural compounds, or irritants in the environment (Patel and Yosipovitch 2007).

Other proposed analgesic mechanisms are based either on the selective activation of κ-opioid receptors, (−)-menthol acting as a weak agonist for them, or the engaging of the supraspinal suprasegmental circuits, leading to the downward inhibition of spinal nociceptors (Klein et al. 2010). Furthermore, menthol produces an increase in skin bloodflow at the site of application, so that local temperature rise through vasodilation mediates, at least in part, the analgesic effect in musculoskeletal pain in a manner similar to heat application therapy. However, the mechanism responsible for this therapeutic effect remains controversial (Patel and Yosipovitch 2007). The stereochemistry of menthol molecule has a profound impact on its biological activity. Of the isomers of menthol, only (−)-menthol exhibited analgesic effects (Kamatou et al. 2013).

7.3.3.1.5 Activity on the respiratory system The effects of menthol at the respiratory level are extremely versatile. It relieves cough, pulmonary congestion, and dyspnoea. The compound inhibits the contraction and responsiveness of smooth alveolar muscles and cough induced by aerosols with citric acid that stimulate airway sensory neurons. It was found that menthol prevents the activation of $Na^+/K^+/Cl^-$ basolateral co-transporter (NKCC1), and it functions as an activator of cystic fibrosis conductivity regulator (CFTR) without affecting bicarbonate-dependent basolateral anionic transporter (NBC1/AE), although all these transporters are commonly cAMP dependent. The regulation of these anion transporters, which are important for mucociliary clearance, takes place possibly via a cAMP-independent/cytoskeleton-dependent modulation. The inhibitory effects on the NKCC1 function may be somewhat beneficial in inflammatory respiratory diseases, as inhibition is believed to counteract the formation of pulmonary edema (Morise et al. 2010). (–)-Menthol inhibits tracheal smooth muscle contractions in guinea pigs. It causes a reduction in bronchoconstriction and respiratory hyperresponsiveness, the effects of which are mediated through the inhibition of the calcium influx. Studies on various experimental models have shown that inhalation is preferable for the medical use of menthol in respiratory conditions (Morise et al. 2010).

7.3.3.1.6 Antitumor activity Dietary addition of (–)-menthol causes a significant inhibition of mammary carcinogenesis induced by 7,12-dimethylbenz[*a*]-anthracene in rats. Menthol has been shown to exhibit anticancer effects against different human cancer cell lines such as prostate, bladder, and oral squamous carcinoma. The antitumor properties of menthol are related to its potential to modulate the TRPM8 receptors and thus to produce cell cycle arrest, mitochondrial membrane depolarization, and inhibition of downstream signaling pathways such as polo-like kinase 1 (Kamatou et al. 2013, Sobral et al. 2014).

7.3.3.1.7 Anti-inflammatory activity (–)-Menthol significantly decreases the production of inflammatory mediators LTB4 (leukotriene B4), IL-1β, and PGE2 (prostaglandin E2) in LPS (lipopolysaccharide)-stimulated monocytes, indicating potential beneficial effects in the therapy of chronic inflammatory diseases such as allergic rhinitis, bronchial asthma, or colitis. It has been found to suppress antigen-induced histamine release from rat peritoneal mast cells and to inhibit passive cutaneous anaphylaxis of guinea pigs (De Cássia da Silveira e Sá et al. 2013). Arakawa and Osawa (2000) showed that peppermint gums enriched with (–)-menthol as well as geraniol or citronellol more effectively alleviate rhinitis symptoms than the nonflavored or normal peppermint-flavored gums.

7.3.3.1.8 Local anesthetic activity Galeotti et al. (2001) have demonstrated that menthol has a strong dose-dependent local anesthetic effect, described in vitro and in vivo. Both enantiomers of menthol, levorotatory and dextrotatory menthol, are anesthetic equiactive. It has also been demonstrated that menthol blocks voltage-gated neuronal sodium channels and those from the skeletal muscle in a concentration-dependent manner, in resting and inactive states (Patel and Yosipovitch 2007).

7.3.3.1.9 Antimicrobial activity Antibacterial and antifungal effects of menthol are expressed on a wide spectrum of microorganisms (*Staphylococcus aureus, Streptococcus pneumoniae, S. pyogenes, Escherichia coli* bacterial strains, and *Candida albicans* and *Trichophyton* sp. fungal pathogens). The compound is used in various antiseptic products or food preservatives. Furthermore, combination of menthol and some antibiotics (erythromicin, oxacillin) or quinolones (norfloxacin) produces synergistic effects. The antimicrobial mechanisms, although incompletely elucidated, are based on the toxic effects on the structure and functionality of the microbial membrane. Thus, it has been hypothesized that the antimicrobial effect of menthol can result, at least in part, from the disruption of the lipid fraction of the microbial membrane and the loss of intracellular components. It has subsequently been shown that menthol, like peppermint essential oil, exhibits antiplasmid activity (Patel and Yosipovitch 2007; Kamatou et al. 2013).

7.3.3.1.10 Skin penetration enhancement As a vehicle in topical and transdermal formulations, menthol can influence both the percutaneous absorption and release of some drugs, by acting as an absorption promoter. The effect can be explained by menthol's ability to destroy the lipid layer of the corneum stratum. The permeation mechanism may be attributed to the preferential hydrogen bonds between menthol and the groups at the ends of the ceramide molecules. In addition, terpenes with polar groups, such as menthol, have been shown to be percutaneous enhancers, especially for the hydrophilic substances. This property, along with other qualities (natural compound, pleasant aroma, efficiency in low concentrations, 0.5–5%), suggest menthol is an ideal vehicle for use in topical combined therapies. Menthol (1–5%) may also be useful for improving the anesthetic effect of tetracaine and other local anesthetics. Although the FDA discourages the use of topical drug combinations, the association of menthol with topical steroids in the treatment of atopic dermatitis and other pruritus symptoms is appealing. In the treatment of pruritus, menthol is already used in combination with camphor and pramoxin (Liu et al. 2005; Patel and Yosipovitch 2007).

7.3.3.1.11 Other effects (+)-Menthol exerts a positive stereoselective and potent modulation on $GABA_A$ receptors with the increase of the GABA flux. Modulation of these receptors exerts profound influences on the neuronal activity, with sedation and anesthesia, in vivo. Menthol's binding site is different from the GABA receptor sites typical for interaction with benzodiazepines, barbiturates, or steroids. For an efficient GABAergic modulation, the presence of the aliphatic chain near to the hydroxyl group from the cyclohexane ring is important. At the same time, menthol can compete for binding sites on $GABA_A$ receptors with propofol, a powerful general anesthetic. Menthol is about 10 times less active than propofol, but also less toxic. This compound may serve as a structural model for the development of new GABAergic modulator drugs (Watt et al. 2008).

7.4 Conclusions

Mono- and sesquiterpenes (both hydrocarbons and oxygenated derivatives) are the main terpene flavors. They have valuable sensory qualities and are major constituents of essential oils of many plants long used in traditional medicine to prevent or treat various conditions. Terpene flavors are of tremendous commercial interest. They are widely used as medicines, flavoring agents, and fragrances in the pharmaceutical, food, confectionery, beverage, and cosmetic industries; in household cleaning products; and in perfumery. Strictly speaking, they are considered non-nutrive dietary components, but these compounds essentially influence human nutrition by enhancing the sensory appeal of foods. Flavors and fragrances can influence human well-being as part of a holistic approach to disease prevention and health promotion. This chapter highlighted the numerous biological properties that have been ascribed to these terpene flavors. These aspects constitute an important foundation for their development of nutraceuticals or even medicines. A better characterization of plant materials, knowledge of the aspects regarding the chemical structure, biological activity, dose-activity relationships, the mechanisms of action, and the interactions with foods and drugs, will enable better understanding of terpenes and additional therapeutic and nutraceutical applications.

References

Abe S, Maruyama N, Hayama K, Ishibashi H, Inoue S, Oshima H, Yamaguchi H (2003). Suppression of tumor necrosis factor-alpha-induced neutrophil adherence responses by essential oils. *Mediators Inflamm.* 12:323–328.

Abdel-Aal E-SM, Akhtar H, Zaheer K, Ali R (2013). Dietary sources of lutein and zeaxanthin carotenoids and their role in eye health. *Nutrients.* 5:1169–1185.

Andrade-Ochoa S, Nevárez-Moorillón GV, Sánchez-Torres LE, Villanueva-García M, Sánchez-Ramírez BE, Rodríguez-Valdez LM, Rivera-Chavira BE (2015). Quantitative structure-activity relationship of molecules constituent of different essential oils with antimycobacterial activity against *Mycobacterium tuberculosis* and *Mycobacterium bovis*. *BMC Complement Altern Med.* 15 (September):332, https://www.ncbi.nlm.nih.gov/pmc/articles/PMC4579641/pdf/12906_2015_Article_858.pdf

Aprotosoaie AC, Gille E, Trifan A, Luca VS, Miron A (2017). Essential oils of *Lavandula* genus: A systematic review of their chemistry. *Phytochem Rev.* 16:761–799.

Aprotosoaie AC, Costache I-I, Miron A (2016). Anethole and its role in chronic diseases. In: Gupta SC, Prasad S, Aggarwal BB (eds). *Drug Discovery From Mother Nature*. Cham: Springer International Publishing; 247–267.

Aprotosoaie AC, Hăncianu M, Costache I-I, Miron A (2014). Linalool: A review on a key odorant molecule with valuable biological properties. *Flavour Fragr J.* 29:193–219.

Arakawa T, Osawa K (2000). Pharmacological study and application to food of mint flavorantibacterial and antiallergenic principles. *Aroma Res.* 1:20–23.

Arana-Sánchez A, Estarrón-Espinosa M, Obledo-Vázquez EN, Padilla-Camberos E, Silva-Vázquez R, Lugo-Cervantes E (2010). Antimicrobial and antioxidant activity of Mexican oregano essential oils (*Lippia graveolens* H.B.K.) with different composition when microencapsulated in β-cyclodextrin. *Lett Appl Microbiol.* 50:585–590.

Bhatia SP, McGinty D, Letizia CS, Api AM (2008). Fragrance material review on menthol racemic. *Food Chem Toxicol.* 46:S228–S233.

Blaschke F, Takata Y, Caglayan E, Law RE, Hsueh WA (2006). Obesity, peroxisome proliferator-activated receptor, and atheroscherosis in type 2 diabetes. *Arterioscler Thromb Vasc Biol.* 26:28–40.

Brahmkshatriya PP, Brahmkshatriya PS (2013). Terpenes: Chemistry, biological role, and therapeutic applications. In: Ramawat KG, Mérillon JM (eds). *Natural Products.* Berlin: Springer-Verlag; 2665–2691.

Breitmaier E (2006). *Terpenes: Flavors, Fragrances, Pharmaca, Pheromones.* Weinheim: Wiley-VCH Verlag.

Brito RG, dos Santos PL, Quintans JSS, de Lucca Júnior W, Araújo AA, Saravanan S, Menezes IR, Coutinho HD, Quintans-Júnior LJ (2015). Citronellol, a natural acyclic monoterpene, attenuates mechanical hyperalgesia response in mice: Evidence of the spinal cord lamina I inhibition. *Chem Biol Interact.* 239:111–117.

Bromm B, Scharein E, Darsow U, Ring J (1995). Effects of menthol and cold on histamine-induced itch and skin reactions in man. *Neurosci Lett.* 187:157–160.

Bruneton J (1995). *Pharmacognosy, Phytochemistry, Medicinal Plants.* Paris: Technique and Documentation.

Buchbauer G, Ilic A (2013). Biological activities of selected mono- and sesquiterpenes: Possible uses in medicine. In: Ramawat KG, Mérillon JM (eds). *Natural Products.* Berlin: Springer-Verlag; 4109–4159.

Burdock GA (2010). *Fenaroli's Handbook of Flavor Ingredients,* 6th ed. Boca Raton, FL: CRC Press.

Burdock GA (1996). *Encyclopedia of Food and Color Additives,* Vol. I. Boca Raton, FL: CRC Press.

Caputi L, Aprea E (2011). Use of terpenoids as natural flavouring compounds in food industry. *Recent Pat Food Nutr Agric.* 3:9–16.

Cardoso Lima T, Mota MM, Barbosa-Filho JM, Viana Dos Santos MR, De Sousa DP (2012). Structural relationships and vasorelaxant activity of monoterpenes. *DARU* 20 (September): 23, https://www.ncbi.nlm.nih.gov/pmc/articles/PMC3555712/pdf/2008-2231-20-23.pdf.

Carnesecchi S, Langley K, Exinger F, Gosse F, Raul F (2002). Geraniol, a component of plant essential oils, sensitizes human colonic cancer cells to 5-fluorouracil treatment. *J Pharmacol Exp Ther.* 301:625–630.

Carnesecchi S, Schneider Y, Ceraline J, Duranton B, Gosse F, Seiler N, Raul F (2001). Geraniol, a component of plant essential oils, inhibits growth and polyamine biosynthesis in human colon cancer cells. *J Pharmacol Exp Ther.* 298:197–200.

Carson CF, Hammer KA, TV Riley (2006). *Melaleuca alternifolia* (tea tree) oil: A review of antimicrobial and other medicinal properties. *Clin Microbiol Rev.* 19:50–62.

Chen W, Viljoen AM (2010). Geraniol: A review of a commercially important fragrance material. *S Afr J Bot.* 76:643–651.

Chen YK, Xie B, Liu XY, King NC, Li C (2006). Crystal structure of nootkatone and its application as tobacco flavorant. *Fine Chem.* 23:980–982.

Chizzola R (2013). Regular monoterpenes and sesquiterpenes (essential oils). In: Ramawat KG, Mérillon JM (eds). *Natural Products.* Berlin: Springer-Verlag; 2973–3008.

Cho M, So I, Chun JN, Jeon J-H (2016). The antitumor effects of geraniol: Modulation of cancer hallmark pathways (review). *Int J Oncol.* 48:1772–1782.

Croteau R (1986). Biochemistry of monoterpenes and sesquiterpenes of the essential oils. In: Cracker LE, Simon JE (eds). *Herbs, Spices, and Medicinal Plants*: *Recent Advance in Botany, Horticulture and Pharmacology.* Phoenix, AZ: Oryx Press; 81–133.

d'Alessio PA, Bisson JF, Béné MC (2014). Anti-stress effects of d-limonene and its metabolite perillyl alcohol. *Rejuvenation Res.* 17:145–149.

Da Silva Almeida JRG, Rocha Souza G, Silva JC, Saraiva SR, Júnior RG, Quintans Jde S, Barreto Rde S, Bonjardim LR, Cavalcanti SC, Quintans LJ Jr. (2013). Borneol, a bicyclic monoterpene alcohol, reduces nociceptive behavior and inflammatory response in mice. *Sci World J.* 2013 (April):808460, https://www.ncbi.nlm.nih.gov/pmc/articles/PMC3654274/pdf/TSWJ2013-808460.pdf.

de Bona da Silva C, Guterres SS, Weisheimer V, Schapoval EES (2008). Antifungal activity of the lemongrass oil and citral against *Candida* spp. *Braz J Infect Dis*. 12:63–66.

de Carvalho CCCR, da Fonseca MMR (2006). Carvone: Why and how should one bother to produce this terpene. *Food Chem*. 95:413–422.

de Carvalho KIM, Bonamin F, Cássia dos Santos R, Périco LL, Beserra FP, de Sousa DP, Filho JM, da Rocha LR, Hiruma-Lima CA (2014). Geraniol: A flavoring agent with multifunctional effects in protecting the gastric and duodenal mucosa. *Naunyn-Schmiedeberg's Arch Pharmacol*. 387:355–365.

De Cássia da Silveira e Sá R, Nalone Andrade L, de Sousa DP (2013). A review on anti-inflammatory activity of monoterpenes. *Molecules*. 18:1227–1254.

De Oliveira Pereira F, Moura Mendes J, Oliveira Lima I, Samara de Lira Mota K, Araújo de Oliveira W, de Oliveira Lima E (2015). Antifungal activity of geraniol and citronellol, two monoterpenes alcohols, against *Tricophyton rubrum* involves inhibition of ergosterol biosynthesis. *Pharm Biol*. 53:228–234.

Do Vale GT, Furtado CE, Santos JG Jr, Viana GSB (2002). Central effects of citral, myrcene and limonene, constituents of essential oil chemotypes from *Lippia alba* (Mill.) N.E. Brown. *Phytomedicine*. 9:709–714.

Eccles R (1994). Menthol and related cooling components. *J Pharm Pharmacol*. 46:618–630.

EFSA Scientific Committee (2014). Scientific opinion on the safety assessment of carvone, considering all sources of exposure. *EFSA J*. 12:3806–3880.

Elsharif SA, Buettner A (2017). Influence of the chemical structure on the odor character of β-citronellol and its oxygenated derivatives. *Food Chem*. 232:704–711.

Elsharif SA, Buettner A (2016). Structure-odor relationship study on geraniol, nerol, and their synthesized oxygenated derivatives. *J Agric Food Chem*. doi: 10.1021/acs.jafc.6b04534

Elsharif SA, Banerjee A, Buettner A (2015). Structure-odor relationships of linalool, linalyl acetate and their corresponding oxygenated derivatives. *Front Chem*. 3 (October):57. https://www.ncbi.nlm.nih.gov/pmc/articles/PMC4594031/pdf/fchem-03-00057.pdf

Erasto P, Viljoen A (2008). Limonene—A review: Biosynthetic, ecological and pharmacological relevance. *Nat Prod Commun*. 3:1193–1202.

Galeotti N, Ghelardini C, Mannelli L, Mazzanti G, Baghiroli L, Bartolini A (2001). Local anaesthetic activity of (+)- and (−)-menthol. *Planta Med*. 67:174–176.

Good Scent Company: http://www.thegoodscentscompany.com/search2.html (accessed June 1, 2017).

Hagvall L, Karlberg AT, Christensson JB (2012). Contact allergy to air-exposed geraniol: Clinical observations and report of 14 cases. *Contact Derm*. 67:20–27.

Kamatou GPP, Vermaak I, Viljoen AM, Lawrence BM (2013). Menthol: A simple monoterpene with remarkable biological properties. *Phytochemistry*. 96:15–25.

Katsukawa M, Nakata R, Takizawa Y, Hori K, Takahashi S, Inoue H (2010). Citral, a component of lemongrass oil, activates PPARα and γ and suppresses COX-2 expression. *Biochim Biophys Acta*. 1801:1214–1220.

Khan AQ, Khan R, Qamar W, Lateef A, Rehman MU, Tahir M, Ali F, Hamiza OO, Hasan SK, Sultana S (2013). Geraniol attenuates 12-O-tetradecanoylphorbol-13-acetate (TPA)-induced oxidative stress and inflammation in mouse skin: Possible role of p38 MAP kinase and NF-κB. *Exp Mol Pathol*. 94:419–429.

Klein AH, Sawyer CM, Carstens MI, Tsagareli MG, Tisklauri N, Carstens E (2010). Topical application of L-menthol induces heat analgesia, mechanical allodynia and a biphasic effect on cold sensitivity in rats. *Behav Brain Res*. 212:179–186.

Kolassa N (2013). Menthol differs from other terpenic essential oils constituents. *Regul Toxicol Pharmacol*. 65:115–118.

Koziol A, Stryjewska A, Librowski T, Sałat K, Gaweł M, Moniczewski A, Lochyński S (2014). An overview of the pharmacological properties and potential applications of natural monoterpenes. *Mini Rev Med Chem*. 14:1156–1168.

Labuda I (2009). Flavor compounds. In: Schaechter M (ed). *Encyclopedia of Microbiology*, 3rd ed. Amsterdam: Academic Press; 305–321.

Lalko J, Api AM (2008). Citral: Identifying a threshold for induction of dermal sensitization. *Regul Toxicol Pharmacol.* 52:62–73.

Lee H, Jeong HS, Kim DJ, Noh YH, Yuk DY, Hong JT (2008). Inhibitory effect of citral on NO production by suppression of iNOS expression and NF-kB activation in RAW264.7 cells. *Arch Pharm Res.* 31:342–349.

Leffingwell J (2006). Chirality and odor perception. Scent of precious woods. *Chim Oggi.* 24:36–38.

Leonhardt RH, Berger RG (2015). Nootkatone. *Adv Biochem Eng Biotechnol.* 148:391–404.

Liu Y, Ye X, Feng X, Zhou G, Rong Z, Fang C, Chen H (2005). Menthol facilitates the skin analgesic effect of tetracaine gel. *Int J Pharm.* 305:31–36.

Lorenzi V, Muselli A, Bernardini AF, Berti L, Pagès JM, Amaral L, Bolla JM (2009). Geraniol restores antibiotic activities against multidrug-resistant isolates from Gram-negative species. *Antimicrob Agents Chemother.* 53:2209–2211.

Ludwiczuk A, Skalicka-Wóżniak K, Georgiev MI (2017). Terpenoids. In: Badal S, Delgoda R, (eds). *Pharmacognosy: Fundamentals, Applications and Strategies*, Amsterdam: Academic Press; 233–266.

Mahapatra SK, Chakraborty SP, Majumdar S, Bag BG, Roy S (2009). Eugenol protects nicotine-induced superoxide mediated oxidative damage in murine peritoneal macrophages in vitro. *Eur J Pharmacol.* 623:132–140.

Mans D, Grivicich I, Peters G, Schwartsmann G (1999). Sequence dependent growth inhibition and DNA damage formation by the irinotecan-5-fluorouracil combination in human colon carcinoma cell lines. *Eur J Cancer.* 35:1851–1861.

Modak T, Mukhopadhaya A (2011). Effects of citral, a naturally occurring antiadipogenic molecule, on an energy-intense diet model of obesity. *Indian J Pharmacol.* 43: 300–305.

Moreira-Lobo DCA, Linhares-Siqueira ED, Cruz GMP, Cruz JS, Carvalho-de-Souza JL, Lahlou S, Coelho-de-Souza AN, Barbosa R, Magalhães PJ, Leal-Cardoso JH (2010). Eugenol modifies the excitability of rat sciatic nerve and superior cervical ganglion neurons. *Neurosci Lett.* 472:220–224.

Morise M, Ito Y, Matsuno T, Hibino Y, Mizutani T, Ito S, Hashimoto N, Kondo M, Imaizumi K, Hasegawa Y (2010). Heterologous regulation of aninon transporters by menthol in human airway epithelial cells. *Eur J Pharmacol.* 635:204–211.

Nagoor Meeran MF, Javed H, Al Taee H, Azimullah S, Ojha S (2017). Pharmacological properties and molecular mechanisms of thymol: Prospects for its therapeutic potential and pharmaceutical development. *Front Pharmacol.* 8 (June):380, https://www.ncbi.nlm.nih.gov/pmc/articles/PMC5483461/pdf/fphar-08-00380.pdf

NPCS Board of Consultants and Engineers (2007). *The Complete Technology Book on Flavours, Fragrances and Perfumes*. New Delhi: NIIR Project Consultancy Services.

Onawunmi GO (1989). Evaluation of the antimicrobial activity of citral. *Lett Appl Microbiol.* 9:105–108.

Ong TP, Heidor R, De Conti A, Dagli MLZ, Moreno FS (2006). Farnesol and geraniol chemopreventive activities during the initial phases of hepatocarcinogenesis involve similar actions on cell proliferation and DNA damage, but distinct actions on apoptosis, plasma cholesterol and HMGCoA reductase. *Carcinogenesis.* 27:1194–1203.

Opdyke DLJ (1979). *Monographs on Fragrance Raw Materials*. Oxford: Pergamon Press.

Ortiz MI, Ramírez-Montiel ML, Ponce-Monter HA, Castañeda-Hernández G, Cariño-Cortés R (2010). The combination of naproxen and citral reduces nociception and gastric damage in rats. *Arch Pharm Res.* 33:1691–1697.

Paduch R, Kandefer-Szerszeń M, Trytek M, Fiedurek J (2007). Terpenes: Substances useful in human healthcare. *Arch Immunol Ther Exp.* 55:315–327.

Patel T, Yosipovitch G (2007). Menthol: A refreshing look at this ancient compound. *J Am Acad Dermatol.* 57:873–877.

Peffley DM, Gayen AK (2003). Plant-derived monoterpenes suppress hamster kidney cell 3-hydroxy-3-methylglutaryl coenzyme A reductase at post-transcriptional level. *J Nutr.* 133:38–44.

Quintans-Júnior LJ, Guimarães AG, de Santana MT, Araújo BES, Moreira FV, Bonjardim LR, Araújo AAS, Siqueira JS, Antoniolli AR, Botelho MA, Almeida JRGS, Santos MRV (2011). Citral reduces nociceptive and inflammatory response in rodents. *Rev Bras Farmacogn.* 21:497–502.

Rana VS, Das M, Blazquez MA (2016). Essential oil yield, chemical composition, and total citral content of nine cultivars of *Cymbopogon* species from western India. *J Herbs Spices Med Plants.* 22:289–299.

Rao AV, Rao LG (2007). Carotenoids and human health. *Pharmacol Res.* 55:207–216.

Reineccius G (2005). *Flavor Chemistry and Technology*, 2nd ed. Boca Raton, FL: CRC Press.

Ribeiro-Filho HV, de Souza Silva CM, de Siqueira RJ, Lahlou S, dos Santos AA, Magalhães PJ (2016). Biphasic cardiovascular and respiratory effects induced by β-citronellol. *Eur J Pharmacol.* 775:96–105.

Riella KR, Marinho RR, Santos JS, Pereira-Filho RN, Cardoso JC, Albuquerque-Junior RLC, Thomazzi SM (2012). Anti-inflammatory and cicatrizing activities of thymol, a monoterpene of the essential oil from *Lippia gracilis* in rodents. *J Ethnopharmacol.* 143:656–663.

RIFM Expert Panel, Belsito D, Bickers D, Bruze M, Calow P, Greim H, Hanifin JM, Rogers AE, Saurat JH, Sipes IG, Tagami H (2007). A toxicologic and dermatologic assessment of ionones when used as fragrance ingredients. *Food Chem Toxicol.* 45:S130–S167.

Saddiq AA, Khayyat SA (2010). Chemical and antimicrobial studies of monoterpene: Citral. *Pest Biochem Physiol.* 98:89–93.

Sell CS (2006). Terpenoids. *Kirk-Othmer Encyclopedia of Chemical Technology.* New York: John Wiley & Sons.

Seo EJ, Lee D-U, Kwak KH, Lee S-M, Kim YS (2011). Antiplatelet effects of *Cyperus rotundus* and its component (+)-nootkatone. *J Ethnopharmacol.* 135:48–54.

Seo K-A, Kim H, Ku H-Y, Ahn HJ, Park SJ, Bae SK, Shin JG, Liu KH (2008). The monoterpenoids citral and geraniol are moderate inhibitors of CYP2B6 hydroxylase activity. *Chem Biol Interact.* 174:141–146.

Seol GH, Kim KY (2016). Eucalyptol and its role in chronic diseases. *Adv Exp Med Biol.* 929:389–398.

Serra S (2015). Recent developments in the synthesis of the flavors and fragrances of terpenoid origin. In: Atta-ur-Rahman (ed). *Studies in Natural Products Chemistry*, Vol. 46. Amsterdam: Elsevier; 201–226.

Shi C, Song K, Zhang X, Sun Y, Sui Y, Chen Y, Jia Z, Sun H, Sun Z, Xia X (2016). Antimicrobial activity and possible mechanism of action of citral against *Cronobacter sakazakii*. *PloS One.* 11 (July):e0159006, http://journals.plos.org/plosone/article/file?id=10.1371/journal.pone.0159006&type=printable

Shin S, Lim S (2004). Antifungal effects of herbal essential oils alone and in combination with ketoconazole against *Trichophyton* spp. *J Appl Microbiol.* 97:1289–1296.

Si L, Chen Y, Han X, Zhan Z, Tian S, Cui Q (2012). Chemical composition of essential oils of *Litsea cubeba* harvested from its distribution areas in China. *Molecules.* 17:7057–7066.

Singh B, Sharma RA (2015). Plant terpenes: Defense responses, phylogenetic analysis, regulation and clinical applications. *3 Biotech.* 5:129–151.

Singh AK, Raina VK, Naqvi AA, Patra NK, Kumar B, Ram P, Khanuja SPS (2005). Essential oil composition and chemoarrays of menthol mint (*Mentha arvensis* L. f. *piperascens* Malinvaud ex. Holmes) cultivars. *Flavour Fragr J.* 20:302–305.

Sobral MV, Xavier AL, Cardosso Lima T, Pergentino de Sousa D (2014). Antitumor activity of monoterpenes found in essential oils. *Sci World J.* 2014 (October):953451, http://dx.doi.org/10.1155/2014/953451

Stahl W, Sies H (2012). β-Carotene and other carotenoids in protection from sunlight. *Am J Clin Nutr.* 96 (Suppl.):1179S–1184S.

Stotz SC, Vriens J, Martyn D, Clardy J, Clapham DE (2008). Citral sensing by transient receptor potential channels in dorsal root ganglion neurons. *Plos One.* 3 (May):e2082, https://www.ncbi.nlm.nih.gov/pmc/articles/PMC2346451/pdf/pone.0002082.pdf

Sultanbawa Y (2016). Lemon myrtle (*Backhousia citriodora*) oils. In: Preedy VR (Ed). *Essential Oils in Food Preservation, Flavor and Safety.* Amsterdam: Academic Press; 517–521.

Surburg H, Panten J (2006). *Common Fragrance and Flavor Materials. Preparation, Properties and Uses,* 5th ed. Weinheim: Wiley-VCH Verlag.

Van der Schaft PH (2007). Chemical conversions of natural precursors. In: Berger RG (ed). *Flavours and Fragrances Chemistry, Bioprocessing and Sustainability.* Berlin: Springer-Verlag; 285–301.

Watt EE, Betts BA, Kotey FO, Humbert DJ, Griffith TN, Kelly EW, Veneskey KC, Gill N, Rowan KC, Jenkins A, Hall AC (2008). Menthol shares general anesthetic activity and sites of action on the $GABA_A$ receptor with the intravenous agent, propofol. *Eur J Pharmacol.* 590:120–126.

Weerawatanakorn M, Wu J-C, Pan M-H, Ho C-T (2015). Reactivity and stability of selected flavor compounds. *J Food Drug Anal.* 23:176–190.

Woyengo TA, Ramprasath VR, Jones PJH (2009). Anticancer effects of phytosterols. *Eur J Clin Nutr.* 63:813–820.

Yang Z, Xi J, Li J, Qu W (2009). Biphasic effect of citral, a flavoring and scenting agent, on spatial learning and memory in rats. *Pharmacol Biochem Behav.* 93:391–396.

Yu SG, Anderson PJ, Elson CE (1995). Efficacy of beta-ionone in the chemoprevention of rat mammary carcinogenesis. *J Agric Food Chem.* 43:2144–2147.

Zachariah TJ (2008). Ajowan. In: Parthasarathy VA, Chempakam B, Zachariah TJ (eds). *Chemistry of Spices.* Cambridge: CAB International; 312–318.

8 The Biopsychology of Flavor Perception and Its Application to Nutraceuticals

Richard J. Stevenson

Contents

8.1 Introduction 201
8.2 Sensory and affective responses to food and drink 202
8.3 Flavor problems with nutraceuticals 206
    8.3.1 Taste 206
    8.3.2 Smell 207
    8.3.3 Somatosensation 208
        8.3.3.1 Texture and astringency 208
        8.3.3.2 Pungency 208
    8.3.4 Appearance 208
8.4 Solutions to flavor-related problems 209
    8.4.1 Masking 209
    8.4.2 Exposure 211
    8.4.3 Expectancy 212
8.5 Conclusions 213
References 214

8.1 Introduction

A number of obstacles must be overcome to convince an adult or a child to consume any novel substance, be it a food, medicine, or nutraceutical. In this chapter we consider the sensory and affective processes involved in food intake. We begin by examining the physiology of the mouth, the senses involved in perceiving food in the mouth, and how these combine and interact to form flavor, the sensory experience associated with eating and drinking (Auvray and Spence, 2008; Lawless, 1996; Prescott, 2004). A significant theme in this section is how experience shapes both the sensory and affective dimensions of flavor. The second and third parts of the chapter have a practical focus. They examine how this knowledge about flavor perception can be used to increase the acceptability of nutraceuticals. The second part focuses on problem tastes, aromas, textures, colors, irritants, and other factors that potentially act to reduce the acceptability of a nutraceutical. The third and final part of the chapter examines potential solutions to the problems identified in the second part, including masking, exposure, and expectancy.

8.2 Sensory and affective responses to food and drink

The sensory experience of food (and drink) is termed flavor (Auvray and Spence, 2008; Lawless, 1996; Prescott, 2004). Flavor has both intrinsic and extrinsic components (Table 8.1). Before we eat we can see, smell, feel, and sometimes hear (e.g., the fizzing of champagne) the food before we place it in our mouths (Spence, 2016). All of these extrinsic components of flavor can potentially alter what we expect a food to be like when we place it in our mouths. More importantly, these extrinsic factors govern whether that food will even make it that far. For example, unappetizing food colors, rotten smells, soggy feel, or a flat drink, are likely to result in rejection (Piqueras-Fiszman and Spence, 2015).

Once a food is in the mouth, three sensory systems are important in forming the intrinsic components of flavor (Stevenson, 2014). The first of these is taste. Taste is detected by receptors located on structures called taste buds, which are mainly located on the surface of the tongue (Smith and Margolskee, 2001). Contrary to popular mythology, different tastants can be detected on any part of the tongue. The tip has the highest concentration of taste buds and is thus most sensitive to all types of tastants. It is generally agreed that there are at least five different sensory categories that can be perceived: sweet, salty, sour, bitter and umami (Bellisle, 1999; McBurney and Gent, 1979).

Sweet tastes seem to be universally liked. Even babies less than an hour old show facial expressions of pleasure—as judged by adults—when provided with sucrose solution (Steiner, 1979). This is not surprising as sweet generally signals the presence of carbohydrates and hence energy. In contrast, bitter tastes seem to be disliked by both babies and adults (Steiner, 1979). This, too, is not surprising, as most plant-based toxins have a bitter taste (e.g., caffeine, quinine, nicotine, solanine) and there is a significant correlation between the degree to which something tastes bitter and its LD_{50} (Lethal Dose, 50%) (Scott and Mark, 1987). Sour tastants are mildly disliked, perhaps because they signal that fruit is unripe or that a food has suffered bacterial spoilage.

Table 8.1 Extrinsic (Outside the Mouth) and Intrinsic (Inside the Mouth) Properties of Flavor

Property	Senses (Major)	Senses (Minor)
Extrinsic	Vision, olfaction	Audition, touch
Intrinsic	Taste, olfaction, touch	Audition

Salty tastes tend to be liked, but this concentration (approximately 0.06–0.18 M), is deprivation dependent. Finally, umami, or protein-like tastes, are generally liked because they signal the presence of protein.

The second intrinsic property of flavor is smell (Mozell et al., 1969). When people eat and drink, the volatile components of the food (or drink) being consumed are released as the food is manipulated in the mouth. These volatile chemicals reach the nose in two ways (Rozin, 1982). First, during chewing bursts of volatile laden air are propelled toward the back of the mouth. When the velopharyngeal flap at the back of the throat opens (as it does briefly during rhythmic chewing) this pushes volatile laden air up the posterior nostrils to the olfactory receptors located deep within the face (Hodgson et al., 2003). A similar process also ensues after swallowing, when air is expelled via the posterior nostrils and up to the olfactory receptors deep within the face (Trelea et al., 2008). Both of these means of transmitting volatile laden air from mouth to nose are termed retronasal olfaction. They stimulate the same olfactory receptors as during sniffing. In this case air passes up the anterior nostrils—the nose—a process termed orthonasal olfaction.

While most people readily acknowledge that taste plays a significant part in the enjoyment of food, the same recognition is not afforded to smell. Most people seem to be unaware that a significant part of the richness of the eating and drinking sensory experience is derived from the olfactory component of the food being consumed (Rozin, 1982). In fact, while there are only a limited number of sensory categories associated with taste (five: sweet, salty, sour, bitter and umami), the number of possible sensory categories for smell seems to be unlimited. This is in part due to the greater complexity of both the olfactory stimulus and, in turn, olfactory perception.

The olfactory mucosa in humans contains 300–400 different types of chemical receptors (Buck, 2000), in contrast to the 10–12 for taste (most of these for bitter tastes). Each olfactory receptor is broadly tuned to detect a different chemical structure/s. The brain uses the pattern of activity across receptor types (in addition to the temporal pattern of their activation) to recognize a particular cherry smell (Wilson and Stevenson, 2006). Note the use of the term *recognize* because the brain has to learn the particular neural input pattern for a particular smell to later detect this pattern against the ever-changing olfactory background. This learning component is very important because most smells are composed of many volatiles. They may vary in composition and degrade over time, but the brain still needs to be able to recognize a smell irrespective of these factors.

The third intrinsic component of flavor comes from the somatosensory system (Christensen, 1984). This contains within it a number of important sensory sub-systems. Texture in food, which is very important for acceptability (think of soggy lettuce or limp potato chips), is perceived both by active exploration of the food using the tongue (touch receptors) and by feedback from the muscles, tendons, and joints in the jaw and face during chewing (proprioceptive feedback from mechanoreceptors). The temperature of food is perceived by sensory receptors located in the mouth, on the tongue, and in the throat. These so-called free-nerve ending receptors, serviced by the trigeminal nerve, not only collate information about temperature but are also sensitive to chemical irritants (Green, 1990). This forms a further subcategory of flavor perception, namely pungency. Humans are perhaps unique among animals in actively seeking out and adding irritants to food. All of these irritants (e.g., chilli, pepper, ginger, mustard) have two primary sensory dimensions: heat (e.g., chilli) or cold (e.g., menthol). They achieve this by activating hot and cold receptors, respectively: the free nerve ending receptors.

What we perceive when we eat and drink is not, however, a disparate set of sensory inputs. Rather, we experience a partially unified sensory experience termed flavor. It is partially unified because there are clearly "parts" that can be discerned (e.g., we know that food is hot or cold irrespective of its other properties), but equally so there are other parts that seem to be difficult to discern. The most notable of these appears to be the failure of most people to recognize that the sense of smell is involved in flavor perception (Rozin, 1982; Stevenson, 2014). One way perceptual psychologists have used to arrive at this understanding (i.e., partial unification) is by studying the interactions between the different intrinsic senses that make up flavor. This has revealed a number of types of interaction that have a purely psychological basis, that is, they do not arise from chemical or other physical interactions among food components (Stevenson, 2009). Understanding these interactions is also practically important because they can be harnessed to alter the flavor of food and hence its acceptability.

Four types of interaction with a psychological basis have been documented among the intrinsic components of flavor. The first and second both concern interactions within a particular sensory system, namely the effect of two different tastes on the perception of each other and likewise for smell. In both these cases, two general principles emerge: (1) the perceived intensity of both components is diminished relative to how each one alone would be perceived at the same concentration (Moskowitz and Barbe, 1977) and (2) this mutual suppression is rarely symmetric, with one component (unpredictably) being reduced in perceived intensity to a greater degree than the other (Schifferstein and Frijters, 1990). The third form of interaction is between smell and touch. It was long believed that increasing viscosity decreased the perceived intensity of a concurrently presented volatile because the viscous substance inhibited release (in various ways) of the volatile agent. It has now been clearly demonstrated that viscosity has a suppressive effect on retronasal olfactory perception and that this has a significant central (i.e., brain-based) component (Bult et al., 2007). The upshot of this is that presenting an odorous substance in a more viscous base will act to decrease its perceived intensity.

The fourth and final form of interaction is between taste and smell. This has been extensively studied, partly because of what it potentially tells us about the unification of these two senses during flavor perception, but also because pragmatically it may be possible to use this knowledge to reduce the amount of sugar and salt in the food we eat. It was known more than 50 years ago that people perceived certain food smells (i.e., sniffed via the nose—orthonasal olfaction) as sweet (Harper et al., 1968). As should now be clear "sweet" is a property of the taste system, not the smell system, and this observation piqued psychologists' interest. It is now known that smells that typically accompany certain tastes come to acquire those same properties (e.g., strawberries smell sweet and sardines smell salty; Stevenson et al., 1995). When an odorant that has such taste-like properties is combined with that taste (e.g., adding tasteless strawberry odorant to an odorless sweet solution) the combination is judged to taste sweeter than the sweet solution alone (with similar effects observed for sour, salty, and bitter; Stevenson et al., 1999). Similarly, the presence of tastes also seems to increase the perceived intensity of concordant odors (e.g., sucrose may increase the perceived fruitiness of strawberry odor; Cliff and Noble, 1990), and this too seems to have a similar basis in experience. While it is highly likely that these types of interactions are the product of learning (i.e., experience) it is not yet clear whether this is responsible for unifying people's experience of taste and smell, as some authors have recently suggested (Lim and Johnson, 2012).

Thus we have focused on the origins of the sensory properties of food and drink, which while clearly important leaves out a significant aspect of this experience—hedonics (i.e., liking/disliking). People generally eat what they like and reject or avoid foods they dislike. Because of their importance to acceptability, the origin of these affective reactions has been even more extensively studied than flavor perception (Yeomans, 2008). While our response to tastes is, at least initially, driven by a probably innate disposition to like (sweet, salty, umami) or dislike (bitter, sour) them, experience can alter this, and experience is even more central in shaping likes and dislikes for whole foods (i.e., combinations of taste, smell, texture, and flavor). Several different experiential mechanisms have been identified (Yeomans, 2008): (1) evaluative conditioning—pleasant or unpleasant tastes becomes associated with the olfactory component of a food or drink; (2) mere exposure—passive experience with a food or drink acts to increase liking for it (this features prominently later in this chapter); (3) flavor–nutrient learning—energy-dense food acts to increase liking for an associated flavor; (4) conditioned taste aversion—experience of illness after a meal becomes associated with food(s)/drink(s) eaten during that meal; (5) observational learning—someone else's reaction to a food comes to shape your reaction to that same food; and (6) drug-related learning—the pharmaceutical effects of food (e.g., caffeine) can increase liking for associated flavors, at least under some conditions. All of these learning-based processes can act to alter the hedonic reaction that will accompany the experience of a particular food or drink.

To summarize, the flavor we experience when we eat and drink is primarily derived from three sensory systems: taste, smell, and touch. These—especially taste and smell—combine to produce flavor. Experience seems to play a role both in our perception of flavor (perhaps acting as the psychological glue between taste and smell) and more crucially in the affective reaction we show to food.

8.3 Flavor problems with nutraceuticals

If all nutraceuticals were as palatable as junk food there would be no issue with acceptability. Rather, the reverse problem would ensue, how to limit intake. However, as with medicines, nutraceuticals can have inherent sensory properties that may initially at least affect acceptability. It is these properties that are the focus of this section (see Table 8.2). Before we discuss them it is important to recognize that several factors moderate whether a particular sensory property will be a problem. First, the physical format of the nutraceutical is clearly the most relevant, whether it is pelletized in some manner (e.g., tablet or spansule) or more like its food-based origins. The former may have fewer sensory/affective problems (assuming that it can be readily swallowed without being chewed) than the latter. Second, the characteristics of the consumer are a further and important consideration (Schiffman, 1992). Children may be more sensitive to certain tastants and less familiar with certain odorants, and also less able to swallow tablets. Pregnant women may become especially sensitive to certain flavors (anything that is mildly disliked may become notably unpleasant, especially thiols). Older adults may be less sensitive to tastes and smells in general, and thus far more tolerant of problem sensory characteristics than any other group. Also, many adults, especially older people, take medications that may impact their sense of taste or smell. Finally, there are some differences in tasting and smelling abilities between children (generally poorer) and adults (generally better). The point here is that the market segment the nutraceutical is targeting needs to be considered when examining flavor-related problems.

8.3.1 Taste

Many medicines taste bitter (Menella et al., 2013). As identified earlier in the chapter, people are born with an innate dislike of bitter tastes, which

Table 8.2 Problem Tastes, Smells, Textures, and Appearances

Sensory System	Potential Problem Characteristics
Taste	Bitterness, high tastant concentration
Smell	Decomposition, fecal, body products, chemical, burnt, oxidized, pungent, fishy, uncooked meat
Touch	Oily or greasy, gritty, dry, difficult to break up or chew, irritant, astringent (tannins)
Vision	Visual texture (greasy/oily), yellow

is not surprising given that most poisons taste bitter. Two factors appear to moderate the acceptability of consuming bitter tasting substances, at least in a food-based context. The first is the large individual variation in bitter taste sensitivity. This was first identified for the bitter-tasting chemical phenylthiocarbamide (PTC) and later for propylthiouracil (PROP). Some individuals were barely able to detect these chemicals at even high concentrations, while others were acutely sensitive, able to detect them at very low concentrations (Glanville and Kaplan, 1965). These differences are genetic in origin, and if one was unlucky enough to recruit a sensory panel to evaluate a new nutraceutical composed of individuals at either end of the sensitivity spectrum, acceptability data would be markedly nonrepresentative (Stevenson, 2017).

A second factor is experience. It has been known since the 1970s that animals exposed to mildly bitter diets can become tolerant to them as may babies fed milk protein hydrolysate formulas. Adults can also develop preferences for bitter-tasting foods and drinks (Moskowitz et al., 1975). Tonic water in Western culture is one, and liking for the bitter and sour tamarind fruit in curries (in the Karnataka region of India) is another. So, while a dislike of bitter tastes may be innate, it is not immutable. There may also be circumstances where dislike for even low levels of bitterness becomes augmented by experience. One documented case is the Aymara Indians of South America, who while equally sensitive to bitter tastes as Westerners, dislike them far more, presumably because their diet is heavily dependent on the potato. Thus they risk cumulative exposure to toxins such as solanine and chaconine in green potatoes.

The other taste categories—sour, sweet, salty and umami—are less likely to feature as problem tastants, unless they are experienced at high concentrations (i.e., high perceived intensity, translating as very approximately greater than 15% sucrose, 7% monosodium glutamate, 1.5% sodium chloride, 0.3% citric acid, and 0.02% quinine). This is because these tastes appear less frequently as the prominent sensory characteristic in naturally occurring nutraceutical compounds.

8.3.2 Smell

Some odor dislikes are shared within particular cultures (and perhaps more widely) while others are idiosyncratic. Odors relating to decomposition, fecal, or bodily products tend to be widely disliked. Although this is believed to result from learning during childhood (Stevenson and Repacholi, 2003), it is uncertain as to whether there may be an innate basis for disliking certain chemicals. Even in cases where disliking may be innately based, the context in which the smell is encountered seems to be important. While 3-methylbutanoic acid may smell pleasant when it is a component in Parmesan cheese, it may smell far less agreeable when it is part of the bouquet arising from dirty feet.

Several other types of volatile are also typically characterized as unpleasant: (1) chemical smells in the context of food are generally not well regarded (sulphurous, petrol-like, papery); (2) burnt smells, at least at higher concentrations; (3) oxidized smells; (4) pungent aromas; (5) fishy smells; (6) uncooked meat; and (7) certain floral smells. Again, while some of these may be appropriate within particular contexts (e.g., rose-flavored Turkish delight), generally, and especially in something not consumed previously, they may tend to engender dislike.

8.3.3 Somatosensation

8.3.3.1 Texture and astringency

People generally have clear expectations about the texture that will accompany a particular food, and any violation of this expectancy is associated with dislike. There are also certain types of texture that seem to be disliked in any context (Christensen, 1984). These include excessively oily or greasy formulations, items that have a gritty or uneven texture, items that are excessively dry and sap the mouth of saliva, or items that are very hard and difficult to macerate. A further relevant characteristic here is astringency, which is where proteins interact with the mucosal surface to make it feel as if it is puckering up (Breslin et al., 1993). This can occur with foods that have high levels of tannins, and this is generally not liked by consumers.

8.3.3.2 Pungency

Where an irritant vehicle is used to dissolve a nutraceutical, or the product itself is or contains an irritant, then the pungency that this presents may be a bar to consumption. As noted earlier, this will depend upon the presentation format of the product, its fat content (which can tend to liberate the irritant and apply it more widely within the mouth), and the concentration of the irritant in terms of the degree to which it will generate pungency.

8.3.4 Appearance

It has been suspected that certain colors may be universally perceived as unpleasant, with moldy green and pus yellow being identified as the least pleasant. Although more recent cross-cultural work suggests that these two colors may not be universally disliked, they do seem to be consistently judged as unpleasant across studies of Western consumers (Taylor et al., 2013). Other aspects of visual appearance may also be important, as they can inadvertently provide possibly misleading cues about the object's texture, notably whether it is shiny (oily/greasy) or whether it is flecked, indicating an uneven texture. All of these factors act to create an expectancy in participants' minds as to what the item may taste like and they may be significant factors in whether it will even be sampled.

8.4 Solutions to flavor-related problems

Three general classes of solution are available to deal with the various problems identified in the preceding section. The first involves perceptual masking, whereby some element is added to the product to eliminate or reduce the impact of the sensory problem (be it taste, aroma, etc). The second is the use of psychological techniques to initially drive consumption to a point where any sensory negatives are negated by the actual or perceived benefits of the product (which then presumably motivates further consumption). The third concerns explicit manipulation of expectancies, based around product design and presentation and the impact these can have on acceptability.

8.4.1 Masking

Masking occurs when a new stimulus (e.g., a tastant) is added to the product, and the effect of this is to reduce or eliminate a particular sensory feature (e.g., bitterness). The terms suppression (i.e., akin to the term masking) and additivity (i.e., making a sensory component more intense) have been used in the academic literature to describe the perceptual effects arising from adding a new element to an existing one (e.g., adding taste A to taste B). Additivity occurs when the combination is perceived to have a greater overall strength, or one of the qualities of the stimulus is perceived to be greater than prior to the stimulus addition. This situation is relatively unusual and is more commonly seen around perceptual thresholds (especially with odors). A far more common effect, and one of central interest here, is suppression, whereby the addition of a new stimulus to a current one results in the participant reporting that one or more features are reduced in intensity (Schifferstein and Frijters, 1990). In effect this amounts to the masking of a problem feature by what should be a more acceptable one (assuming that a suitable agent is selected). It is this phenomenon that is of interest here.

Suppression can be used to mask bitter tastes (Mennella et al., 2013). This is usually accomplished by the addition of sweet or sour tastants, with the amount required dependent upon the intensity of the bitter taste (Figure 8.1; Bartoshuk, 1975). The use of artificial sweeteners to achieve this can be potentially problematic, because if they have bitter side tastes (i.e., after the sweetness sensation has passed, the artificial sweetener then tastes bitter), these may act additively with the bitterness that the sweetener is trying to mask. This taps into a broader potential problem because bitter tastes seem to have a different time course than sweet tastes, with peak bitterness developing more slowly and lasting longer than peak sweetness. Formulation approaches may be used to deal with this, by trying to ensure that sweetness is delivered for longer (by including slowly dissolving particles containing the sweetener), so that the time course of delivery exceeds that of the bitterness.

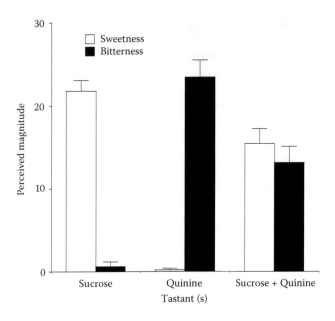

Figure 8.1 Sweetness and bitterness ratings for 11% sucrose alone, 0.4% quinine alone, and the two as a mixture at the same concentrations, demonstrating how sucrose sweetness suppresses quinine bitterness and vice versa. Data from Bartoshuk, 1975.

Suppression can also be used to successfully mask problem odors (Moskowitz and Barbe, 1977). The usual strategy is to use an odorant that is familiar and generally preferred (e.g., a citrus-based formulation), with the concentration adjusted so that the unpleasant odor is either difficult to detect or effectively blended in with the masking agent. It may be particularly useful to add a tastant congruent to the masking odorant, especially if the formulation itself does not have much of a taste. This is because the presence of a congruent tastant appears to increase the perceived intensity of its matching odor (Lim and Johnson, 2012), seemingly with lesser effect on any problem odorant. A further means of altering the olfactory component, this time without adding any additional tastants or odorants, is to increase the viscosity of the product (Bult et al., 2007). As noted earlier, more viscous formulations reduce perceived odor intensity. Moreover, there would be no reason in principle why this could not be combined with the previous strategy of adding a masking odor and a congruent taste.

Texture masking does not appear to have been explored in the academic literature. It would appear easier to alter the formulation by additional milling or by using tasteless and odorant viscous-enhancing products (e.g., carboxymethylcellulose). For masking irritant aspects of a stimulus, increasing viscosity is one means of reducing the apparent pungency, and a similar effect may also be obtained by adding sucrose (Green, 1990). Certain tastants (and odorants) are themselves irritants, and may act additively with any irritant in the product to increase pungency.

8.4.2 Exposure

Novice alcohol drinkers ultimately make the transition from consuming ethanol-free beverages to, in the case of spirit drinkers, drinking 40% ethanol, something that would be highly aversive to a novice drinker. This transition from novice to alcohol consumer is probably aided by the availability of drink products that ease the beginning drinker into alcohol consumption, but without the unpleasant pungency that may put the novice off (Copeland et al., 2007). Presumably, the intoxicating effects of alcohol reinforce consumption, and this then propels further consumption, but this clearly occurs hand-in-hand with alterations in what the drinker enjoys. In other words, the harsh pungent effects of the alcohol come to be liked. Research on chilli and fizzy drink consumption (Rozin, 1990) suggest that this process cannot be wholly reliant on the presence of the drug ethanol. Both of these are initially aversive, like alcohol, and yet through the process of exposure both come to be liked (Rozin, 1990). The lesson here is that understanding these type of processes may be important for designing ways to get people to consume an initially unappealing product, and even to come to strongly like it. The question is how.

Several different theories have been advanced to explain people's liking for chilli pepper. These include: (1) its ability to trigger salivation, thus making a bland diet more acceptable; (2) its high vitamin C content, which may be reinforcing in a diet lacking this vitamin; and (3) the suggestion that chilli may trigger release of endogenous opioids. None of these suggestions has been sustained. Instead, two factors seem to be important. The first concerns a person's initial motivation to consume. In chilli-eating cultures this generally occurs during childhood, where chilli is available at every meal and the child (not under any pressure to eat it) sees all of his older siblings and family members consuming it and enjoying it. This is known to be a powerful impetus to eat something new, especially if it involves imitating the actions of highly valued others. The second concerns a person's continued motivation to consume. In this case it appears that people come to like the burning sensation of chilli. One part of this appears to involve knowing that the burning sensation is "safe." Another concerns our general preference for sensory stimulation, be it hot or cold showers, massage of varying degrees of vigor, and loud music. Finally, the process of exposure alone has been consistently shown to increase liking for drinks, foods, and chilli (Figure 8.2; Stevenson and Yeomans, 1995).

The upshot of this is that if a suitably motivating rationale can be generated for initially consuming the product or a suitably supportive context can be created, then the process of exposure itself may be sufficient to enable the person to come to like the flavor of the nutraceutical, even if the flavor has

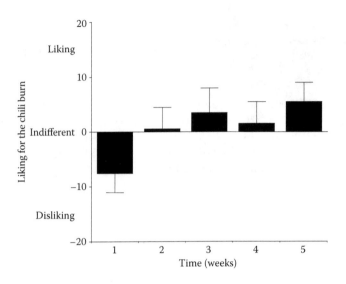

Figure 8.2 *Individual ratings of liking for chilli. Participants were asked to eat a meal containing chilli once a week for 5 weeks and on each occasion rate their like/dislike of its burn. Data from Stevenson & Yeomans, 1995.*

problem features. In the case of nutraceuticals a powerful initial motivation is usually the perceived health-related benefits of consumption. These are likely to enable intake of what might be rejected if it were a food. Assuming the problem flavor component is not too overwhelming, the process of exposure may result in it becoming liked. If it generates more profound levels of bitterness, pungency, or peculiar texture, these unusual sensory attributes may also become enjoyed, in much the same way that people can develop a preference for the chilli burn, tonic water, and blue cheese.

8.4.3 Expectancy

When adults view a food or drink, they form expectancies of what that food or drink will taste like (Stevenson, 2009). When the food or drink is placed in the mouth, any mismatch between the expectancy and the product's actual flavor will determine whether it is accepted or rejected. Two psychological processes are known to operate in this situation. The first is termed assimilation. It occurs when the difference between expectation and actual flavor is sufficiently small for this gap not to matter (i.e., both representations are "assimilated"). In certain circumstances, the gap can be quite considerable. One classic example of this is white wine that has been colored red. Participants report an odor and taste profile for it that matches a typical red wine, not a white wine (Morrot et al., 2001). A further example illustrates both assimilation and its opposite, contrast. In this instance, participants were presented with an ice cream with a pink color (Yeomans et al., 2008). For all intents and purposes it looked like strawberry ice cream, but when participants sampled it, it was in fact salmon-flavored mousse. The contrast here between people's expectations and the actual taste generated highly negative evaluation of the ice cream. Another group of participants who were provided information beforehand about what the ice cream actually contained were far more favorably inclined when they tasted it. Presumably, in this case expectation and reality were considerably closer and participants assimilated the difference between them.

Several factors are influential in dictating product expectancies. The first sampling of any product is usually with the eyes, and so the product packaging and information will form an important part of expectation building. In this light, any connection with health-food products and medicines will likely lead to a low expectation of enjoyment. Indeed, it seems likely that low palatability is to some extent part of the experience for deriving a health benefit, at least in Protestant countries (U.K., U.S.A.) where morality is closely entwined with hedonics (i.e., if pleasant then immoral, if unpleasant then moral—and healthy), and thus a disagreeable experience is regarded as morally beneficial (Brandt and Rozin, 1997). Once the packaging is open, the appearance of the product itself again sets up expectations as to its likely flavor. Its color and visual texture and smell are likely to remind adult participants of something they have consumed before, and this will then inform their expectations of its flavor. All of these aspects of a product are relatively easy to manipulate, but the goal to bear in mind needs to be the type of expectancy to be generated in the consumers mind—medicine or food?

8.5 Conclusions

Flavor has both extrinsic and intrinsic components. The extrinsic components are important for forming an expectancy of what a product will be like when it is consumed. Once the product is in the mouth, taste, smell, and somatosensation (touch) combine to generate flavor, with this being compared to the expected flavor, leading to rejection or acceptance. In the mouth, flavor can have various perceptual problems, with bitterness, unpleasant aromas, and texture the most prominent. The first two can be ameliorated to some degree by masking, using other tastants and odorants, or by altering the viscosity of the product. Importantly, people's rationales for consuming a product are important considerations in terms of what they will be prepared to tolerate, although exposure may, in the end, even with an initially quite unpleasant formulation, lead to liking. Finally, it is important to evaluate the product formulation on a sensory panel akin to the consumers who will likely eat the product. This simple recommendation may result in considerable cost savings in the long run.

References

Auvray M, Spence C (2008). The multisensory perception of flavor. *Consciousness Cogn.* 17:1016–1031.
Bartoshuk L (1975). Taste mixtures: Is mixture suppression related to compression? *Physiol Behav.* 14:643–649.
Bellisle F (1999). Glutamate and the UMAMI taste: Sensory, metabolic, nutritional and behavioural considerations. A review of the literature published in the last 10 years. *Neurosci Biobehav Rev.* 23:423–438.
Brandt A, Rozin P (1997). *Morality and Health.* New York: Routledge.
Breslin PAS, Gilmore MM, Beauchamp GK, Green BG (1993). Psychophysical evidence that oral astringency is a tactile sensation. *Chem Senses.* 18:405–417.
Buck LB (2000). The molecular architecture of odor and pheromone sensing in mammals. *Cell.* 100:611–618.
Bult JHF, de Wijk RA, Hummel T (2007). Investigations on multimodal sensory integration: Texture, taste, and ortho- and retronasal olfactory stimuli in concert. *Neurosci Lett.* 411:6–10.
Christensen CM (1984). Food texture perception. In: Mark E (ed). *Advances in Food Research.* New York: Academic Press; 159–199.
Cliff M, Noble AC (1990). Time-intensity evaluation of sweetness and fruitiness and their interaction in a model solution. *J Food Sci.* 55:450–455.
Copeland J, Stevenson RJ, Gates PW, Dillon P (2007). Young people and alcohol: The palatability of ready to drink (RTD) alcoholic beverages among 12 to 30 year olds. *Addiction.* 102:1740–1749.
Glanville EV, Kaplan AR (1965). Food preference and sensitivity of taste for bitter compounds. *Nature.* 205:851–853.
Green BG (1990). Effects of thermal, mechanical, and chemical stimulation on the perception of oral irritation. In: Green BG, Mason JR, Kare MR (eds). *Chem Senses: Irritation,* Vol. 2. New York: Marcel Dekker; 171–192.
Harper R, Land DG, Griffiths NM, Bate-Smith EC (1968). Odour qualities: A glossary of usage. *Br J Psychol.* 59:231–252.
Hodgson M, Linforth RST, Taylor A (2003). Simultaneous real-time measurements of mastication, swallowing, nasal airflow and aroma release. *J Agric Food Chem.* 51:5052–5057.
Lawless HT (1996). Flavor. In: Friedman M, Carterette E (eds). *Handbook of Perception and Cognition: Cognitive Ecology,* Vol. 16. San Diego: Academic Press; 325–380.
Lim J, Johnson MB (2012). The role of congruency in retronasal odor referral to the mouth. *Chem Senses.* 37:515–522.
McBurney DH, Gent JF (1979). On the nature of taste qualities. *Psychol Bull.* 86:151–167.
Menella J, Spector A, Reed D, Coldwell S (2013). The bad taste of medicine: Overview of basic research on bitter taste. *Clin Ther.* 35:1225–1246.
Morrot G, Brochet F, Dubourdieu D (2001). The color of odors. *Brain Lang.* 79:309–320.
Moskowitz HR, Barbe CD (1977). Profiling of odor components and their mixtures. *Sens Process.* 1:212–226.
Moskowitz HR, Kumaraiah V, Sharma KN, Jacobs HL, Sharma SD (1975). Cross-cultural differences in simple taste preferences. *Science.* 190:1217–1218.
Mozell M, Smith B, Smith P, Sullivan R Swender P (1969). Nasal chemoreception in flavor identification. *Arch Otolarynogol.* 90:367–373.
Piqueras-Fiszman B, Spence C (2015). Introduction. In: Piqueras-Fiszman B, Spence C (eds). *Multisensory Flavor Perception.* Cambridge: Woodhead Publishing; 1–8.
Prescott J (2004). Psychological processes in flavour perception. In A. Taylor and D. Roberts (eds). *Flavour Perception.* Oxford: Blackwell; 256–277.
Rozin P (1982). "Taste-smell confusions" and the duality of the olfactory sense. *Percept Psychophys.* 31:397–401.

Rozin P (1990). Getting to like the burn of chili pepper. In: Green B, Mason R, Kare M (eds). *Chem Senses, Vol. 2: Irritation*. New York: Marcel Dekker; 231–269.

Schifferstein HNJ, Frijters JER (1990). Sensory integration in citric acid/sucrose mixtures. *Chem Senses*. 15:87–109.

Schiffman SS (1992). Olfaction in aging and medical disorders. In: Serby MJ, Chobor KL (eds). *Science of Olfaction*. New York: Springer-Verlag; 500–25.

Scott TR, Mark GP (1987). The taste system encodes stimulus toxicity. *Brain Res*. 414: 197–203.

Smith DV, Margolskee RF (2001). Making sense of taste. *Sci Am*. March:26–33.

Spence C (2016). Sound: The forgotten flavor sense. In: Piqueras-Fiszman B, Spence C (eds). *Multisensory Flavor Perception*. Cambridge: Woodhead Publishing; 81–100.

Steiner J.E (1979). Human facial expressions in response to taste and smell stimulation. In: Reese H, Lipsitt L (eds). *Advances in Child Development and Behavior*, Vol. 13. New York: Academic Press; 257–295.

Stevenson RJ (2009). *The Psychology of Flavour*. Oxford: Oxford University Press.

Stevenson RJ (2014). Flavor binding: Its nature and cause. *Psychol Bull*. 140:487–510.

Stevenson RJ (2017). Psychological correlates of habitual diet. *Psychol Bull*. 143:53–90.

Stevenson RJ, Prescott J, Boakes RA (1995). The acquisition of taste properties by odors. *Learn Motiv*. 26:433–455.

Stevenson RJ, Prescott J, Boakes RA (1999). Confusing tastes and smells: How odours can influence the perception of sweet and sour tastes. *Chem Senses*. 24:627–635.

Stevenson RJ, Repacholi BM (2003). Age related changes in children's hedonic response to male body odor. *Dev Psychol*. 39:670–679.

Stevenson RJ, Yeomans M (1995). Does exposure enhance liking for the chilli burn? *Appetite*. 24:107–120.

Taylor C, Clifford A, Franklin A (2013). Color preferences are not universal. *J Exp Psychol: Gen*. 142:1015–1027.

Trelea IC, Atlan S, Deleris I, Saint-Eve A, Marin M, Souchon I (2008). Mechanistic mathematical model for in vivo aroma release during eating of semiliquid foods. *Chem Senses*. 33:181–192.

Wilson DA Stevenson RJ (2006). *Learning to Smell: Olfactory Perception From Neurobiology to Behavior*. Baltimore, MD: John Hopkins University Press.

Yeomans MR (2008). Learning and hedonic contributions to human obesity. In: Blass E (ed). *Obesity*. Sunderland, MA: Sinauer Associates; 211–236.

Yeomans MR, Chambers L, Blumenthal H, Blake A (2008). The role of expectancy in sensory and hedonic evaluation: The case of smoked salmon ice-cream. *Food Qual Prefer*. 19:565–573.

9

Flavor Nanotechnology: Recent Trends and Applications

Komal Parmar, Jayvadan Patel, and Navin Sheth

Contents

9.1 Introduction 217
9.2 Nanotechnology in flavors 222
9.3 Techniques for nanoencapsulation of flavor components 223
    9.3.1 Spray drying 223
    9.3.2 Freeze drying 224
    9.3.3 Coacervation phase separation 224
    9.3.4 Extrusion emulsion 225
    9.3.5 Supercritical fluid technology 226
    9.3.6 Emulsion method 227
    9.3.7 Electrospray 227
9.4 Applications of nanoencapsulation in flavor compounds 228
9.5 Evaluation of flavor products 229
9.6 Conclusion 230
References 230

9.1 Introduction

Flavor is an important consideration that influences our like and dislike of food products. It influences consumer decisions about the food quality in their selection, acceptance and ingestion of food products. Flavor is the sensory notion of food products that is ascertained by the chemical senses of the taste and smell buds. Aroma, another aspect of flavor, is defined as a substance that causes the reaction of receptors in the nose (Zuidam and Heinrich, 2010). The British Standards Institution defines flavor as the combination of organoleptic properties such as odor and taste of food components that is perceived by the respective aroma and taste receptors when consumed and that may be influenced by various sensations such as tactile, pain, heat, and cold (Astray et al., 2007). Aristotle identified only two categories of flavor: sweet and bitter. Today we distinguish five basic categories: sweet, salty, sour, bitter, and savory. Flavor results from combinations of hundreds or even thousands of organic compounds present in the food ingredients. Research carried out on extraction of these organic compounds led to extraction of mainly hydrocarbons, esters, aldehydes, ketones, and alcohols. Generally, these compounds are volatile in nature or have low boiling points. They exist in solid, liquid, or gas form. Table 9.1 lists examples of some flavor compounds used in the food industry. Table 9.2 lists other aroma compounds classified on the basis of their functional groups.

Flavors are classified into two major groups based on source: natural flavors and synthetic flavors. Natural flavors are extracted from natural sources such as flowers, fruits, and spices and purified by various processes. Generally,

Table 9.1 Common Flavors Used in the Food Industry

Compound	Formula	Source	Perception	Application in Food Industry
Monosodium glutamate	$C_5H_8NO_4Na$	Gluten, bacterial fermentation	Umami taste	Protein-rich food product containing fish, meat, or milk
Disodium guanylate	$C_{10}H_{12}N_5Na_2O_8P$	Dried fish or dried seaweed	Umami taste	Instant noodles, snacks, potato chips, cured meat, packaged soups. and thinned vegetables
Acetic acid	$C_2H_4O_2$	Fermentation, oxidation of ethanol	Sour	Cooking ingredient, beverages, pickling, herbicide
Vanillin	$C_8H_8O_3$	Chemical synthesis	Sweet vanilla	Milk products, beverages, candies
Limonene	$C_{10}H_{16}$	Chemical synthesis	Orange taste	Beverages, candies
Raspberry ketone	$C_{10}H_{12}O_2$	Chemical synthesis	Raspberry taste	Beverages, candies
Ethyl vanillin	$C_9H_{10}O_3$	Chemical synthesis	Intense vanilla odor	Beverages, chocolates, candies
Diacetyl	$C_4H_6O_2$	Byproduct of fermentation, chemical synthesis	Buttery flavor	Beverages, artificial butter,
Camphor	$C_{10}H_{16}O$	Natural source	Camphor taste	Cooking ingredient in Asia
Menthol	$C_{10}H_{20}O$	Chemical synthesis, Natural source	Mint taste	Beverages, candies
Maltol	$C_6H_6O_3$	Natural source	Coke and butterscotch taste	Beverages, wine
Ethyl maltol	$C_7H_8O_3$	Chemical synthesis	Fragrant aroma	Beverages, wine, meat products
Geranyl acetate	$C_{12}H_{20}O_2$	Chemical synthesis	Rose flavor	Beverages, candies
Ethyl hexanoate	$C_8H_{16}O_2$	Fermentation of yeasts	Fruity flavor	Beverages
Isoamyl acetate	$C_7H_{14}O_2$	Chemical synthesis	Banana flavor	Beverages, candies
Citral	$C_{10}H_{16}O$	Natural source	Lemon flavor	Beverages, candies

Table 9.2 Classification of Flavor Compounds Based on Functional Group Present

Group	Compound	Flavor
Alcohol	Menthol	Peppermint
	Furaneol	Strawberry
	1-Hexanol	Woody
Aldehyde	Cuminaldehyde	Spicy
	Isovaleraldehyde	Fruity
	Vanillin	Vanilla
Ketones	Raspberry ketone	Raspberry
	2-acetyl-1-pyrroline	Jasmine rice
	Cyclopentadecanone	Musk fragrance
Esters	Fructone	Fruity
	Hexyl acetate	Fruity
	Ethyl methylphenylglycidate	Strawberry
Lactones	Gamma-undecalactone	Peach
	Gamma-dodecalactone	Milky peach
	Gamma-decalactone	Buttery sweet
	Gamma-nonalactone	Coconut
	Gamma-valeroactone	Hay-like
	Gamma-hexalactone	Vanilla
Thiols	Benzyl mercaptan	Garlic
	Grapefruit mercaptan	Grapefruit

natural flavors are isolated from the essential oils using the extraction process. A single essential oil can contain a number of organic compounds with different properties, For example, lemon oil consists of organic compounds such as α-pinene, camphene, β-pinene, sabinene, myrcene, α-terpinene, linalool, β-bisabolene, limonene, trans-a-bergamotene, nerol, and neral (Lota et al., 2002, Hamdan et al., 2010). Synthetic flavors are obtained from chemical synthesis based on the conversion of a precursor organic compound to flavor through one or many steps involved in the process. Raspberry ketone is synthesized by crossed-aldol condensation of 4-hydroxybenzaldehyde with acetone, which is further hydrogenated over rhodium on alumina to form raspberry ketone (Smith, 1996). Flavor enhancers are also utilized in the food industry for the purpose of enhancing food taste. Examples of flavor enhancers include monosodium glutamate, disodium guanylate, calcium diglutamate, inosinic acid, maltol, ethyl maltol, and leucine. A 2016 global market research report suggests rapid growth of the flavor market in global food industry, predicted to reach approximately US$ 15.60 billion by 2021, growing at a rate of 7.5% between 2016 and 2021 (Zion Market Research, 2016).

Flavor is an essential additive in the food product. The amount of flavor must be optimized. Consumers are cautioned that too much salt leads to unpalatable food. A high concentration of flavor can lead to nausea, vomiting, headaches, intolerance, and even severe damage to body organs (Jinap and Hajeb, 2010; Husarova and Ostatnikova, 2013). Another important parameter that is a vital attribute of food product quality is the stability of flavor with respect to time (Weerawatanakorn et al., 2015; Given, 2009).

Nanotechnology is the science that deals with manipulation of organic and inorganic matter at the nano range. Nanotechnology has opened up an extremely important area of research and development in the functional food industry. Food technologists modify our food by utilizing nanotechnology-enabled innovations to the utmost. The potentially favorable effects on the food that result also have a positive effect on our health. A nanofood is defined as food prepared from/by using nanotechnology techniques or tools during cultivation, production, processing, and/or packaging. Nanotechnology has a prospective impact on several aspects of the food industry, such as use of agrochemicals in the agrifood sector; use of nanosensors/nanobiosensors in crop protection; and use of nanodevices in genetic manoeuvring, diagnosis of plant disease, precision processes, and smart packaging (Chellaram et al., 2014; Sekhon, 2014; Parisi et al., 2015).

Nanoscale food research uses controlled release of nanostructured food components or direct nutrients to provide a quality food product (Qureshi et al., 2012). Extending shelf life and elegance of food products to achieve food stability is also an emerging area of interest in nanotechnology studies (He and Hwang, 2016). Various applications of nanotechnology in the food sector currently being researched involve modification of the structure of natural food components to craft them into suitable nanosize carriers of bioactive components (Sekhon, 2010). Food packaging and preservation incorporates utilization of nanocomposites with unique characteristics. For example, packaging material intended to replace the polystyrene clamshell used in fast food restaurants is made up of potato starch and calcium carbonate which is lightweight, biodegradable, and has good thermal insulation (Garcia et al., 2010). Recent developments in the use of nanopesticides has provided reliable performance for crop protection, thus enabling future potential in the agri-food sector in determining environmental fate of crop (Ali et al., 2014; Kah and Hofmann, 2014). Nanomaterials are used in food processing methods, including include incorporation of nutraceuticals for smart delivery, nanoencapsulation of nutrients, color and flavor in food systems, bioseparation of proteins, solubilization, and sampling of contaminants (Ravichandran, 2010; Huang et al., 2010; Weiss et al., 2006). Use of nanomaterials in food processing methods provides an enhanced enzyme support system to protect them from environmental factors and thus can be employed to enhance flavor, nutritional value, and health benefits (Rashidi and Khosravi-Darani, 2011).

The global nanotechnology market continues to grow and achieve impressive results. Reports, reviews, patents, and research publications underscore the potential impact of nanotechnology on the food sector (Chen et al., 2006; Sekhon, 2010; GuhanNath et al., 2014; Wesley et al., 2014; Feng et al., 2009). Table 9.3 lists potential applications of nanotechnology in the food sector. In recent years, many industries have focused on flavor nanotechnology for enhancement of the taste of marketed food products (Heller, 2006).

Table 9.3 Applications of Nanotechnology in the Food Sector

Agriculture	Processed Food Products	Packed Food Products	Nutraceutical Supplements
Smart delivery of pesticides, herbicides, and fertilizers, e.g., macronutrient fertilizer coated with zinc oxide nanoparticles (Milani et al., 2015)	Enhances stability and provides protection against oxidation, e.g., essential oils loaded in nanosystems (Bilia et al., 2014)	Metal-based nanosensors to detect the toxins, e.g., aflatoxin B_1 in milk (Meetoo, 2011)	Nanoparticles that improve bioavailability of nutrients and nutraceuticals (Acosta, 2009)
As diagnostic device to monitor environmental conditions for betterment of crops, e.g., monitoring of organophosphorus pesticides (Srilatha, 2011)	Enhancement in bioavailability and efficacy of nutritional elements (McClements et al., 2015)	Detecting pesticides on the surface of fruits and vegetables, e.g., single walled carbon nanotubes are used (Sozer and Kokini, 2009)	Nanoscale delivery for enhancement of stability, biocompatibility, and permeability of nutraceuticals (Neves et al., 2016)
For genetic modification of plants for further improvement, e.g., mesoporous silica nanostructure to transport DNA to transform plant cells (Torney et al., 2007)	To enhance flavor of the food product, e.g., Slim Shake Chocolate uses nanocluster technology to produce cocoa clusters to enhance the flavor of cocoa (Shahidi et al., 2006)	For manufacture of nanomaterial-based containers, e.g., use of silicon dioxide to reduce leakage of moisture from container (Coma, 2008)	Nanotechnology for some functional ingredients, such as rendering hydrophilic substance to fat soluble and vice versa (Das et al., 2009)
Evaluate enzyme substrate interactions by single molecule detection (Tan et al., 2016)	Allows better selection of raw materials (Ravinchandran, 2010)	Nanoencapsulation to coat food products, e.g., nanolaminates (Flanagan and Singh, 2006)	Cellulose nanocrystals for nutrients (Sunasee et al., 2016)
Recycling of agricultural waste, e.g., nano-engineered enzymes produce cellulose from waste plant materials (Dhewa, 2015)	Helps improve appearance of food products by changing viscosity and gelation (Ray and Okamoto, 2003)	Removes unwanted odor, e.g., NanoCream-PAC (powdered activated carbon) absorbs components with unpleasant odor (Burdo, 2005)	Nanochelates for nutrient delivery without affecting organoleptic properties (Bhosale et al., 2013)
Improvement of soil by use of nanomaterials, e.g., soil enhancer product (Mukhopadhyay et al., 2014)	Delivery of minerals and nutrients in the food (Klaine et al., 2008)	Acts as an antimicrobial agent, e.g., silver-coated nanocomposites (Moraru et al., 2003)	

9.2 Nanotechnology in flavors

Flavor ingredients typically have stability issues because many factors affect their chemical components. Flavor stability is an important parameter to be evaluated. It is directly or indirectly related to quality and acceptability of the food. Therefore, an interest in controlling the stability and food quality has developed recently. Physicochemical properties of flavor components and their interactions with other food additives affect the quality of the food. For example, some flavor components are more stable in water-soluble carbohydrates (Fathi et al., 2014) while others are stable in lipid coatings (Fathi et al., 2012). During the manufacturing process of any food product, loss of aroma may occur due to heating or stirring methods involved. Loss of aroma may even occur during intake. Synthetic flavors that provide the aroma in the finished product have been explored. However, these synthetic flavor ingredients are also sensitive to oxidation, heat, moisture, and light (Garwood et al., 1995). Encapsulation of flavor components in various carrier materials at the nanoscale may help protect, target, modify, and/or enhance their functional properties (Singh, 2016).

In recent years, numerous nanotechnology strategies for nanoencapsulation of ingredients have been investigated. These include nanoemulsion, nanocrystals, nanotubes, nanopowder, nanocomplexes, liposomes, niosomes, and nanoparticles, as shown in Figure 9.1. Techniques involved in the preparation of such nano-based systems include coacervation, spray drying, homogenization, freeze drying, and extrusion methods. Materials utilized for the formation of nano-based systems should be inert; have the ability to form the wall and seal the core, withstand stress, and protect the flavor component from environmental factors; have a palatable taste; and be economically viable. Table 9.4 lists some of the carrier materials suitable for nanoencapsulation.

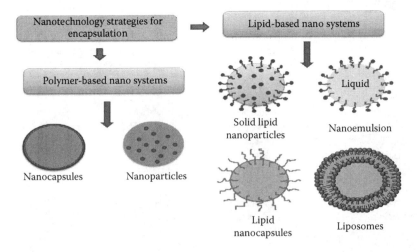

Figure 9.1 *Nanoencapsulation strategies.*

Table 9.4 Classification of Polymers Used for Nanoencapsulation

Natural	Synthetic/Semisynthetic	Inorganic
Carbohydrate based, e.g., cellulose derivative, starch	Polylactic acid	Silica
Plant based, e.g., gum karaya, gum arabic	Polyglycolic acid	Calcium phosphates
Marine based, e.g., carrageenan, alginate	Poly ε-caprolactone	Titanium
Animal/microbial based, e.g., xanthan, chitosan	Polyanhydrides	
Protein based, e.g., albumin, gluten	Polyamides	
Lipid based, e.g., phospholipids, waxes		

9.3 Techniques for nanoencapsulation of flavor components

Nanotechnology's place in the food sector extends far beyond mere utilization of various techniques for nanoencapsulation. Novel techniques that may improve the quality of nanoencapsulation of flavor ingredients are also being explored. This section describes some of the important methods utilized in nanoencapsulation.

9.3.1 Spray drying

Spray drying is the most common method employed in the food industry for nanoencapsulation of food components (Reineccius, 2004). Spray drying is the most frequently used technique in large-scale nanoencapsulation of flavors (Turchiuli et al., 2013). It is an established method that works on the principle of atomizing the suspensions or solutions, followed by a drying process to generate nanosize particles with the active ingredient encapsulated in the carrier material. The major drawback of this method is that it is not applicable for volatile material (Jafari et al., 2008). Various parameters influence the size of particles formed and the encapsulation efficiency of the active ingredients.

Properties of the wall materials affect the encapsulation efficiency of flavor components. Choice of the wall material is decisive since it determines the emulsion properties, retention of volatile flavor components, and shelf life of the nanocapsule after drying. Carbohydrates, including modified starch, cellulose derivatives, gums, and cyclodextrin, constitute the list of choice wall materials for nanoencapsulation by the spray drying method. However, carbohydrates may undergo gelation due to increase in temperature during the process. Other factors include viscosity of solution. The higher the viscosity, the greater the risk of clogging the orifice with atomizer remains (Jafari et al., 2008). Spray dryer features affect the size of the particles formed. The nano spray dryer, with the help of a piezoelectric-driven actuator, moves the mesh upward and downward, which further generates millions of fine droplets. Figure 9.2 illustrates the formation of nanocapsules by the spray drying method.

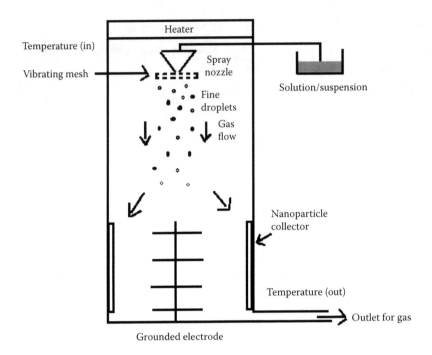

Figure 9.2 Spray drying method for nanoparticles.

9.3.2 Freeze drying

Freeze drying or lyophilization is a widely used process for preparation of nanoparticles. The process involves drying of the product. The solvent, usually water, is crystallized at very low temperature and then sublimated directly into the vapor state. The process is highly suitable for thermolabile materials. Compared to other drying techniques, the materials dried with this technique have long shelf life and maintain the shape of the nanocapsule due to fixation of shape during the process (Ciurzynska and Lenart, 2011). The method is ideal for most flavor components that contain volatile oils. Figure 9.3 shows a schematic diagram of the process of freeze drying.

9.3.3 Coacervation phase separation

Coacervation phase separation is an older method employed in the encapsulation process. It is a colloidal phenomenon. The technique involves separation of coacervates from the mixture of colloidal liquids after agglomeration. Figure 9.4 describes the working principle for this method.

Many factors affect the coacervation process, including change in temperature, change in pH, increase in the amount of micromolecular substance or another macromolecular substance, type of biopolymeric material, surface charge, rate of agitation, and solubilities of flavors and biopolymer in solvent.

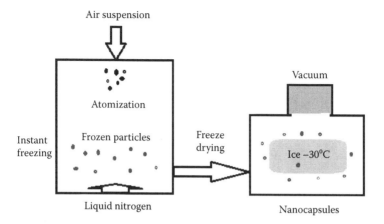

Figure 9.3 Freeze drying process for preparation of nanocapsules.

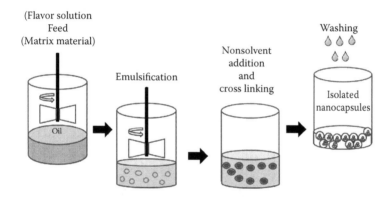

Figure 9.4 Coacervation phase separation steps for nanocapsule formation.

Based on the number of polymers utilized in the process, the method is classified into two broad categories: simple and complex (Feng et al., 2009). The phenomenon of agglomeration is based on the electrostatic force of attraction between oppositely charged molecules. The food flavor components become entrapped within the particle formed by complex formation of positively charged and negatively charged polymeric materials and are thereby encapsulated in the nanoparticles.

9.3.4 Extrusion emulsion

The extrusion microemulsion technique is reproducible and suitable for small volumes. The method is free of organic solvents. Emulsion at the micro or nano level is produced by shaking organic solutions of the micro- or nanoparticles with water. Droplets are formed with large size distribution ranging from

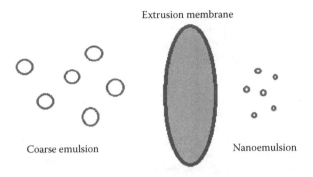

Figure 9.5 *Emulsion extrusion method converting coarse emulsion to nanoemulsion.*

10 to 200 nm in diameter. In this method, nanoglobules containing flavor components are passed through the membrane to control the size of nanoparticles. Passing of globules through such membranes reduces the size of nanoparticle-coated droplets due to breaking of the droplets during the extrusion process (Tangirala et al., 2007; De-Jesus et al., 2013). Encapsulation of flavor components via this method is suitable for volatile and unstable flavors (Shimoni, 2000). Factors that affect the quality of nanocapsules formed include type of emulsifier, amounts of emulsifier and flavor component, and pressure applied against the membrane during the extrusion process (Raoa and Geckeler, 2011). Figure 9.5 shows a schematic diagram of the emulsion extrusion method.

9.3.5 Supercritical fluid technology

In recent years, supercritical fluids have been used for encapsulation of thermolabile components in a method similar to the spray drying technique. Supercritical carbon dioxide is the supercritical fluid that is most widely explored in nanoencapsulation process. Supercritical carbon dioxide possesses several requisite characteristics, including inertness, low toxicity, economical (compared to organic solvents used), high volatility (hence easy removal), and nonflammability, compared to other compounds that can be used in the process, such as propane or nitrogen. Supercritical fluid technology consists of various subtechniques classified on the basis of the function of supercritical fluid in the process. Supercritical fluid can act as a solvent, an antisolvent, a cosolvent, a solute, and an extractant (Silva et al., 2014; Ezhilarasi et al., 2013). Among the most widely used methods is the rapid expansion of supercritical solutions (RESS), where supercritical fluid acts as a solvent to solubilize the thermolabile component and polymer. The solution is made to expand in the area with low pressure, so a sudden drop in high pressure will precipitate out the solute from the supercritical fluid, resulting in formation of nanoparticles. Other methods, such as the supercritical antisolvent (SAS) method, employs supercritical fluid as antisolvent where the thermolabile component and polymer are solubilized in a suitable organic solvent. The system is then passed through the antisolvent, which decreases

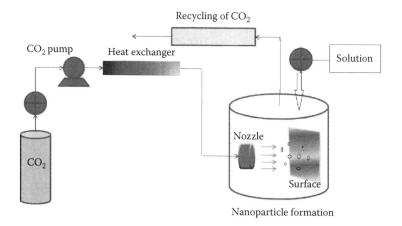

Figure 9.6 *Schematic representation of nanoparticle formation by supercritical fluid technology.*

the solubility of solute in the solvent and precipitates to form nanoparticles. Factors to be determined include temperature, pressure, solute-to-polymer ratio, and the rate of solution flow. Figure 9.6 shows the primary processes of supercritical fluid technology involved in nanoparticle formation.

9.3.6 Emulsion method

Emulsion technology is widely applied in the encapsulation of bioactive materials via the formulation of nanoemulsion. This technique is very effective for encapsulating both lipid- and water-soluble food components. Flavor volatile oils can themselves be emulsified at nano level and thereby encapsulated further (Rao and McClements, 2011; Ezhilarasi et al., 2013). The method is simple with high encapsulation efficiency, less stress during processing, and narrow size distribution. It is easy to scale up. The process involves a two-step procedure: emulsification and removal of organic solvent by evaporation. Nanoemulsion formed by mechanical shearing is stirred continuously for an optimal time to harden the globules to nanocapsules. Any organic solvent utilized is evaporated. Factors affecting encapsulation of flavor components depend on the polarity of the ingredient into aqueous or organic phase (Yadav et al., 2015). The organic solvent used must be volatile enough to evaporate from the mixture. The size of nanocapsule formed depends on the shear rate, temperature, chemical composition of organic solvent, emulsifier, amount of polymer, emulsifier, and polymer-to-component ratio. Figure 9.7 shows the steps involved in the formation of nanospheres by the emulsion evaporation method.

9.3.7 Electrospray

Electrospray is a one-step novel technique used to produce nanoparticles containing bioactive compounds. It is a simple and low-cost process to encapsulate

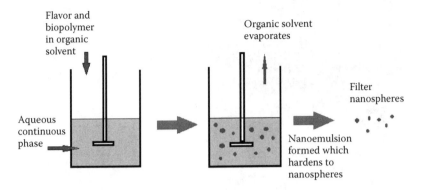

Figure 9.7 *Emulsion evaporation method to produce nanospheres.*

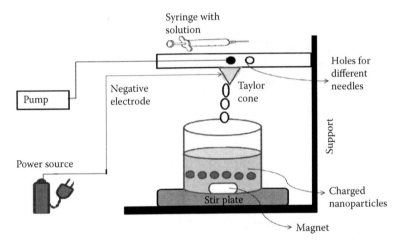

Figure 9.8 *Electrospray method for nanoparticle preparation.*

flavor components. The process works on the principle of deformation of the droplet interface by the electric field, which produces droplets in the micro- or nanometer range. The electric field applied to the droplet generates an electrostatic charge on the surface of the droplet which is capable of overcoming the cohesive force of the droplet, thereby helping to form nanospheres (Tapia-Hernandez et al., 2015). Figure 9.8 describes how the electrospray technique works to form nanoparticles.

9.4 Applications of nanoencapsulation in flavor compounds

Nanoencapsulation has proved to be a potential drug delivery system for many active ingredients to date. Nanoencapsulation of food ingredients provides improved bioavailability, solubility, and stability, and increased potency. This section describes applications of nanoencapsulation for flavor components.

Preservation of food flavors in food and beverages is most important with respect to the compounds prone to oxidation, heat, light, and other components acting as impurities. Encapsulation provides these ingredients with a barrier against such environmental factors. Tian and coworkers studied the instability of 2,5-dimethyl pyrazine, a major flavor component in seafood. They successfully embedded the aroma constituent into hydroxypropyl-beta-cyclodextrin nanocapsules (Tian et al., 2016). Stability of coffee aroma was improved by encapsulating into gelatine based nano-emulsions protecting from unwanted effects in the presence of environmental conditions (Balassa et al., 1970). Limonene was investigated for encapsulation in alginate and polyvinyl alcohol matrix by the freeze-thawing method in order to protect the liquid aroma compounds from oxidative degradation (Levi et al., 2011). Citral, a widely used additive in beverages, undergoes acid-catalyzed cyclization and oxidation in low pH conditions. Stability enhancement of citral was achieved by encapsulating it in nanoemulsion (Zhao et al., 2013). Protection of eugenol against light oxidation was achieved by encapsulation into hydroxypropyl-beta-cyclodextrin and polycaprolactone prepared by the inclusion and emulsion diffusion methods, respectively (Choi et al., 2009). Thermal stability of eugenol was improved by embedding into chitosan nanoparticles. The studies revealed 2.7-fold greater radical scavenging activity of encapsulated eugenol nanoparticles (Woranuch and Yoksan, 2013).

Nanoencapsulation of flavor components has been investigated for taste enhancement to increase the organoleptic properties of the product. As discussed earlier, Slim Shake Chocolate uses nanocluster technology to produce cocoa clusters to enhance the flavor of cocoa (Shahidi et al., 2006). Nanosalt claims to reduce the salt content of the product because of enhanced surface area. This enables the consumer to experience the product with a similar taste as before but with substantially reduced salt content (Fabra et al., 2012). Masking of undesirable flavor is also an important parameter to be considered in the manufacturing of food products. Self assembly of food proteins can result in protection of nutraceuticals and masking of undesirable aromas of compounds (Des Rieux et al., 2006; Luecha et al., 2010).

9.5 Evaluation of flavor products

The consumer evaluates food product by the quality and type of aroma. Thus analysis of aroma in the context of its content, presence, and composition is important. The traditional approach of evaluating the aroma of any product involves trained assessors. The result generated is in form of scores given to the aroma. This is a successful technique but it suffers from certain limitations. Thus, instrumental analysis of organoleptic properties is a good alternative for evaluation purposes. This method includes gas chromatography for the assessment of content and composition of aroma. Unpredictable interactions of flavor components can also be determined with the help of gas chromatography (Azarnia et al., 2012).

A replacement for the human tongue has been developed in the form of an electronic-tongue, also known as the e-tongue. The e-tongue has different sensors that work differently for different compounds (Del Valle, 2010). It is used in the analysis of liquids, thus preventing the exposure of human beings to harmful substances or awkward tastes. Investigators have successfully utilized the e-tongue as a tool to study the taste behavior of flavor compounds (Sliwinska et al., 2014).

9.6 Conclusion

Food additives have many limitations, which can be overcome by efficient encapsulation. Nanotechnology provides successful and effective encapsulation of such components. The choice of approach in nanotechnology depends on various factors associated with the properties of the active ingredient to be encapsulated and the type of biopolymer utilized for encapsulation. Ultimate product cost also influences the choice of method to be employed. Nanoparticles used for encapsulation of such flavor components provide many advantages that contribute to high acceptability of the final food product. In the end, the purpose of nanotechnology in any sector of food science will always focus on food safety, efficiency, healthiness of product, and well-being of the community.

References

Acosta E (2009). Bioavailability of nanoparticles in nutrient and nutraceutical delivery. *Curr Opin Colloid Interface Sci.* 14:3–15.

Ali MA, Rehman I, Iqbal A, Din S, Rao AQ, Latif A, Samiullah TR, Azam S, Husnain T (2014). Nanotechnology, a new frontier in agriculture. *Int J Adv Life Sci.* 1:129–138.

Astray G, Garcia-Riob L, Mejuto JC, Pastranac L (2007). Chemistry in food: Flavors. *Electron J Environ Agric Food Chem.* 6(2):1742–1763.

Azarnia S, Boye JI, Warkentin T, Malcolmson L (2012). Application of gas chromatography in the analysis of flavor compounds in field peas. In: Salih B, Çelikbiçak O (eds), *Gas Chromatography in Plant Science, Wine Technology, Toxicology and Some Specific Applications.* InTech, doi: 10.5772/34053, 1–17.

Balassa LL, Tomahawk LT, Grove B (1970). Encapsulation of aromas and flavors. *United States Patent 3,495,988*:1–6.

Bhosale RR, Ghodake PP, Mane AN, Ghadge AA (2013). Nanocochleates: A novel carrier for drug transfer. *J Sci Innov Res.* 2(5):964–969.

Bilia AR, Guccione C, Isacchi B, Righeschi C, Firenzuoli F, Bergonzi MC (2014). Essential oils loaded in nanosystems: A developing strategy for a successful therapeutic approach. *Evid Based Complementary Altern Med.* Article ID 651593.

Burdo OG (2005). Nanoscale effects in food-production technologies. *J Eng Phys Thermophys.* 78(1):90–96.

Chellaram C, Murugaboopathib G, Johna AA, Sivakumarc R, Ganesand S, Krithikae S, Priyae G (2014). Significance of nanotechnology in the food industry. *APCBEE Procedia.* 8:109–113.

Chen LY, Remondetto GE, Subirade M (2006). Food protein based materials as nutraceutical delivery systems. *Trends Food Sci Technol.* 17:272–283.

Choi M, Soottitantawat A, Nuchuchua O, Min S, Ruktanonchai U. (2009). Physical and light oxidative properties of eugenol encapsulated by molecular inclusion and emulsion-diffusion method. *Food Res Int.* 42(1):148–156.

Ciurzynska A, Lenart A (2011). Freeze-drying—Application in food processing and biotechnology—A review. *Pol J Food Nutr Sci.* 61(3):165–171.

Coma V (2008). Bioactive packaging technologies for extended shelf life of meat-based products. *Meat Sci.* 78(2):90–103.

Das M, Saxena N, Dwiwedi P (2009). Emerging trends of nanoparticles application in food technology: Safety paradigms. *Nanotoxicology.* 3(1):1018.

De Jesus MB, Allan R, Inge SZ, Eneida P (2013). Microemulsion extrusion technique: A new method to produce lipid nanoparticles. *J Nanopart Res.* 15:1960–1963.

Del Valle M (2010). Electronic tongues employing electrochemical sensors. *Electroanalysis.* 22(14):1539–1555.

Des Rieux A, Fievez V, Garinot M, Schneider YJ, Preat V (2006). Nanoparticles as potential oral delivery systems of proteins and vaccines: A mechanistic approach. *J Control Release.* 116:1–27.

Dhewa T (2015). Nanotechnology applications in agriculture: An update. *Octa J Environ Res.* 3(2):204–211.

Ezhilarasi PN, Karthik P, Chhanwal N, Anandharamakrishnan C (2013). Nanoencapsulation techniques for food bioactive components: A review. *Food Bioproc Tech.* 6(3):628–647.

Fabra MJ, Chambin O, Voilleyc A, Gayd J, Debeaufort F (2012). Influence of temperature and NaCl on the release in aqueous liquid media of aroma compounds encapsulated in edible films. *J Food Eng.* 108(1):30–36.

Fathi M, Martin A, McClements DJ (2014). Nanoencapsulation of food ingredients using carbohydrate based delivery systems. *Trends Food Sci Technol.* 39(1):18–39.

Fathi M, Mozafari MR, Mohebbi M (2012). Nanoencapsulation of food ingredients using lipid based delivery systems. *Trends Food Sci Technol.* 23:13–27.

Feng T, Xiao Z, Tian H (2009). Recent patents in flavor microencapsulation. *Recent Pat Food Nutr Agric.* 1(3):193–202.

Flanagan J, Singh H (2006). Microemulsions: A potential delivery system for bioactives in food. *Crit Rev Food Sci Nutr.* 46(3):221–237.

Garcia M, Forbe T, Gonzalez E (2010). Potential applications of nanotechnology in the agro-food sector. *Cienc Tecnol de Alimen.* 30(3):573–581.

Garwood RE, Mandralis ZI, Scott AW (1995). Encapsulation of volatile aroma compounds. U.S. Patent No. US005,399,368A.

Given PS (2009). Encapsulation of flavors in emulsions for beverages. *Curr Opin Colloid Interface Sci.* 14:43–47.

GuhanNath S, Sam AI, Allwyn AS, Ranganathan TV (2014). Recent innovations in nanotechnology in food processing and its various applications—A review. *Int J Pharm Sci Rev Res.* 29(2):116–124.

Hamdan D, El-Readi MZ, Nibret E, Sporer F, Farrag N, El-Shazly A, Wink M (2010). Chemical composition of the essential oils of two citrus species and their biological activities. *Pharmazie.* 65(2):141–147.

He X, Hwang H (2016). Nanotechnology in food science: Functionality, applicability, and safety assessment. *J Food Drug Anal.* 24:671–681.

Heller L (2006). Flavor firm uses nanotechnology for new ingredient solutions. http://www.foodnavigator-usa.com/Suppliers2/Flavor-firm-uses-nanotechnology-for-new-ingredient-solutions. Accessed 2 March 2017.

Huang Q, Yu H, Ru Q (2010). Bioavailability and delivery of nutraceuticals using nanotechnology. *J Food Sci.* 75(1):R50–R56.

Husarova V, Ostatnikova D (2013). Monosodium glutamate toxic effects and their implications for human intake: A review. *J Med Res.* 2013: Article ID 608765.

Jafari SM, Assadpoor E, He Y, Bhandari B (2008). Encapsulation efficiency of food flavors and oils during spray drying. *Drying Technol.* 26:816–835.

Jinap S, Hajeb P (2010). Glutamate: Its applications in food and contribution to health. *Appetite.* 55:1–10.

Kah M, Hofmann T (2014). Nanopesticide research: Current trends and future priorities. *Environ Int.* 63:224–235.

Klaine SJ, Alvarez PJJ, Batley GE, Fernandes TF, Handy RD, Lyon DY, Mahendra S, McLaughlin MJ, Lead JR (2008). Nanomaterials in the environment: Behavior, fate, bioavailability, and effects. *Environ Toxicol Chem.* 27(9):1825–1851.

Levi S, Raca V, Manojloviu V, Rakiu V, Bugarskib B, Flockc T, Krzyczmonikd KE, Nedoviu V (2011). Limonene encapsulation in alginate/poly (vinyl alcohol). *Procedia Food Sci.* 1:1816–1820.

Lota ML, de Rocca SD, Tomi F, Jacquemond C, Casanova J (2002). Volatile components of peel and leaf oils of lemon and lime species. *J Agric Food Chem.* 50(4):796–805.

Luecha J, Sozer N, Kokini JL (2010). Synthesis and properties of corn zein/montmorillonite nanocomposite films. *J Mater Sci.* 45(13):3529–3537.

McClements DJ, Zou L, Zhang R, Salvia-Trujillo L, Kumosani T, Xiao H (2015). Enhancing nutraceutical performance using excipient foods: Designing food structures and compositions to increase bioavailability. *Compr Rev Food Sci Food Saf.* (14):824–847.

Meetoo DD (2011). Nanotechnology and the food sector: From the farm to the table. *Emirates J Food Agric.* 23(5):387–407.

Milani N, Hettiarachchi GM, Kirby JK, Beak DB, Stacey SP, McLaughlin MJ (2015). Fate of zinc oxide nanoparticles coated onto macronutrient fertilizers in an alkaline calcareous soil. *PLoS One.* 10(5):e0126275.

Moraru CI, Panchapakesan CP, Huang Q, Takhistov P, Liu S, Kokini JL (2003). Nanotechnology: A new frontier in food science. *Food Technol.* 57(12):24–29.

Mukhopadhyay SS (2014). Nanotechnology in agriculture: Prospects and constraints. *Nanotechnol Sci Appl.* 7:63–71.

Neves AR, Martins S, Segundo MA, Reis S (2016). Nanoscale delivery of resveratrol towards enhancement of supplements and nutraceuticals. *Nutrients.* 8(3):131.

Parisi C, Vigani M, Rodríguez-Cerezo E (2015). Agricultural nanotechnologies: What are the current possibilities? *Nano Today* 10:124–127.

Qureshi AM, Karthikeyan S, Karthikeyan P, Khan PA, Uprit S, Mishra UK (2012). Application of nanotechnology in food and dairy processing: an overview. *Pak J Food Sci.* 22:23–31.

Rao J, McClements DJ (2011). Formation of flavor oil microemulsions, nanoemulsions and emulsions: Influence of composition and preparation method. *J Agric Food Chem.* 59(9):5026–5035.

Raoa JP, Geckeler KE (2011). Polymer nanoparticles: Preparation techniques and size-control parameters. *Prog Polym Sci.* 36:887–913.

Rashidi L, Khosravi-Darani K (2011). The applications of nanotechnology in food industry. *Crit Rev Food Sci Nutr.* 51:723–730.

Ravichandran R (2010). Nanotechnology applications in food and food processing: Innovative green approaches, opportunities and uncertainties for global market. *Int J Green Nanotechnol Phys Chem.* 1(2):P72–P96.

Ray SS, Okamoto M (2003). Polymer/layered silicate nanocomposites: A review from preparation to processing. *Prog Polym Sci.* 28(11):1539–1641.

Reineccius GA (2004). The spray drying of food flavors. *Drying Technol.* 22(6):1289–1324.

Sekhon BS (2010). Food nanotechnology—An overview. *Nanotechnol Sci Appl.* 3:1–15.

Sekhon BS (2014). Nanotechnology in agri-food production: An overview. *Nanotechnol Sci Appl.* 7:31–53.

Shahidi F, Weiss J, Chen H (2006). Nanotechnology in nutraceuticals and functional foods. *Food Technol.* 60(3):30–36.

Shimoni E (2000). Nanoencapsulation of food ingredients: From macromolecular nanostructuring to smart delivery systems. *Lab Funct Foods Nutraceut Food Nanosci.* 43:317–326.

Silva EK, Angela M, Meireles A (2014). Encapsulation of food compounds using supercritical technologies: Applications of supercritical carbon dioxide as an antisolvent. *Food Public Health* 4 (5):247–258.

Singh H (2016). Nanotechnology applications in functional foods: Opportunities and challenges. *Prev Nutr Food Sci.* 21(1):1–8.

Sliwinska M, Wisniewska P, Dymerski T, Namiesnik J, Wardenck W (2014). Food analysis using artificial senses. *J Agric Food Chem.* 62:1423–1448.

Smith LR (1996). Rheosmin ("raspberry ketone") and zingerone, and their preparation by crossed aldol-catalytic hydrogenation sequences. *Chem Educ.* 1(3):1–18.

Sozer N, Kokini JL (2009). Nanotechnology and its applications in the food sector. *Trends Biotechnol.* 27(2):82–89.

Srilatha B (2011). Nanotechnology in agriculture. *Nanomed Nanotechnol.* 2:123.

Sunasee R, Hemraz UD, Ckless K (2016). Cellulose nanocrystals: A versatile nanoplatform for emerging biomedical applications. *Expert Opin Drug Deliv.* 13(9):1243–1256.

Tan S, Gu D, Liu H, Liu Q (2016). Detection of a single enzyme molecule based on a solid-state nanopore sensor. *Nanotechnology.* 27:155502.

Tangirala R, Revanur R, Russell TP, Emrick T (2007). Sizing nanoparticle-covered droplets by extrusion through track-etch membranes. *Langmuir.* 23:965–969.

Tapia-Hernandez JA, Torres-Chavez PI, Ramírez-Wong B, Rascon-Chu A, Plascencia-Jatomea M, Barreras-Urbina CG, Rangel-Vazquez NA, Rodríguez-Felix F (2015). Micro- and nanoparticles by electrospray: Advances and applications in foods. *J Agric Food Chem.* 63:4699–4707.

Tian H, Xu T, Dou Y, Li F, Yu H, Ma X (2016). Optimization and characterization of shrimp flavor nanocapsules containing 2,5-dimethylpyrazine using an inclusion approach. *J Food Process Preserv.* DOI: 10.1111/jfpp.13015.

Torney F, Trewyn BG, Lin VSY, Wang K (2007). Mesoporous silica nanoparticles deliver DNA and chemicals into plants. *Nat Nanotechnol.* 2:295–300.

Turchiuli C, Cuvelier M-E, Giampaoli P, Dumoulin E (2013). Aroma encapsulation in powder by spray drying, and fluid bed agglomeration and coating. In: Yanniotis S, Taoukis P, Stoforos NG and Karathanos VT (eds), *Advances in Food Process Engineering Research and Applications.* New York: Food Engineering Series, Springer; 255–265.

Weiss J, Takhistov P, McClements DJ (2006). Functional materials in food nanotechnology. *J Food Sci.* 71(9):R107–R116.

Weerawatanakorn M, Wu JC, Pan MH, Ho CT (2015). Reactivity and stability of selected flavour compounds. *J Food Drug Anal.* 23(2):176–190.

Wesley SJ, Raja P, Raj AA, Tiroutchelvamae D (2014). Review on nanotechnology applications in food packaging and safety. *Int J Eng Res.* 3(11):645–651.

Woranuch S, Yoksan R (2013). Eugenol-loaded chitosan nanoparticles: I. Thermal stability improvement of eugenol through encapsulation. *Carbohydr Polym.* 96(2):578–585.

Yadav V, Sharma A, Singh SK (2015). Microencapsulation techniques applicable to food flavors research and development: A comprehensive review. *Int J Food Sci Nutr.* 4:119–124.

Zhao Q, Ho C, Huang Q (2013). Effect of ubiquinol-10 on citral stability and off-flavor formation in oil-in-water (o/w) nanoemulsions. *J Agric Food Chem.* 61:7462–7469.

Zion Market Research (2016). Flavors Market by Type (Natural and Synthetic) for Beverages, Bakery, Confectionery, Dairy, Savory & Snacks and Others Applications: Global Industry Perspective, Comprehensive Analysis, Size, Share, Growth, Segment, Trends and Forecast, 2015–2021. Report available at https://www.zionmarketresearch.com/report/flavors-market. Accessed 25 Feb 2017.

Zuidam JN, Heinrich E (2010). Encapsulation of aroma. In: Zuidam NJ, Nedovic VA (eds). *Encapsulation Technologies for Food Active Ingredients and Food Processing.* Dordrecht: Springer; 127–160.

10

Nanoencapsulation of Flavors
Advantages and Challenges

Farhath Khanum, Syeda Juveriya Fathima, N. Ilaiyaraja, T. Anand, Mahantesh M. Patil, Dongzagin Singsit, and Gopal Kumar Sharma

Contents

10.1 Introduction 236
    10.1.1 What is flavor? 236
    10.1.2 Production of natural flavoring substances 236
        10.1.2.1 Extraction processes 237
        10.1.2.2 Distillation process 237
        10.1.2.3 Solvent extraction 238
        10.1.2.4 Supercritical fluid extraction 238
        10.1.2.5 Microwave extraction 238
        10.1.2.6 Pressurized liquid extraction 239
    10.1.3 Biotechnological production processes 239
    10.1.4 Flavor compounds 239
10.2 Perception of flavor 240
10.3 Flavor encapsulation 242
    10.3.1 Flavor encapsulation challenges 242
    10.3.2 Applications of encapsulation 242
10.4 Methods of nanoencapsulation of flavors 244
    10.4.1 Physical/mechanical methods 245
        10.4.1.1 Spray drying 245
        10.4.1.2 Spray cooling/spray chilling or spray congealing 246
        10.4.1.3 Fluid bed spray coating 248
        10.4.1.4 Extrusion 248
        10.4.1.5 Freeze drying 250
        10.4.1.6 Co-crystallization 251
    10.4.2 Physico-chemical methods 251
        10.4.2.1 Coacervation 251
        10.4.2.2 Liposome entrapment 252
    10.4.3 Chemical methods 256
        10.4.3.1 Molecular inclusion 256
        10.4.3.2 Interfacial polymerization 258

10.5	Release mechanisms		258
	10.5.1	Release rates	260
	10.5.2	Controlled flavor release	260
		10.5.2.1 Flavor release by diffusion	262
		10.5.2.2 Flavor release by degradation	262
		10.5.2.3 Flavor release by swelling	262
		10.5.2.4 Flavor release by melting	263
		10.5.2.5 Stability of encapsulated flavors	263
		10.5.2.6 Characterization of nanocapsules	263
10.6	Conclusion		264
References			265

10.1 Introduction

Encapsulation is a method whereby substances in liquid, gas, or solid form are packed in micro-/nanocapsules to protect substance (the core) within a sealed capsule. Barrett K. Green of the National Cash Register Company discovered the encapsulation process accidentally in the 1950s while attempting to create a carbonless copy paper that would provide multiple copies. The pharmaceutical industry later adopted the technology and improved the methodology further to safeguard medications release and target delivery. These improvements enabled selection of the specific location where the medication needs to be applied and facilitated timed or all-at-once release in response to manufacturer demands and requirements. Following this trend, the food industry emphasized microencapsulation of flavors/aromas to prevent flavor loss during processing. Intense research is underway to evaluate new materials and methods of encapsulation to avoid degradation that results in loss of food quality and appeal. Research has reduced the capsule size to "nano" to improve its efficiency and quantitative requirements. Encapsulation also facilitates aggregation of incompatible substances in a formulation.

10.1.1 What is flavor?

Flavor is one of the most valuable ingredients in any food product. Flavor plays an important role in consumer satisfaction. Flavor influences further consumption of food. Flavor is experienced through taste, smell, and feel. Flavor develops naturally in foods such as fruits, vegetables, meats, and grains throughout their growth and ripening and during cooking and fermentation. The extreme heating conditions used in food processing to eliminate microorganisms volatilize these flavor components, leaving behind a bland product. Storage of foods for longer duration also leads to loss of aroma/flavor (see Figure 10.1).

10.1.2 Production of natural flavoring substances

The flavoring substances present naturally in plant and animal sources must be extracted through distillation processes that separate specific substances from a

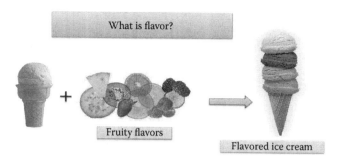

Figure 10.1 Flavor.

Table 10.1 Types of Flavor

Flavor	Description
Natural flavoring substances	Substances that are extracted from natural sources such as vegetable or animal materials and are not further chemically modified or changed
Nature-identical flavoring substances	Substances that are chemically similar to natural substances but that are obtained by chemical synthesis or by chemical modification of other natural substances
Artificial flavoring substances	Substances obtained by chemical synthesis or chemical modification of natural substances, but which are not present in natural products
Thermal process flavoring	Flavoring that is industrially produced by the controlled heating of several components (e.g., Maillard reaction); during the process, intense flavors develop; the basic materials for this reaction are amino acids and reducing sugars
Smoke flavoring	Fresh smoke is obtained by controlled burning of hardwood, which happens mostly in the absence of air; the smoke generated is condensed and mixed with solvents such as cooking oil or carriers such as table salt
Other flavoring preparations	This group includes essential oils, such as clove and eucalyptus oil; the oils are obtained from plants such as spices, herbs, fruit or blossoms and represent the odoriferous and/or tasty "essence" of the plant

natural mixture. Based on the flavoring capacity these substances are isolated, identified, and encapsulated either in natural or purified form (see Table 10.1).

10.1.2.1 Extraction processes

Extraction processes used include steam distillation, solvent extraction, supercritical fluid extraction, microwave extraction, and pressurized liquid extractions (Dima et al., 2014; Mhemdi et al., 2011; IOFI 2012; Zhanga et al., 2011).

10.1.2.2 Distillation process

During distillation, mixtures of liquids are separated by heating to different preset temperatures depending upon the boiling points of the liquids present. The steam is collected and cooled. During distillation in the presence of water (hydrodistillation), the essential oil components form an azeotropic mixture (i.e., mixtures of two different liquids) with water. Most of the essential oils do not mix well with water in the liquid phase so after condensation, they form two layers and are separated by decantation. The time required for the distillation varies from 30 to 60 minutes or longer depending on the quantity of the material. Duration influences not only the yield but also the composition of the extract. There are two primary methods of distillation: Clevenger distillation, where the material to be extracted is immersed in water, which is then boiled, and steam distillation, where steam is passed through a bed of the material to be extracted. In both methods, the vaporized volatile components are carried by the steam to a condenser. Upon condensation, oil- and water-rich layers are formed, which are separated by decantation or pipetting out the oil layer. During both types of distillation, the sample is exposed to temperatures close to 100°C, which can lead to changes in hermolabile components. Prolonged heating in water can lead to hydrolysis of esters, polymerization of aldehydes, or decomposition of other components. These techniques are used with plants whose essential oils are difficult to extract. Due to the high temperatures used, the danger of decomposition is high.

10.1.2.3 Solvent extraction

Solvent extraction is used to extract essential oils/flavors/aromas from plants that cannot withstand the high heat used in steam distillation. Very delicate fragrances such as jasmine and linden blossom cannot survive the process of distillation because the essential oils are unusable and unstable due to the heat. To extract their magical aromas, the process of solvent extraction is used. Solvents used for extraction include alcohol, hexane, ethanol, ether, methanol, and petroleum ether.

10.1.2.4 Supercritical fluid extraction

Supercritical fluids above their critical point exhibit liquid-like (solvent power, negligible surface tension) as well as gas-like (transport) properties. These properties are used in the supercritical fluid extraction of flavors/essential oils. A number of solvents such as CO_2, propane, butane, and ethylene are used for this purpose. Supercritical fluid extraction helps in the processing of plant material at low temperatures, reducing thermal degradation, and it avoids the use of toxic solvents.

10.1.2.5 Microwave extraction

In solvent-free microwave extraction (SFME) source material is exposed to microwave radiation via a microwave applicator. Microwaves increase the vibrational energy in water molecules. This continued increase in energy causes the water to boil and change phase. SFME been shown to remove up to 65% of available oils and take anywhere from 1.5 to 45 min. In comparison to steam distillation SFME can yield up to three times the amount of oil. The shorter processing time reduces the cost of energy. One main difference between SFME and steam distillation is the source of the steam for the extraction. In SFME the steam is generated from water remaining in the leaves of the plants. This causes greater disruption of the surface of the leaves and is one likely reason more oil is extracted.

10.1.2.6 Pressurized liquid extraction

Pressurized liquid extraction (PLE) uses elevated temperatures and pressures to drastically improve the speed of the extraction process. Increasing the temperature increases the diffusion rates of the extractives, the solubility of the extractives and their mass transfer, and decreases the viscosity and surface tension of the solvents. These changes improve the contact of the extractives/analytes with the solvent and enhance the extraction efficiency. The principle of PLE is simple. The sample is placed in the extraction cell and extracted with a solvent at a temperature ranging from room temperature to 200°C and at a relatively high pressure (from 4 to 20 MPa). The selected solvent is pumped to fill the extraction cell containing the sample, which is kept for a specified time at the selected pressure and temperature. Then the extracted solvent is transferred to a collection vial. The sample and all the connective tubings are then rinsed with a known amount of solvent. Recently an additional nitrogen purge has been included to guarantee the complete removal of the solvent from the PLE system. These steps can be repeated several times if necessary. The total extraction time is typically 15–45 min. PLE has significant advantages over other techniques. The matrix components that are not dissolved in the extraction solvent may be retained inside the sample extraction cell, thereby reducing the requirement of an additional filtration step.

10.1.3 Biotechnological production processes

The basis of biotechnological production processes is that very closely matching flavoring substances can be produced by microorganisms. The bacteria used are acetic acid bacteria or enzymes acting as biocatalysts. The flavoring industry uses specific microorganisms other than fungi and fungi suitable for food. Use of isolated and purified enzymes in place of microorganisms is increasing. After production, the flavoring substances are isolated by extraction or distillation. This is a cost-effective technique of obtaining flavors in bulk (Krings and Berger, 1998; Kim, 2005; Gupta et al., 2015).

10.1.4 Flavor compounds

Flavor and aroma are inextricably tied together, creating the ability to detect and distinguish specific aroma and flavor components, which is crucial for sensory analysis. Aroma components, also known as odorant, fragrance, or flavor components, are basically chemical compounds that have smell, fragrance, or odor. For a chemical compound to have a smell or odor it should be sufficiently volatile to be transported to the olfactory system in the upper part of the nose. Generally, molecules falling into this category have molecular weights of <300–400 Da. Flavors involve both taste and smell, whereas fragrances involve only smell. Flavors tend to be naturally occurring, fragrances tend to be synthetic. The aroma chemicals play a significant role in the production of flavorings, which the food industry strives to improve, and which increase the appeal of their products. Arguably, then, these aroma compounds could be considered the most important constituents of food.

Based on structure, the aroma compounds can be classified as esters, linear terpenes, cyclic terpenes, aromatics, and amines. Alcohols, aldehydes, ketones, lactones, acids, neutral compounds and thiols serve as aroma compounds (Naknean and Meenune, 2010; Fisher and Scott, 1997; Wu et al., 2016).

10.2 Perception of flavor

The perception of flavor is the most multisensory experience of daily life (Spence, 2015). Flavor perception is one of the most complex processes of human behavior (Dalton et al., 2000). It encompasses almost all of the senses, particularly the sense of smell, which is involved through odor images generated in the olfactory pathway. In the human brain, the perceptual systems are closely linked to systems for learning, memory, emotion, and language. Thus distributed neural mechanisms contribute to food preference and food cravings (Shepherd, 2006).

It is important to distinguish between orthonasal smell when we sniff, which tells us about the aroma of food or the bouquet of the wine, and the retronasal smell when air is pulsed out from the back of the nose as we swallow. While the distinction between these two senses of smell has been recognized for more than a century (Shepherd, 2012), only recently have researchers been able to provide empirical support for the claim that different neural substrates may actually be involved in processing these two kinds of olfactory information (Small and Prescott, 2005). It is the retronasal aromas that are combined with gustatory cues to give rise to flavors. Trigeminal input also contributes to flavor perception (Spence, 2015).

Tastes and smells are the perceptions of chemicals in the air or in our food. Separate senses with their own receptor organs, taste and smell are nonetheless intimately entwined. Flavor is defined as a perception that includes gustatory, oral-somatosensory, and retronasal olfactory signals that arise from the mouth as foods and beverages are consumed. During eating, foods are subjected to two main oral processes. Chewing, including biting and crushing with the teeth, and progressive impregnation by saliva result in the formation of a cohesive bolus and swallowing of the bolus. During this complex mouth process, flavor compounds are progressively released from the food matrix. The core sensory signals combine in the central nervous system of humans. It is proposed that oral-somato sensory and olfactory inputs are first integrated in the anterior ventral insula. The core flavor perception is then conveyed to upstream regions in the brainstem and thalamus as well as downstream regions in the amygdala, orbitofrontal cortex, and anterior cingulate cortex to produce the rich flavorful experiences that guide our feeding behavior (Salles et al., 2011).

Airborne odor molecules, called odorants, are detected by specialized sensory neurons located in a small patch of mucous membrane lining the roof of the nose. Axons of these sensory cells pass through perforations in the overlying bone and enter two elongated olfactory bulbs lying against the underside of the frontal lobe of the brain. Odorants stimulate receptor proteins found on hair-like cilia at the tips of the sensory cells, a process that initiates a neural response. An odorant acts on more than one receptor but does so to varying degrees. Similarly, a single receptor interacts with more than one different odorant, also to varying degrees. Therefore, each odorant has its own pattern of activity, which is set up in the sensory neurons. This pattern of activity is then sent to the olfactory bulb, where other neurons are activated to form a spatial map of the odor. Neural activity created by this stimulation passes to the primary olfactory cortex at the back of the underside, or orbital, part of the frontal lobe. Olfactory information then passes to adjacent parts of the orbital cortex, where the combination of odor and taste information helps create the perception of flavor (see Figure 10.2). Psychophysical, neuro-imaging, and neuro-physiological studies on cross-modal sensory interactions involved in flavor perception have started to provide an understanding of the integrated activity of sensory systems that generate such unitary perceptions, and hence the mechanisms by which these signals are functionally united when anatomically separated. The flavor perception depends upon neural processes occurring in chemosensory regions of the brain, including the anterior insula, frontal operculum, orbito frontal cortex, and anterior cingulate cortex, as well as upon the interaction of this chemosensory "flavor network" with other heteromodal regions, including the posterior parietal cortex and possibly the ventral lateral prefrontal cortex.

The olfactory cells lining the nose bind a variety of molecules responsible for aroma and relay electrical signals to a specialized area of the brain called the

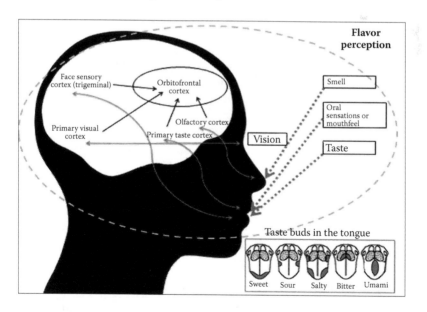

Figure 10.2 *Perception of flavor. (From Redondo, N. et al., Food & Function, 5, 1686–1694, 2014.)*

olfactory bulb, which in turn shuttles messages to the brain's smell recognition center. Detection of aromas is highly specific; each aroma molecule activates a specific receptor on a particular membrane cell of the nasal mucosa. When an aroma molecule binds to a receptor, it sets off a sequence of events involving special signal proteins, called G proteins, which controls the opening or closing of channels in the cell membrane. Buck and Axel (1991) showed the receptor-bearing cells send projections directly to the olfactory bulb. This helps explain why we are so immediately sensitive to odors, and why separating taste from odor in food is next to impossible. The sensory lexicons developed by Drake and Civille (2003) not only provide a standardized way of describing flavors but also link these flavors to chemical compounds.

10.3 Flavor encapsulation

The sensory characteristics of food, in particular the taste and flavor, as well as the fascinating aroma of freshly prepared delicacies, have a very specific effect on the consumer's choice of food. Retention of these mostly volatile compounds has been a challenge to the food industry.

Encapsulation of flavor consists of protecting a flavor compound or a mixture of compounds with an exclusive envelope. The main purpose of encapsulation is to limit the degradation or loss of flavor during processing and storage of foods.

10.3.1 Flavor encapsulation challenges

Some of the challenges to encapsulating flavors are (Campanile, 2007; Rowe and Winkel, 2009; Kwak, 2014):

- Flavors are mainly available as liquids.
- Flavors are usually a mixture of many components with different physico-chemical properties; they are usually relatively hydrophobic and relatively small and volatile.
- The sensory threshold may vary dramatically.
- Flavors may be unstable in certain conditions or may interact with the food matrix.
- Flavors generally have single-release kinetics.
- Stabilization in water-continuous systems is problematic.
- Controlled release must be maintained.
- Prevention of off-flavor formation in shelf life must be prevented.
- Loss of flavor must be avoided.

10.3.2 Applications of encapsulation

Applications of encapsulation include the following:

- Reduce the reactivity of the core material with the outside environment, for example oxygen and water.

- Decrease the rate of evaporation or transfer of the core material to the outside environment.
- Regulate the release of the core content so as to achieve the proper delay in release until the right stimulus is sensed.
- Mask the taste of the core material wherever desired.
- Regulate the quantity of the core material to be released when it is only used in very small amounts, but at the same time achieve uniform dispersion in the host material.
- Create a distinct barrier between two or more incompatible substances.
- Provide value additions to the product such as increased shelf life, better performance, and/or increased sensory appeal of the product.
- Increase stability of the product at high temperatures.
- Help achieve cost effectiveness, improve quality, and add value to the products and processes.
- Make the handling of the core substances easier by
  - Preventing lump formations,
  - Spreading the core material more uniformly,
  - Converting a liquid to a solid form, and
  - Enabling easy mixing of the core material (see Figure 10.3).

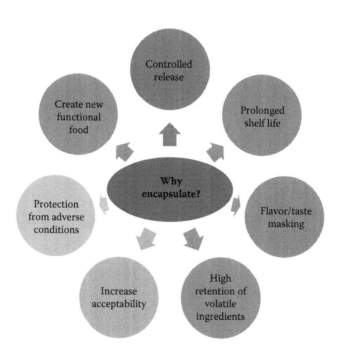

Figure 10.3 Uses of encapsulation.

10.4 Methods of nanoencapsulation of flavors

Fresh food flavor is key to consumer acceptability and product recognition. A significantly large part of the current literature on the encapsulation of essential oils, a major natural source of flavor, deals with micrometric-size capsules, which are used for the protection of the active compounds against environmental factors (e.g., oxygen, light, moisture, and pH), to decrease oil volatility, and to transform the oil into a powder (Ishwarya et al., 2015; Hernandez-Sanchez and Gutierrez-Lopez, 2015; Rai et al., 2015; Kwak, 2014; Gibbs et al., 1999). Encapsulation in nanometric particles is an alternative for overcoming these problems that, in addition, due to the subcellular size, may increase the cellular absorption mechanisms and bio-efficacy (see Figure 10.4).

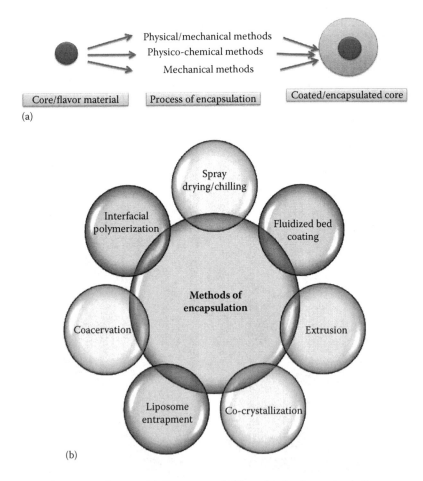

Figure 10.4 Encapsulation. (a) Process and (b) methods of encapsulation.

10.4.1 Physical/mechanical methods

10.4.1.1 Spray drying

Spray drying is a process that involves the atomization of a liquid into a spray of fine droplets, carried to further contact with a hot drying gas to evaporate the moisture to form the solid product, which is finally recovered usually via a cyclone unit. Spray-drying technology has undergone constant evolution in the past years. The synthesis of polymeric nano-sized particles obtained in a one-step procedure by spray drying a polymer solution has emerged recently.

Spray drying is the commercial process that is widely used in large-scale production of encapsulated flavors and volatiles (Jafari et al., 2006; Turchiuli et al., 2013). Teixeira et al. (2004) have shown that retention of aroma/flavor compounds is high in spray-drying technique. Many heat-labile (low-boiling-point) materials can be encapsulated by spray drying because the core material does not reach high temperatures (Sharma and Tiwari, 2001). In this process the substance to be encapsulated is dispersed in a carrier material, followed by atomization and spraying of the mixture into a hot chamber (Watanabe et al., 2002), where encapsulation occurs. The capsules are then transported to a cyclone separator from where they are recovered. The chemical and physical properties of the wall and core materials, solid content of the dryer, processing temperature, and the nature and performance of the encapsulating support, i.e., emulsion-stabilizing capabilities, film-forming ability, and low viscosity at a high concentration determine the retention of volatile core material during encapsulation by spray drying (Rosenberg et al., 1990; Goubet et al., 1998). Optimal wall materials for spray drying have high water solubility, low viscosity at high concentration, effective emulsifying properties, film-forming abilities, and efficient drying properties (Reineccius, 1988; Re-MI, 1998). Microstructures of spray-dried particles/capsules have been shown to be affected by different parameters such as wall composition and properties, flavor-to-wall ratio, atomization and drying parameters, uneven shrinkage at the early stages of drying, the effect of a surface tension-driven viscous flow, and storage conditions. The disadvantage of spray drying is that the low-boiling point aromatics can be lost during spray drying and the core material may also be on the surface of the capsule. This leads to oxidation and possible change and/or loss in the flavor of the encapsulated product (Desobry et al., 1997). The color of β-carotene was found to decrease with increase in temperature (Desobry et al., 1997). In another study, where rosemary essential oils were encapsulated by spray drying, moisture content, hygroscopicity, and wettability have been shown to be significantly affected by the wall material concentration, inlet air temperature, and feed flow rate. Bulk density was positively influenced by the wall material concentration and negatively influenced by the inlet air temperature. Particle density was shown to be influenced by the wall material concentration and the inlet air temperature in a negative manner (Fernandes et al., 2013).

Table 10.2 Advantages and Disadvantages of Spray Drying

Advantages	Disadvantages
Economical	Uniformity of capsules is low
Good quality of capsules in good yield	Limitation in the choice of wall material
Good solubility of the capsules	Low viscosity at relatively high concentration
Small size	Produces very fine powder that needs further processing
Good stability capsules	Less suitable for heat-sensitive material

Table 10.2 outlines advantages and disadvantages of the spray-drying technique.

Li Xiang et al. (2010) described preparation of different types of polymeric nanoparticles using polymers such as arabic gum, whey protein, polyvinyl alcohol, modified starch, and maltodextrin based on a nano spray dryer. Compared to the traditional spray dryer, the nano spray dryer is characterized by a vibration mesh spray technology producing tiny droplets in a range of a smaller order of magnitude than the traditional devices. The production of droplets depends on a piezoelectric actuator speed-up at an ultrasonic frequency (60 kHz) controlling vibration of a slim perforated membrane with micron-sized holes that may vary from 4 to 7 μm in diameter. The final size and standard deviation of the nanoparticles obtained depend on several parameters such as the nature and concentration of the polymer, the spray mesh size, the operating conditions (drying temperature, feed rate, drying gas flow rate), or the concentration of surfactant, if present in the formulation.

10.4.1.2 Spray cooling/spray chilling or spray congealing

Spray cooling and spray chilling are comparatively less-expensive encapsulation techniques and are therefore routinely used for encapsulation of flavor compounds to improve heat stability, delay release in wet environments, and/or convert liquid flavor into free-flowing powders (Gouin, 2004; Sillick and Gregson, 2012). Technically, these technologies are similar to spray drying where the flavor compounds or the core material is dispersed in a liquefied wall material and atomized. There is generally no water to be evaporated. Emulsification of the flavor compounds into molten wall materials is followed by atomization to disperse droplets from the feedstock. Once the droplets are mixed with a cooling medium they are solidified into powder form (Risch, 1995).

10.4.1.2.1 Spray chilling In this technique the wall material is melted and atomized through a pneumatic nozzle into a vessel that typically contains a carbon dioxide ice bath temperature (50°C) such as a holt-melt fluidized bed (Augustin et al., 2001). Spray chilling is a lipid-based system that involves the addition of the compound of interest to a liquefied lipid carrier, and the resulting mixture is fed through an atomizer nozzle. When the nebulized material is put into contact with the environment, which is cooled below the melting point of the matrix material, the vehicle solidifies (due to interaction of the molten material and cold air), and solid lipid particles are formed at the same time. Since this technology is lipid based, carriers, such as wax, and oils, such as palm oil, beeswax, cocoa butter, and kernel oil, can also be used. In addition to allowing the controlled release of these ingredients, this encapsulation technique can potentially change the functionality, reduce the hygroscopicity, mask taste or odor, change solubility, and provide physical protection. Spray-chilled flavors are finding applications in functional foods, baked food products, soup mixes, and foods containing a high level of fat.

10.4.1.2.2 Spray cooling Spray cooling, also called spray congealing, is similar to spray chilling. The only difference is the temperature of the reactor in which the coating material is sprayed. The flavor compounds are mixed with the molten matrix material and spray cooled. Vegetable oil can be used. The optimum melting point is 45–122°C (Risch, 1995). The disadvantage of spray chilling and spray cooling is the special care needed in handling and storage conditions (Taylor, 1983).

10.4.1.2.2.1 Advantages The use of spray chilling has increased due to the numerous advantages associated with the technological process, such as speed, performance, and relatively low cost. Once this technique does not require the use of water or organic solvents, the elimination of residual solvents will not be required. Moreover, it is a fast, safe and reproducible physical process because it is associated with an easy adjustment of particle size (Albertini et al., 2004). In the last two decades, spray chilling has been considered an environmentally friendly technique compared to other procedures such as spray drying. Spray chilling also reduces energy use and time of operation (Passerini et al., 2010). Another positive aspect is the ease of scaling up production because this technique can be operated continuously with the elimination of some manufacturing steps (Emas and Nyqvist, 2000). Another advantage of spray chilling is the use of low temperature in the process; heat is not required. Moreover, there is usually controlled release of the solid lipid microparticle (SLM) contents around the melting point of the carrier and by digestion of the carrier in the intestine. The SLMs are almost perfect spheres, which are structures that give free-flowing powders.

10.4.1.2.2.2 Disadvantages Although SLMs offer some advantages, there are technological disadvantages, including low encapsulation capacity of the active material; possibility of expulsion of the active ingredient by the matrix during the particle shelf life; and the degradation of lipid carrier applied for the dispersion development, which may affect the active ingredient material, thereby influencing its stability and/or its release profile (Turton and Cheng, 2007). Another limitation to be considered is the proper choice of the material to be encapsulated. The encapsulated material should be stable at the carrier melting temperature. For many thermolabile or unstable food ingredients, the degradation temperature is recurrently low, thereby reducing the possibility of using a suitable carrier. The use of molten mixtures should also be considered, which requires operational attention to avoid agglomeration and solidification of such materials in the production line, thus impairing the efficiency of the atomization process. The method has a drawback related to fast cooling rates that sometimes crystallize the lipid matrix into a polymorphic form (an unstable arrangement), which leads to the development of disordered chains and/or undesirable orientation, thus promoting lower barrier properties due to the tendency of more stable arrangements, resulting in the release of the active ingredient (Eldem et al., 1991; Emas and Nyqvist, 2000). Other possible disadvantages include the effect of particles on food texture, which depends on the particle size, as well as the possibility of particles to float in liquid systems. In addition, SLMs prepared by spray chilling are insoluble in water due to the lipid carrier, which can limit or not allow some applications. Moreover, the matrix structure of the particles obtained by spray chilling leads to the dispersion of the part of the active material that is not protected on the surface of the particles.

10.4.1.3 Fluid bed spray coating

Fluid bed spray coating processes can be grouped into three steps. First, the flavor compounds or any particle to be coated are fluidized in the hot atmosphere of the coating chamber. Then the coating material is sprayed through a nozzle onto the particles in the form of a film. Finally, there is a cycle of wetting and drying stages. The small droplets of the sprayed liquid spread onto the particle surface and coalesce. The solvent or the mixtures is then evaporated by the hot air and the coating material adheres to the particles (Jacquot and Pernetti, 2003). In short, this technique is based on a nozzle spraying of coating material into a fluidized bed of aroma particles in a hot milieu (Panda et al., 2001).

The fluid bed process is widely employed in the pharmaceutical and cosmetic industries; its use in the food industry to encapsulate flavors has also been studied (Turchiuli et al., 2013; Dezarn, 1998; Lee and Krochta, 2002; Chua and Chou, 2003). This seems to be the most suitable method for encapsulating flavors because the wall materials used in flavor systems, which are readily dissolved and form strong inter-particle bridges on redrying (Buffo et al., 2002). This technology also allows specific particle size distribution and low porosities to be designed into the product (Uhlemann and Morl, 2000).

Other advantages of the fluidized bed method (Mujumdar and Devahastin, 2000) include high drying rates due to good gas–particle contact, which leads to optimal heat and mass transfer rates; a smaller flow area; lower capital and maintenance costs; and ease of handling and control.

10.4.1.4 Extrusion

Encapsulation of flavors via extrusion has been used for volatile and unstable flavors in glassy carbohydrate matrices (Manojlovic et al., 2008; Reineccius, 1991; Gunning et al., 1999; Zeller et al., 1999; Benczedi and Bouquerand, 2001).

The main advantage of extrusion technology is the protection of flavors from oxidation. Carbohydrate matrices in the glassy state have very good barrier properties and extrusion is a convenient process enabling the encapsulation of flavors in such matrices (Gouin, 2004). However, diffusion of flavor from extruded carbohydrates is enhanced by physical/structural defects such as cracks, very thin walls, or pores formed during or after processing (Villota and Hawkes, 1994). Extrusion of polymer solutions through nozzles to produce beads or capsules has been used thus far on a laboratory scale (Heinzen, 2002). During processing, high temperature, high pressure, and shear result in extensive physicochemical changes to the ingredient materials. These harsh conditions can cause a considerable loss of pre-added flavor because of thermal degradation, vaporization, oxidation, or flashing-off at the die (Yuliani et al., 2006). A higher retention (up to 100%) of flavor compounds (cinnamaldehyde, eugenol, nonanoic acid and 3-octanone) complexed with β-cyclodextrin in corn starch extrusion was reported by Kollengode and Hanna (1997) at an extrusion temperature of 100°C. A recent study demonstrated a very high retention of d-limonene encapsulated in β-cyclodextrin (average of 92.2%) at extrusion temperatures up to 167°C (Yuliani et al., 2005). Another technique that could be used to protect flavor during extrusion is a novel protein precipitation technique based on sodium caseinate. However, the presence of added protein in the feed material can influence the rheological behavior of the melt and the product properties (Fernandez-Gutierrez et al., 2004).

10.4.1.4.1 Simple extrusion The volatile compound to be extruded is dispersed in a matrix polymer at 110°C. This mixture of polymer and the volatile compounds is then forced through a die. The filaments obtained are dropped quickly into a desiccant liquid, which hardens the extruded mass and traps the active substances (Rizvi et al., 1995; Crouzet, 1998). Isopropyl alcohol is the most common liquid used for the dehydration and hardening process. The strands or filaments of hardened material are shredded into small pieces, separated, and dried (Risch, 1995).

10.4.1.4.2 Double-capillary extrusion devices

10.4.1.4.2.1 Coaxial double capillary device In a coaxial double capillary device, the flavor compounds/core substance and the carrier material are fed through the inner and outer openings of a coaxial double capillary. The core is usually a liquid and the polymer can be applied as a solution or as a melt (the core and the coat fluids must be immiscible). At the tip of the coaxial nozzle the two fluids form a unified jet flow, which breaks up to form the corresponding droplets.

10.4.1.4.2.2 Centrifugal extrusion device A centrifugal extrusion device has a central rotating cylinder with nozzles located on its outer circumference. The liquid flavor is passed through the inner orifice and the liquid shell material through the outer orifice forming a coextruded rod of flavor components surrounded by wall material. As the cylinder rotates, the extruded rods break into droplets, which form capsules (Schalmeus, 1995).

10.4.1.4.2.3 Recycling centrifugal extrusion In recycling centrifugal extrusion the technology of rotating cylinder extrusion is combined with a system for recycling of the excess coating fluid. The flavor or core material is dispersed in the carrier material. The suspension is extruded through the rotating cylinder in such a way that the excess coating fluid is atomized and separated from the coated particles. Excess coating fluid is then collected and recycled, while the resulting capsules are hardened by cooling or solvent extraction. Optimizing the heating temperature, pressure, level of the emulsifier, residence time, and extrusion vessel, the production of encapsulated flavoring with a high flavor load can be achieved. But, as mentioned previously, the diffusion of flavors out of extruded carbohydrates is enhanced by structural defects such as cracks, thin walls, or pores formed during or after processing (Miller and Mutka, 1986; Wampler, 1992).

Extrusion provides flavors outstanding protection from oxidation, offers exceptional shelf life, and produces visible pieces of flavoring; however, extrusion is commercially more expensive than spray drying. The flavor load is 8–12% compared to 20% in spray drying. Another disadvantage is that extrusion is a high temperature process; flavoring must be able to withstand 110–1200°C for substantial time periods because this is a batch process that subjects the flavor to significant time periods at elevated temperatures (Reinecius, 2013).

10.4.1.5 Freeze drying

The freeze-drying technique, also known as lyophilization, is one of the most useful processes for drying heat-sensitive substances that are unstable in aqueous solutions. In this technique, once the water crystallizes, the nonfrozen solution is viscous and the diffusion of flavors is prevented (Chranioti and Tzia, 2014). Once the freeze drying starts, the surface of the solution becomes an amorphous solid in which selective diffusion is possible (Karel and Langer, 1988). A comparison of spray drying, tray drying, drum drying, and freeze drying to encapsulate cold-pressed orange oil with gum acacia and modified food starch was carried out by Buffo and Reineccius (2001). These authors concluded that freeze drying is the process that gives the most desirable properties to the end product. Minemoto et al. (1997) have compared oxidation of methyl linoleate when encapsulated with gum arabic by hot air drying and freeze drying and showed that freeze drying was better than hot air drying. It has been shown that the freeze-drying process maintained the shape of the microcapsules because of fixation by freezing (Nagata, 1996). However this freeze-drying technique has certain drawbacks such as high costs, expensive storage and transport of particles, and long processing time (Jacquot and Pernetti, 2003).

10.4.1.6 Co-crystallization

Co-crystallization is an economical and flexible alternative to various flavor encapsulation processes, as the procedure is relatively simple (Chen, 1994). Very few studies have been published that report on the use of co-crystallization to encapsulate the required compounds (Chen et al., 1988; Beristain and Vernon-Carter, 1994; Beristain et al., 1996). The technique has been used to encapsulate fruit juices, essential oils, flavors, and sugars (Chen et al., 1988).

In this process, the crystal structure of sucrose is modified from a perfect crystal to a conglomerate. This structure provides a porous configuration that can accept the addition of a second ingredient (Beristain et al., 1996).

For example, crystallization of supersaturated sucrose syrup has been achieved at high temperature (above 120°C), low moisture, and TSS (95–97°Brix), and flavor compounds can be added to the sucrose syrup at the time of spontaneous crystallization (Bhandari et al., 1998). The crystal structure of sucrose can be modified to form aggregates of very small crystals that incorporate the flavors either by inclusion within the crystals or by entrapment, which also enhance flavor stability (Mullin, 1972; Chen et al., 1988). The granular product has a low hygroscopicity, free-flow ability, and dispersion characteristics (LaBell, 1991; Quellet et al., 2001). But, during this process, the liquid flavor converts into dry granules, which leads to degradation of heat-sensitive compounds (Bhandari et al., 1998). Beristain et al. (1996) encapsulated orange peel oil using a co-crystallization process and showed that retention of volatile oil by the co-crystallization process was similar to spray-dried and extruded products. Although the product had a free-flowing property, oxidation of flavors could not be avoided, and the addition of strong antioxidant was necessary.

10.4.2 Physico-chemical methods

10.4.2.1 Coacervation

Coacervation, also known as phase separation, is an electrostatically driven liquid–liquid phase separation technique, resulting from the association of oppositely charged macro-ions. The term coacervate refers to spherical aggregates of colloidal droplets held together by hydrophobic forces. It is a phenomenon occurring in colloidal solutions. It is often regarded as the original method of encapsulation developed by the National Cash Register Co. in the 1950s (Risch, 1995; Green and Scheicher, 1955; Nakagawa et al., 2004). In simple terms coacervation consists of dissolving gelatin, a polymer. Examples of anionic biopolymers used are gelatin B, sodium alginate, arabic gum, pectin, xanthan gum, nucleic acids, or proteins. The most widely used polycations are gelatine A as a biopolymer, chitosane as a modified biopolymer, and polyethylene imin as a synthetic polymer in water, heated to solubilize it. The flavor compound or any desired core material is mixed in this gelatin solution. The rate of mixing and stirring can be altered to obtain the desired droplet size. Upon alteration of the thermodynamic conditions the mixture separates into two liquid phases: one rich in colloid, i.e., the coacervate, and the other not containing colloid (Korus, 2001). Generally, the core material used in the coacervation should be compatible with the polymer and be insoluble (or scarcely soluble) in the coacervation media. Tolstuguzov and Rivier (1997) described a process for encapsulating solid particles within a protein. In this process, the additive in a protein solution was mixed with a polysaccharide and maintained at a pH greater than the isoelectric point of the protein. A mixture with two phases was formed, one of which was the heavier phase containing the encapsulated material. Coacervation can be classified as simple or complex, based on the polymer used. In simple coacervation only one type of polymer, along with strongly hydrophilic agents, is used. In complex coacervation two or more types of polymers are used. The flavor material should be dissolved in the mixture as the coacervation proceeds and the coacervate nuclei are then adsorbed onto the surface of the volatile compounds. Flavor may also be added during or after phase separation. In all cases, the coacervation mixture must be continuously stirred. The addition of suitable droplet stabilizer may also be necessary to avoid coagulation of the resulting microcapsules (Arshady, 1999; King, 1995). This technology has not been very popular in the food industry because it is laborious and expensive. Optimization of wall material concentration in the emulsification and coacervation process is problematic because the concentration needed to obtain a fine emulsion may be different from what is needed to increase the yield of microcapsules. Other limitations of flavor encapsulation by coacervation include evaporation of volatiles, dissolution of active compound into the processing solvent, and oxidation of product (Flores et al., 1992). The complex coacervates are highly unstable and toxic chemical agents, such as glutaraldehyde, are required to stabilize them (Sanchez and Renard, 2002).

10.4.2.2 Liposome entrapment

Nanoliposomes are microscopic vesicles that consist of phospholipid bilayers entrapping one or more aqueous compartments. Their unique properties have triggered numerous applications in several scientific and technological fields. Nanoliposomes can provide controlled release of various bioactive agents, including food ingredients and nutraceuticals, at the right place and the right time. Therefore, they increase the effectiveness and cellular uptake of the encapsulated material. Reactive, sensitive, or volatile additives can be turned into stable ingredients using nanoliposomes. They can be prepared using completely natural ingredients or indigenous molecules found in our bodies; thus they are biocompatible and acceptable for human consumption. The permeability of liposomes and nanoliposomes can be influenced by their phospholipid content, their method of preparation, and their number of lamellae. Liposome size and morphology have a strong influence on their distribution within food systems. In addition, liposomes that will leak and release their contents at a particular temperature or pH have been designed, and they can be used very conveniently in food systems. The encapsulated material is released under the influence of a specific stimulus at a specified stage. For example, flavors and nutrients may be released upon consumption, whereas sweeteners that are susceptible to heat may be released toward the end of baking, thus preventing undesirable caramelization in the baked product (Dziezak, 1988; Desai and Park, 2005). Liposomes can trap both hydrophobic and hydrophilic compounds, prevent decomposition of the entrapped core, and release the entrapped core material at designated targets.

10.4.2.2.1 Liposome and nanoliposome manufacture All the methods of preparing the liposomes involve four basic stages (Akbarzadeh et al., 2013):

1. Drying out lipids from organic solvent.
2. Dispersing the lipid in an aqueous medium.
3. Separating and purifying the resultant liposome.
4. Analysis of the final product.

10.4.2.2.2 Methods of liposome preparation and core material loading Generally two methods are used for liposome preparation: passive loading techniques and active loading techniques.

Passive loading techniques can be classified into three different methods: mechanical dispersion method; solvent dispersion method; and detergent removal method (removal of non-encapsulated material).

10.4.2.2.2.1 Mechanical dispersion methods Types of mechanical dispersion methods include: sonication; French pressure cell; freeze-thaw liposomes; lipid film hydration by hand shaking, nonhand shaking, or freeze drying; microemulsification; membrane extrusion; and dried reconstituted vesicles (Riaz, 1996; Anwekar et al., 2011).

10.4.2.2.2.1.1 Sonication Sonication is the most extensively used technique for the preparation of small unilamellar vesicles (SUV). In this method, multilamellar vesicles (MLVs) are sonicated (bath or probe type) under a passive atmosphere. The main drawbacks of this method are very low internal volume and therefore encapsulation efficacy, possible degradation of phospholipids and compounds encapsulated, elimination of large molecules, metal pollution from probe tip, and the presence of MLV along with SUV (Riaz, 1996).

There are two types of sonication techniques. In probe sonication the tip of a sonicating probe is dipped directly into the liposome dispersion. In bath sonication, the liposome dispersion in a cylinder is placed in a bath sonicator. Temperature control of the lipid dispersion is easier in this method, in contrast to sonication by dispersal directly using the probe tip. The material to be sonicated can be kept in a sterile vessel, unlike the probe units (Kataria et al., 2011).

10.4.2.2.2.1.2 French pressure cell The French pressure cell technique is named after its inventor, Charles Stacy French (Akbarzadeh et al., 2013). the French pressure cell method involves the extrusion of MLV through a small opening. Here large vesicles are converted to small vesicles using high pressure. From this method one can get unilamellar or oligolamellar vesicles 30–80 nm in diameter, depending on the pressure applied. Moreover, liposomes obtained by this method are more stable compared to those from sonication (Mayer et al., 1986; Song et al., 2011; Zhang, 2011; Mozafari, 2005a). A limitation of the method is that the working volumes are comparatively small (about 50 ml as the maximum) (Anwekar et al., 2011).

10.4.2.2.2.1.3 Freeze-thawed liposomes SUVs are mixed with the material to be entrapped and rapidly frozen and thawed slowly. They are sonicated for 15–30 sec which disperses aggregated materials to LUV. This type of synthesis is strongly inhibited by increasing the phospholipid concentration and the ionic strength of the medium. Encapsulation efficacies of 20%–30% have been obtained.

10.4.2.2.2.1.4 Solvent dispersion methods

10.4.2.2.2.1.4.1 Ether injection (solvent vaporization) Lipids are dissolved in diethyl ether or ether-methanol. This mixture is slowly injected into an aqueous solution of the material to be encapsulated at 60°C–65°C or under reduced pressure followed by removal of ether under vacuum, which leads to the creation of liposomes. The main disadvantages of the technique are that the population is heterogeneous (70–200 nm) and the exposure of compounds to be encapsulated to organic solvents at high temperature (Schieren et al., 1978).

10.4.2.2.2.1.4.2 Ethanol injection A lipid is dissolved in ethanol and this solution is taken in a syringe and rapidly injected into a huge excess of aqueous phase or buffer. The SLVs are formed immediately. The disadvantages of the method are that the population of vesicles obtained is heterogeneous (30–110 nm), liposomes are very dilute, the removal of all ethyl alcohol is difficult because it forms an azeotrope with water, and the probability of the various biologically active macromolecules to denature and inactivate in the presence of even low amounts of ethanol is high.

10.4.2.2.2.1.4.3 Reverse phase evaporation method The liposomes prepared by the reverse phase evaporation method have a high aqueous space-to-lipid ratio and are capable of entrapping a large percentage of the aqueous material. Reverse-phase evaporation is based on the creation of inverted micelles. These inverted micelles are further sonicated in a mixture of a buffered aqueous phase, which contains the water-soluble molecules to be encapsulated into the liposomes and an organic phase in which the amphiphilic molecules are solubilized. The slow evaporation of the organic solvent leads to the conversion of these inverted micelles into viscous state and gel structures. Liposomes prepared by reverse phase evaporation technique can be made from numerous lipid formulations and have aqueous volume-to-lipid ratios up to four times higher than hand-shaken liposomes or multilamellar liposomes (Anwekar et al., 2011, Kataria et al., 2011).

10.4.2.2.2.1.4.4 Detergent removal method The detergents are used to solubilize lipids at their critical micelle concentrations (CMC). As the detergent is detached, the micelles become increasingly surface tension-free in phospholipids and finally combine to form LUVs. The detergents can be removed by dialysis (Daemen et al., 1995; Shaheen et al., 2006), detergent absorption using beaded organic polystyrene absorbers such as XAD-2 beads and Bio-beads SM2, gel-permeation chromatography using Sephadex G-50, Sephadex G-100, Sepharose 2B-6B, and Sephacryl S200-S1000, or by dilution.

10.4.2.2.2.1.4.5 Freeze-protectant for liposomes (Lyophilization) Chemical compounds extracted from natural material be it from plant or animal source usually are degraded because of oxidation and other chemical reactions before they are delivered to the target site. A majority of these products are lyophilized from simple aqueous solutions. Generally, water is the only solvent that must be detached from the solution using the freeze-drying/lyophilizing process, but there are still a number of examples where pharmaceutical products are manufactured via processes that require freeze-drying from organic co-solvent systems (Hemanthkumar and Spandana, 2011). Freeze-drying (lyophiliszation) involves the removal of water from products in the frozen state at tremendously low pressures, which is a sublimation process. The process is normally used to dry products that are thermolabile as they could be degraded by heat-drying (Mozafari and Mortazavi, 2005). During this type of drying, trehalose is being used as an excellent cryoprotectant (freeze-protectant) for liposomes.

Until recently, large-scale production of liposomes and nanoliposomes was limited by poor encapsulation efficiencies, lack of a continuous production process, and the use of organic solvents. These solvents not only affect the structure and stability of the entrapped substance but will also remain in the final encapsulation formulation, thus contributing to toxicity and influencing the stability of the liposomal system (Vemuri and Rhodes, 1995; Cortesi et al., 1999). Another issue in liposomal encapsulation for the food industry is the scaling up of the preparation method at acceptable levels and costs. These problems can be addressed by employing a new preparation method known as the heating method by which liposomes and nanoliposomes (in addition to some other carrier systems) can be prepared in one step in less than an hour using a single apparatus in the absence of potentially toxic solvents (Mozafari, 2005b; Mozafari et al., 2007; Mortazavi et al, 2007). This method is economical and capable of manufacturing bioactive carriers, including liposomes and nanoliposomes, with a superior monodispersity and storage stability using a simple protocol. Another important feature of the method is that it can be adapted from small to industrial scales. The heating method is obviously most suitable for production of carrier systems for different in vitro and in vivo applications and involves heating and stirring (less than 1000 rpm) the carrier ingredients, in the presence of a plyol, at 40–120°C, based on the properties of the ingredients, presence or absence of cholesterol, and type of material to be entrapped (Mozafari, 2005; Mozafari et al., 2007a,b; Mortazavi et al., 2007). Recently Mozafari and his team showed that nanoliposomes prepared by the heating method are completely nontoxic toward cultured cells while nanoliposomes prepared by a conventional method using volatile solvents showed significant levels of cytotoxicity (Mortazavi et al., 2007). A further improved version of the heating method, called the Mozafari method, has recently been employed for the encapsulation and targeted delivery of the food-grade antimicrobial nisin. Another method of liposome production without using toxic solvents is the microfluidization technique (Vemuri et al., 1990; Zheng et al., 1999) using a microfluidizer equipment. This equipment has traditionally been used in the pharmaceutical industry to make liposomal formulations (Vemuri et al., 1990) and pharmaceutical emulsions. More recently, Jafari et al. (2006) employed the microfluidizer to produce flavor emulsions for homogenized milk (Jafari et al., 2006). Microfluidization works on the principle of dividing a pressure stream into two parts, passing each part through a fine orifice, and directing the flows at each other inside the of microfluidizer chamber (Jafari et al., 2006). Within the interaction chamber, cavitation, along with shear and impact, reduces size of the particles.

10.4.2.2.2.1.4.6 Solid lipid nanoparticles Solid lipid nanoparticles (SLN) (Bilia et al., 2014, Fathi et al., 2012) are nanoscale-size particles prepared using lipids that remain solid at room temperature (or/and body temperature). The lipid component may consist of a broad range of lipid and lipid-like molecules such as triacylglycerols or waxes (Mehnert and Mäder, 2001). The diameter of such lipid particles can also be quite small, that is, in the range of 50nm–1 µm. Active ingredients can be solubilized homogeneously either in the core of the SLNs or in the outside. The advantage of SLNs as a delivery system for lipophilic active components is reported to lie in the immobilization of active elements by the solid particle structure leading to an increased chemical protection, less leakage, and sustained release (Weiss et al., 2008). This physical property allows a better control of both the physical (against recrystallization) and chemical (against degradation) stability of the delivered constituents.

10.4.3 Chemical methods

10.4.3.1 Molecular inclusion

Cyclodextrins are a group of cyclic oligosachharides. Cyclodextrins are enzymatically modified starch molecules, which can be made by the action of cyclodextrin glucosyl transferase upon starch. After cleavage of starch by the enzyme, the ends are joined to form a circular molecule of glucopyranose units with α (1–4) linkage. The characteristics of cyclodextrin and their use as encapsulating material have been extensively explained (Hedges and McBride, 1999). The commonly found cyclodextrins have 6, 7, or 8 glucopyranose units in the cycle and are called α-cyclodextrins, β-cyclodextrins, and γ-cyclodextrins, respectively. The pyranose units are so aligned as to form central cavities that are hydrophobic in nature (Figure 10.5). These cavities are the places where the guest molecules sit. The inclusion complexes are defined as the result of interactions between guest molecules and the surrounding lattice (Godshall, 1997). A typical application is the protection of unstable flavor chemicals (Uhlemann et al., 2002). In the food industry, flavors have been encapsulated within cyclodextrins (Reineccius and Risch, 1986; Loftsson and Kristmundsdottir, 1993; Reineccius et al., 2002). The inner hydrophobic cavity of β-cyclodextrin is toroid shaped, and its molecular dimensions allow total or partial inclusion of a wide range of aroma/flavor compounds. While the central cavity of the molecule creates a relatively hydrophobic environment, its external surface has a hydrophilic character. This supramolecular conformation is largely responsible for the characteristic physico-chemical properties of cyclodextrins (Steinbock et al., 2001). According to Goubet et al. (1998), the retention of aroma compounds can be influenced to a large

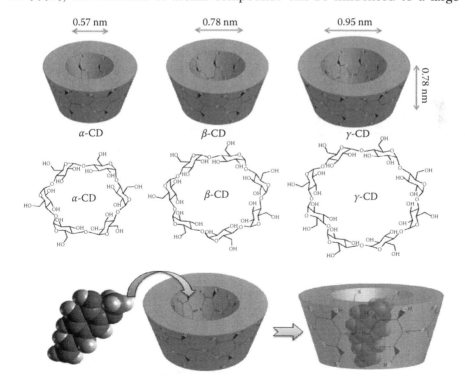

Figure 10.5 *Molecular inclusion complexes with β-cyclodextrin. Source: http://unam .bilkent.edu.tr/~uyar/Research.html.*

extent by the molecular weight of the compound, shape, steric hindrance, functionality, polarity, and volatility of the core material. Once the complex is formed, liberating guest molecules requires the presence of water or high temperature (Reineccius et al., 2002). Pagington (1986) and Bhandari et al. (1999) accounted for several methods for complexing β-cyclodextrin with flavor compounds. The most used are shaking a cyclodextrin with flavor compound in aqueous solution and filtering the precipitated complex; mixing solid cyclodextrin with guest molecules in a powerful blender, and bubbling the flavors, as vapors, through a solution of cyclodextrin; and making a paste of the flavor substance with the cyclo-dextrin-water.

Qi and Hedges (1995) provide a detailed experimental protocol of the coprecipitation method suitable for laboratory evaluation. These researchers claim that the paste method was easier for large-scale production, as less water would, subsequently, have to be removed during drying. The addition of maltodextrins and β-cyclodextrin as stabilizing or thickening agents could retain some aroma compounds in food matrices during thermal processes (cooking and pasteurization) (Jouquand et al., 2004). α,α-Trehalose(α-D-glucopyranosyl α-D-glucopyranoside), a nonreducing carbohydrate (a disaccharide), is also used to prepare inclusion complexes.

10.4.3.2 Interfacial polymerization

One of the advantages of these polymers is their very rapid polymerization—occurring within seconds—initiated by ions present in the medium water. An advantage of interfacial polymerization techniques is high efficiency of encapsulation (e.g., insulin up to 95%). In addition, the advantage of obtaining nanocapsules by this method is that the polymer is formed in situ, allowing the polymer membrane to follow the contours of the inner phase of an oil/water or water/oil emulsion. In this case, the main disadvantage is the use of organic solvents required for the external phase. Washing of solvents and replacement by water represents a time-consuming and difficult procedure. Methods of encapsulation and their processing steps are shown in Table 10.3.

10.5 Release mechanisms

Encapsulation enables controlled release of the active ingredient over a longer period of time. In response to specific stimuli, the particle releases its active ingredient in a specific environment. A possible release mechanism is through the increase of the temperature above the melting point of the carrier. The particle type, geometry, and composition define the release mechanism of the active ingredient. The processes that employ hydrophilic carriers generally trigger a more rapid release of the active ingredient in relation to those agents that have a lipid carrier (fats and waxes), thus retarding the release of

Table 10.3 Methods of Encapsulation and Their Process Steps

Method	Process Steps
Spray drying	1. Dissolve active ingredients in aqueous coating solution. 2. Atomize. 3. Dehydrate.
Fluid bed coating	1. Fluidize active ingredients powder. 2. Spray coating. 3. Dehydrate or cool.
Spray chilling/cooling	1. Dissolve active ingredients in heated lipid solution. 2. Atomize. 3. Cool.
Melt injection	1. Melt the coating. 2. Disperse or dissolve active ingredients in the coating. 3. Extrude through filter. 4. Cool and dehydrate.
Melt extrusion	1. Melt the coating. 2. Dissolve active ingredients in the coating. 3. Extrude. 4. Cool.
Emulsification	1. Disperse active ingredients and emulsifiers in water or oil phase. 2. Mix oil and water phases under shear.
Coacervation	1. Prepare o/w emulsions with lipophilic active ingredients in oil phase. 2. Mix under strong shaking. 3. Induce three immiscible phases. 4. Cool. 5. Crosslink.
Liposome entrapment	1. Disperse lipid molecules in water, with active ingredients in lipid or water phase. 2. Reduce size by high shear or extrusion. 3. Remove free active ingredients.
Encapsulation by rapid expansion of supercritical fluid (RES)	1. Create a dispersion of active ingredients and dissolved or swollen shell material in supercritical fluid. 2. Release the fluid to precipitate the shell onto the active ingredients.
Freeze drying	1. Disperse active ingredients and carrier material in water. 2. Freeze the sample. 3. Drying under low pressure 4. Grind (optional).
Preparation of microspheres via emulsification	1. Emulsify water with biopolymer in oil phase. 2. Add gelling agent under shear.
Preparation of microspheres via extrusion	1. Disperse active in alginate solution. 2. Drop into gelling bath.
Co-extrusion	1. Dissolve active ingredients in oil. 2. Prepare aqueous or fat coating. 3. Use concentric nozzle, and press simultaneously the oil phase through the inner nozzle and the water phase through the outer one. 4. Drop into gelling or cooling bath.

Source: With kind permission from Springer Science+Business Media: Encapsulation Technologies for Active Food Ingredients and Food Processing, 2010, Zuidam, N.J., Shimoni, E.

the active compound (Whorton, 1995) The release of active ingredients from particles occurs via erosion, leaching of the matrix, and pH of the saliva. Depending on the type and concentration, some surfactants can affect the dissolution rate of the matrix.

Different types of release mechanisms have been put forth for microcapsules. They are listed below.

1. A compressive force breaks open the capsule by mechanical means.
2. Shear force breaks open the capsule (the force as in Waring blender or a Z-blade type mixer).
3. The wall may be dissolved, leaving the core as such.
4. The wall may be melted, releasing the core to the environment such as that occurring during baking.
5. High temperature or high water level makes the core diffuse through the wall at a slow rate.

Exact studies on release mechanisms/kinetics for nanoencapsulated flavors are very scanty (Soottitantawata et al., 2005).

10.5.1 Release rates

The release rates that are obtainable from a single microcapsule are generally zero order, half order, or first-order order. Zero order occurs when the core is a pure material and releases through the wall of a reservoir microcapsule as a pure material. Half-order release occurs with matrix particles. First-order release occurs when the core material is actually a solution. As the solute material releases from the capsule the concentration of solute material in the solvent decreases and a first-order release is achieved. It should be noted that these types of release rates occur from a given single microcapsule. A mixture of microcapsules will include a distribution of capsules varying in size and wall thickness. The effect, therefore, is to produce a release rate different from the above-mentioned orders because of the ensemble of microcapsules. It is therefore essential to carefully study on an experimental basis the release rate from an ensemble of microcapsules and to recognize that the deviation from theory is due to the distribution in size and wall thickness (Shahidi and Han, 1993).

10.5.2 Controlled flavor release

Controlled release may be described as a method by which one or more active agents or ingredients are made available at a desired location/site, time, and at a definite rate (Pothakamury and Barbosa-Canovas, 1995). Release of flavor from complex matrices is an important aspect for the food industries as well as the perception (Guichard, 2000). Release of encapsulated volatile compounds depends on several interrelated processes such as diffusion of

Table 10.4 The Flavor-Release Profile

Encapsulation Technique	Controlled Release Mechanism
Simple coacervation	Prolonged release
Complex coacervation	Prolonged release (diffusion) and triggered release (pH, dehydration, effect mechanical, dissolution or enzymatic effect)
Spray drying	Prolonged release and triggered release
Fluid bed drying	Triggered release (pH or heat treatment)
Extrusion	Prolonged release

Source: Richard, J., Benoît, J.P.,. Microencapsulation. In: Techniques de l'ingenieur. J. 2210, 1–20. Paris, Techniques de l'ingenieur, 2000..

the volatile compound through the matrix, type and geometry of the particle, transfer from the matrix to the environment, and degradation and/or dissolution of the matrix material (Pothakamury and Barbosa-Canovas, 1995). De Roos (2000) showed that two factors control the rate of flavor release from products: the thermodynamic factor, i.e., the comparative volatility of the flavor compounds in the food matrix and air phases under equilibrium conditions, and the kinetic factor, i.e., the resistance to mass transport from product to air. The mechanism of release for the capsule may also be based on solvent effects, such as melting, diffusion, degradation, or particle fracture, as described later in this chapter (Table 10.4).

- Triggered release—Release occurs due to a change in the local environment, such as pH, temperature, moisture, pressure, and electromagnetic. This is used to achieve immediate, delayed, or pulsatile release profiles.
- Sustained release—Release occurs for an extended period of time (the zero order). This can be used to achieve constant active ingredient exposure for a fixed time period.
- Burst release
- Combination release profiles

The stability of the encapsulated flavor in the matrix is dependent on the type of food ingredient and the physico-chemical properties of the flavor compound. Retention will clearly bring a conspicuous decrease in flavor perception. Generally, flavor release decreases with increasing lipid level in the food matrix, with the exception of hydrophilic compounds (Guichard, 2002). The presence of salts increases the volatility of the flavor compound. Boland et al. (2004) investigated the release of 11 flavor compounds from gelatin, starch, and pectin gels. These authors demonstrated that flavor release was significantly affected by the texture of the gels. The gelatin gel showed enhanced flavor release in the presence of saliva, while the starch and pectin gels showed a reduction in flavor release under these conditions. Interaction between proteins and aroma compounds has been extensively studied (Lubbers et al., 1998), showing that covalent binding, hydrogen binding and hydrophobic interactions are all present and detectable.

10.5.2.1 Flavor release by diffusion

Solubility of a compound in the matrix that establishes a concentration in the matrix and the permeability of the compound through the matrix control the diffusion as per the concentration gradient. Diffusion is important in food as it is the predominant mechanism (in the absence of cracks, pinholes, or other defects) in controlled release from encapsulation matrices due to their huge surface area per unit weight and thin walls of the capsules that function as semipermeable membranes (Crank, 1975; Cussler, 1997). The vapor pressure of a volatile substance on each side of the matrix is the driving force influencing diffusion (Gibbs et al., 1999). There are three principal steps in the release of a flavor compound from the matrix system: diffusion of the compound to the surface of the matrix, partition of the volatile compound between the matrix and the surrounding food, and transport away from the matrix surface (Fan and Singh, 1989). It is obvious that if the food component is not soluble in the matrix, then it will not enter the matrix, and so diffusion cannot take place irrespective of the pore size of the matrix (Reineccius, 1995). Two distinct mechanisms of diffusion may apply. One mechanism is molecular or static diffusion, caused by the random movement of the molecules in the stagnant/sluggish fluid. The rate of molecular diffusion remains almost constant with flavor type. The second mechanism is eddy or convective diffusion, which transports particles of the fluid from one location to another, carrying with them the dissolved flavor compound. The rate of eddy diffusion is usually much higher than the rate of molecular diffusion and is independent of the flavor type (Roos, 2003).

10.5.2.2 Flavor release by degradation

Flavor can be released from the capsule by degradation/erosion of the capsule or the coating material. This happens, for example, in lipid or protein coating by the use of specific enzymes. Homogeneous and heterogeneous degradation are both detectable. Heterogeneous degradation occurs when degradation is confined to a thin layer at the surface of the delivery system, whereas homogenous degradation is a result of degradation occurring at a uniform rate throughout the polymer matrix (Pothakamury and Barbosa-Canovas, 1995).

10.5.2.3 Flavor release by swelling

When the capsules are placed in a thermodynamically compatible medium, the matrix polymer swells because of absorption of fluid from the medium. The aroma compounds in the swollen matrix then diffuse out (Fan and Singh, 1989).

The degree of swelling is controlled by water absorption or presence of solvents such as glycerin or propylene glycerol (Gibbs et al., 1999b).

10.5.2.4 Flavor release by melting

The coating material can be dissolved/solubilized using appropriate solvents or by employing thermal treatments. The coating melts away, releasing the core contents or the active ingredients in the environment. Generally the lipid coatings are melted by heat, water-soluble coatings are melted by increasing the moisture content, and water-insoluble coatings are melted by appropriate solvents.

10.5.2.5 Stability of encapsulated flavors

Shahidi and Han (1993) have shown 99–100% flavor retention over a period of 10 years under normal room temperature storage conditions in β-cyclodextrin inclusion complexes. Xiao et al. (2016) have shown that nanocapsules loaded with styrallyl acetate were stable over 8 months. Thermal stability analyzed by thermogravimetric analysis (TGA) showed that the styrallyl acetate nanocapsules exhibited excellent thermal stability compared to the pure oil. Ahmed et al. (2017) have reported that nanoencapsulation of rosemary essential oils increases the thermal stability of the essential oils.

10.5.2.6 Characterization of nanocapsules

Characterization of flavor capsules can be carried out by physical, chemical, and organoleptic methods.

10.5.2.6.1 Dynamic light scattering
Dynamic light scattering is caused by Brownian movement of particles. Movement/diffusion of small isometric particles in liquids is fast, causing faster fluctuations in the intensity of scattered light compared to big particles that diffuse more slowly. These intensity fluctuations are recorded.

10.5.2.6.2 Electron microscopy
The size distribution of the capsules is usually carried out by electron microscopy, generally by negative staining and freeze-fracture techniques. The surface morphology and size distribution are carried out by transmission electron microscopy (TEM), while the organization of the core material within the capsule is studied using scanning electron microscopy (SEM) in secondary electron imaging mode (SEI) (Rosenberg et al., 1984).

10.5.2.6.3 Electron spin resonance
Electron spin resonance (ESR) experiments facilitate the assessment of the nano environment, including the polarity and viscosity, of the oily phase in nanocapsules.

10.5.2.6.4 Nuclear magnetic resonance Nuclear magnetic resonance (NMR) relaxation techniques are a means of obtaining rapid, sensitive, and nondestructed information about molecular dynamics and intermolecular interactions in the capsules or any complex food systems.

10.5.2.6.5 Small angle neutron scattering Small angle neutron scattering (SANS) is a technique whereby cold neutrons permeate materials. When they hit upon nano-sized structures, they are scattered to small angles. The structures can be reconstructed from the scattering image. This technique allows the characterization of structures or objects in the nanometer scale, typically in the range of 1–200 nm. The information that can be obtained from SANS is primarily the average size, size distribution, and spatial correlation of nanoscale structures, as well as shape and internal structure.

10.6 Conclusion

Retention of high value ingredients, such as flavors, the presence of which improves functionality and appeal of foods, is of utmost importance to the food industry. Several developments have been made in the area of encapsulation of food flavors. The choice of a suitable technique of encapsulation depends on several features, including the properties of the flavor compounds, the degree of stability required during storage and processing, the properties of the food matrix, the specific release properties required, the maximum obtainable flavor load in the powder, and production cost. Each encapsulation technique presents advantages and disadvantages. Encapsulation by spray drying is the most economical and flexible way that the food industry can employ to encapsulate flavor ingredients. This technology is now becoming available to satisfy the increasingly specialized needs of the industry. The fluid-bed process is also a promising encapsulation technique for large-scale production of flavor powders. These technologies of encapsulation are being modified to reduce the size of the capsule to nanometers in at least one dimension to increase the efficiency of encapsulation. Nanoencapsulation technology for flavor encapsulation seems to be more viable and promising.

The major challenge faced by the food industry is to limit the loss of flavor. The loss of flavor is mainly due to the volatile nature of the compounds themselves, processing conditions, and the occurrence of several undesirable reactions such as oxidation and chemical degradation. Employment of encapsulation techniques has significantly resolved these issues. Food is best appreciated for its appealing flavors. It is therefore essential that the flavor be released and felt while in the mouth. This challenge for the encapsulated flavors has been resolved by encapsulating in a starch matrix, which shields them during thermal processing. The complexation was found to increase with increase in amylose content. The helical structure of amylase has a hydrophobic cavity in which the volatiles and essential oils get trapped. The amylase–flavor complexes become hydrolyzed during chewing in the presence of the enzyme α-amylase and aroma is completely released.

Another challenge is the additional cost. The consumer will be burdened because the cost of the final product will increase. Stability challenges of encapsulates during processing, storage, or cooking of the food product are other matters of concern. If an encapsulate leaks prematurely, it cannot deliver its value. Stability is especially challenging when water is present. Increased complexity of production process and/or supply chain is another challenge. This might vary from simply ordering and dosing an additional component to post-dosing encapsulates after the product preparation process and prior packaging. Undesirable consumer notice (visual or feel) of encapsulates in food products could be an additional challenge. This will, however, depend on, for example, particle size and contrasts between encapsulate and food product in the refractive index, color, and texture. In liquid foods, particles below 50 µm are in general not noticed by eye or touch. In dry food products, the particle size should be larger so it will be noticed because the contrasts, in general, are smaller.

References

Ahmed, MS, Hussein, Kamil MM, Lotfy SN, Mahmoud KF, Mehaya FM, Mohammad AA (2017). Influence of nano-encapsulation on chemical composition, antioxidant activity and thermal stability of rosemary essential iil. *Am J Food Technol.* 12:170–177.

Akbarzadeh, Rezaei-Sadabady R, Davaran S, Joo SW, Zarghami N, Hanifehpour Y, Samiei M, Kouhi M, Nejati-Koshki1 K (2013). Liposome: Classification, preparation, and applications. *Nanoscale Res Lett.* 8:102.

Albertini, Passerini N, González-Rodríguez ML, Perissutti B, Rodriguez L (2004). Effect of Aerosil® on the properties of lipid controlled release microparticles. *J Control Release.* 100:233–246.

Anwekar H, Patel S, Singhai AK (2011). Liposomes as drug carriers. *Int J Pharm Life Sci.* 2(7):945–951.

Arshady R (1999). Microspheres, microcapsules and liposomes. In: Arshady R (ed). *Preparation and Chemical Applications*, Vol. I. London, UK: Citus Books; 279–322.

Augustin MA, Sanguansri L, Margetts C, Young B (2001). Microencapsulation of food ingredients. *Food Aust.* 53:220–223.

Benczedi D, Bouquerand PE (2001). Process for the Preparation of Granules for the Controlled Release of Volatile Compounds. 01/17372 A1, March 15.

Beristain CI, Vernon-Carter EJ (1994). Utilization of mesquite (Prosopis julijlora) gum as emulsion stabilizing agent for spray-dried encapsulated orange peel oil. *Drying Technol.* 12 (7):1727–1733.

Beristain CI, Vazquez A, Garcia HS, Vernon-Carter EJ (1996). Encapsulation of orange peel oil by co-crystallization. *Lebensm Wiss Technol.* 29:645–647.

Bhandari BR, Datta N, D'Arcy BR, Rintoul GB (1998). Co-crystallization of honey with sucrose. *Lebensm Wiss Technol.* 31:138–142.

Bhandari BR, D'Arcy BR, Padukka I. (1999). Encapsulation of lemon oil by paste method using b-cyclodextrin: encapsulation efficiency and profile of oil volatiles. *J Agric Food Chem.* 47:5194–5197.

Bilia AR, Guccione C, Isacchi B, Righeschi C, Firenzuoli F, Bergonzi MC (2014). Essential oils loaded in nanosystems: a developing strategy for a successful therapeutic approach. *Evid Based Complement Altern Med.* http://dx.doi.org/10.1155/2014/651593.

Boland AB, Buhr K, Giannouli P, Ruth SMV (2004). Influence of gelatine, starch, pectin and artificial saliva on the release of 11 flavour vomounds from model gel systems. *Food Chem.* 86:401-411.

Buffo RA, Reineccius GA (2001). Comparison among assorted drying processes for the encapsulation of flavors. *Perfum Flavor.* 26:58-67.

Buffo RA, Probst K, Zehentbauer G, Luo Z, Reineccius GA (2002). Effects of agglomeration on the properties of spray-dried encapsulated flavours. *Flavour Fragr J.* 17:292-299.

Buck L, Axel R (1991). A novel multigene family may encode odorant receptors: A molecular basis for odor recognition. *Cell.* 65:175-187.

Campanile F (2007). Micrencapsulation and controlled release of flavours. Chemsource Symposium, 27th-28th June, RAI, Amsterdam.

Chen AC (1994). Ingredient technology by the sugar co-crystallization process. *Int Sugar J.* 96:493-494.

Chen AC, Veigra MF, Anthony BR (1988). Cocrystallization: An encapsulation process. *Food Technol.* 42:87-90.

Chua KJ, Chou SK (2003). Low-cost drying methods for developing countries. *Trends Food Sci Technol.* 14:519-528.

Chranioti C, Tzia C (2014). Arabic gum mixtures as encapsulating agents of freeze-dried fennel oleoresin products. *Food Bioproc Technol.* 7:1057-1065.

Cortesi R, Esposito E, Gambarin S, Telloli P, Menegatti E, Nastruzzi C (1999). Preparation of liposomes by reverse-phase evaporation using alternative organic solvents. *J Microencapsul.* 16:251-256.

Crank J (1975). *The Mathematics of Diffusion*, 2nd edn. Oxford: Oxford University Press.

Crouzet J (1998). Arôˆmes alimentaires. In: *Techniques del'ingenieur.* Paris: Agroalimentaire F 4100; 1-16.

Cussler EL (1997). *Diffusion, Mass Transfer in Fluid Systems*, 2nd edn. Cambridge: Cambridge University Press.

Daemen T, Hofstede G, Ten Kate MT, Bakker-Woudenberg IA, Scherphof GL. (1995). Liposomal doxorubicin-induced toxicity: Depletion and impairment of phagocytic activity of liver macrophages. *Int J Cancer.* 61:761-721.

Dalton P, Doolittle N, Nagata H, Breslin PAS (2000). The merging of the senses: integration of subthreshold taste and smell. *Nat Neurosci.* 3:431-432.

de Barros Fernandes RV, Vilela Borges S, Botrel DA (2013). Influence of spray drying operating conditions on microencapsulated rosemary essential oil properties. *Food Science and Technology.* 33(1):171-178.

De Roos KB (2000). Physicochemical models of flavour release from foods. In: Roberts DD, Taylor AJ (eds). *Flavour Release.* Washington, DC: American Chemical Society; 126-141.

Desai KGH, Park HJ (2005). Recent developments in microencapsulation of food ingredients. *Drying Technol.* 23(7):1361-1394.

Desobry S, Netto FM, Labuza TP (1997). Comparison of spray-drying, drum-drying and freeze-drying for b-carotene encapsulation and preservation. *J Food Sci.* 62:1158-1162.

Dezarn TJ (1998). Food ingredient encapsulation. In: Risch SJ, Reineccius GA (eds). *Encapsulation and Controlled Release of Food Ingredients.* ASC Symposium Series 590. Washington, DC: American Chemical Society; 74-86.

Dima C, Cotarlet M, Tiberius B, Bahrim G, Alexe P, Dima S (2014). Encapsulation of coriander essential oil in β-cyclodextrin: Antioxidant and antimicrobial properties evaluation. *Rom Biotechnol Lett.* 19(2):9128-9140.

Drake MA, Civille GV (2003). Flavour lexicons. *Compr Rev Food Sci Food Saf.* 2:33-40.

Dziezak JD (1988). Microencapsulation and encapsulated ingredients. *Food Technol.* 42(4):136-151.

Eldem P, Speiser H (1991). Polymorphic behavior of sprayed lipid micropellets and its evaluation by differential scanning calorimetry and scanning electron microscopy. *Pharm Res.* 8:178-184.

Emas M, Nyqvist H (2000). Methods of studying aging and stabilization of spray-congealed solid dispersions with carnauba wax. 1. Microcalorimetric investigation. *Int J Pharm.* 197:117–127.

Fan LT, Singh SK (1989). *Controlled Release: A Quantitative Treatment.* Berlin: Springer-Verlag.

Fathi M, Mozafari MR, Mohebbi M (2012). Nanoencapsulation of food ingredients using lipid based delivery systems. *Trends Food Sci Technol.* 23(1):13–27.

FRV de Barros, Borges SV, Botrel DA (2013). Influence of spray drying operating conditions on microencapsulated rosemary essential oil properties. *Food Science and Technology.* 33(1).

Fernandez-Gutierrez JA, Martin-Martinez ES, Martinez-Bustos F, Cruz-Orea A (2004). Physicochemical properties of casein–starch interaction obtained by extrusion process. *Starch/Stärke.* 56:190–198.

Fisher C, Scott TR (1997). *Food Flavours: Biology and Chemistry.* The Royal Society of Chemistry, Cambridge.

Flores RJ, Wall MD, Carnahan DW, Orofino TA (1992). An investigation of internal phase losses during the microencapsulation of fragrances. *J Micro-encapsul.* 3:287–307.

Gibbs BF, Kermasha S, Alli I, Mulligan CN (1999). Encapsulation in the food industry. *Int J Food Sci Nutr.* 50:213–224.

Godshall MA (1997). How carbohydrates influence food flavor. *J Food Technol.* 51:63–67.

Goubet I, Le Quere JL, Voilley A (1998). Retention of aroma compounds by carbohydrates: Influence of their physicochemical characteristics and of their physical state. *J Agric Food Chem.* 48:1981–1990.

Gouin S (2004). Microencapsulation: Industrial appraisal of existing technologies and trends. *Trends Food Sci Technol.* 15:330–347.

Green BK, Scheicher L (1955). Pressure Sensitive Record Materials. US Patent no. 2, 217, 507, Ncr C.

Guichard E (2000). Interaction of food matrix with small ligands influencing flavour and texture. *Food Res Int.* COST Action, 96:187–190.

Guichard E (2002). Interactions between flavor compounds and food ingredients and their influence on flavour perception. *Food Rev Int.* 18:49–70.

Gunning YM, Gunning PA, Kemsley EK, Parker R, Ring SG, Wilson RH, Blake A (1999). Factors affecting the release of flavour encapsulated in carbohydrate matrixes. *J Agric Food Chem.* 47:5198–5205.

Gupta C, Prakash D, Gupta S (2015). A biotechnological approach to microbial based perfumes and flavours. *J Microbiol Exp.* 2(1):00034.

Hedges A, Mcbride C (1999). Utilization of β-cyclodextrin in food. *Cereal Foods World.* 44, 700–702, 704.

Heinzen C (2002). Microencapsulation solves time dependent problems for food makers. *Eur Food Drink Rev.* 3:27–30.

Hemanthkumar M, Spandana V (2011). Liposomal encapsulation technology a novel drug delivery system designed for ayurvedic drug preparation. *Int Res J Pharm.* 2(10):4–6.

Hernández-Sánchez H, Gutierrez-Lopez GF (2015). *Food Nanoscience and Nanotechnology.* Springer, New york.

IOFI: International Organization of the Flavor Industry; Code of Practice 2012. www.iofi .org/datastream.aspx?type=doc/120306_IOFI_Code_V1_3_GA.

Ishwarya SP, Anandharamakrishnan C, Stapley AGF (2015) Spray-freeze-drying: A novel process for the drying of foods and bioproducts. *Trends Food Sci Technol.* 41(2):161–181.

Jafari SM, He YH, Bhandari B. (2006). Nano-emulsion production by sonication and microfluidization – a comparison. *Int J Food Prop.* 9(3):475–485.

Jacquot M, Pernetti M (2003). Spray coating and drying processes. In: Nedovic U, Willaert R (eds). *Cell Immobilization Biotechnology.* Series: Focus on biotechnology. Dordrecht: Kluwer Academic Publishers; 343–356.

Jouquand C, Ducruet V, Giampaoli P (2004). Partition coefficients of aroma compounds in polysaccharide solutions by the phase ratio variation method. *Food Chem.* 85:467–474.

Karel M, Langer R (1988). Controlled release of food additives. In: Risch SJ, Reineccius GA (eds). *Flavour Encapsulation.* ACS Symposium Series 370. Washington, DC: American Chemical Society; 177–191.

Kataria S, Sandhu P, Bilandi A, Akanksha M, Kapoor B (2011). Stealth liposomes: A review. *Int J Res Ayurveda Pharm.* 2(5):1534–1538.

King AK (1995). Encapsulation of food ingredients: A review of available technology, focusing on hydrocolloids. In: Risch SJ, Reineccius GA (eds). *Encapsulation and Controlled Release of Food Ingredients.* Washington, DC: American Chemical Society; 26–41.

Kim AY (2005). *Application of Biotechnology to the Production of Natural Flavor and Fragrance Chemicals.* Natural Flavors and Fragrances ACS Symposium Series, Vol. 908, Chapter 4; 60–75.

Kollengode ANR and Hanna MA (1997). Cyclodextrin comlexed flavor retention in extruded starches. *J Food Sci.* 62(5):1057–1060.

Korus J (2001). Microencapsulation of flavours in starch matrix by coacervation method. *Pol J Food Nutr Sci.* 10(51):17–23.

Krings U, Berger RG (1998). Biotechnological production of flavours and fragrances. *Appl Microbiol Biotechnol.* 49:1–8.

Kwak H-S (2014). *Nano- and Microencapsulation for Foods.* New York: John Wiley & Sons.

LaBell F (1991). Co-crystallization process aids dispersion and solubility. *Food Process.* 52:60–63.

Lee SY and Krochta JM (2002). Accelerated shelf life testing of whey protein coated pea nuts analyzed by static gas chromatography. *J Agric Food Chem.* 50:2022–2028.

Li Xiang, Anton N, Arpagaus C, Belleteix F, Vandamme TF (2010). Nanoparticles by spray drying using innovative new technology: The Büchi Nano Spray Dryer B-90. *J Control Release.* 147(2):304–310.

Loftsson T, Kristmundsdottir T (1993). Microcapsules containing water-soluble cyclodextrin inclusion complexes of water-insoluble drugs. In: El-Nokaly MA, Piatt DM, Charpentier BA (eds). *Polymeric Delivery Systems.* Washington, DC: American Chemical Society; 168–189.

Lubbers S, Landy P, Voilley A (1998). Retention and release of aroma compounds in food containing proteins. *J Food Technol.* 52:68–74.

Manojlovic V, Rajic N, Djonlagic J, Obradovic B, Nedovic V, Bugarski B (2008). Application of electrostatic extrusion—Flavour encapsulation and controlled release. *Sensors* 8:1488–1496.

Mayer LD, Bally MB, Hope MJ, Cullis PR (1986). Techniques for encapsulating bioactive agents in to liposomes. *Chem Phys Lipids.* 40:333–345.

Mehnert W, Mäder K (2001). Solid lipid nanoparticles: production, characterization and applications.*Adv Drug Deliv Rev.* 47(2–3):165–196.

Miller DH, Mutka JR (1986). US Patent, 4, 610–890. Sunkist Growers Inc., Sherman Oaks, CA.

Minemoto Y, Adachi S, Matsuno R (1997). Comparison of oxidation of menthyl linoleate encapsulated with gum arabic by hot-air-drying and freeze drying. *J Agric Food Chem.* 45(12):4530–4534.

Mhemdi H, Rodier E, Kechau N, Fages J (2011). A supercritical tunable process for the selective extraction of fats and essential oils from coriander seeds. *J Food Eng.* 105:609–616.

Mozafari MR, Mortazavi SM (2005). *Nanoliposomes: From Fundamentals to Recent Developments.* Oxford, UK: Trafford Publishing. Ltd. ISBN 1-4120-5545-8.

Mozafari MR (2005a). Liposomes: An overview of manufacturing techniques. *Cell Mol Biol Lett.* 10:711–719.

Mozafari MR (2005b). Method and apparatus for producing carrier complexes. UK Patent No. GB 0404993.8, Int. Appl. No. PCT/GB05/000825 (03/03/2005).

Mozafari MR, Reed CJ, Rostron C (2007). Cytotoxicity evaluation of anionic nanoliposomes and nanolipoplexes prepared by the heating method without employing volatile solvents and detergents. *Pharmazie*. 62(3):205–209.

Mortazavi SM, Mohammadabadi MR, Khosravi-Darani K, Mozafari MR (2007). Preparation of liposomal gene therapy vectors by a scalable method without using volatile solvents or detergents. *J Biotechnol*. 129:604–613.

Mullin JW (1972). *Crystallization Kinetics. Crystallization*. Cleveland, OH: CRC Press, Inc; 213–216.

Mujumdar AS, Devahastin S (2000). Fluidized bed drying. In: Mujumdar AS, Suvachittanont S (eds). *Developments in Drying Vol. 1: Food Dehydration*. Bangkok: Kasetsart University Press; 59–111.

Nagata T (1996). Techniques and application of electron microscopic radioautography. *J Electro Microsc*. 45:258–274.

Nakagawa K, Iwamoto S, Nakajima M, Shono A, Satoh K (2004). Micro channel emulsification using gelatine and surfactant-free coacervate microencapsulation. *J Colloids Interface Sci*. 278:198–205.

Naknean P, Meenune M (2010). Factors affecting retention and release of flavour compounds in food carbohydrates. *Int Food Res J*. 17;23–34.

Pagington JS (1986). Beta-cyclodextrin. *Perfum Flavor*. 11:49–58.

Panda RC, Zank J, Martin H (2001). Modelling the droplet deposition behaviour on a single particle in fluidized bed spray granulation process. *Powder Technol*. 115:51–57.

Passerini N, Qi S, Albertini B, Grassi M, Rodriguez L, Craig DQ (2010). Solid lipid microparticles produced by spray congealing: Influence of the atomizer on microparticle characteristics and mathematical modeling of the drug release. *J Pharm Sci*. 99:916–931.

Pothakamury UR, Barbosa-Canovas GV (1995). Fundamental aspects of controlled release in foods. *Trends Food Sci Technol*. 6:397–406.

Qi ZH, Hedges AR (1995). Use of cyclodextrins for flavors. In: Ho CT, Tong CH (eds). *Flavor Technology: Physical Chemistry, Modification and Process*. ACS Symposium Series 610. Washington DC: American Chemical Society; 231–243.

Quellet C, Schudel M, Ringgenberg R (2001). Flavors and fragrance delivery systems. *Chimia*. 55:421–428.

Rai M, Ribeiro C, Mattoso L, Duran N. (2015). *Nanotechnologies in Food and Agriculture*. Switzerland: Springer. doi:10.1007/978-3-319-14024-7. ISBN: 978-3-319-14023-0 (Print). 978-3-319-14024-7 (Online).

Redondo N, Gomez-Mrtinez S, Marcos A (2014). Sensory attributes of soft drinks and their influence on consumers' preferences. *Food Funct*. 5:1686–1694.

Reineccius GA (1988). Spray-drying of food flavors. In: Risch SJ, Reineccius GA (eds). *Flavor Encapsulation*. ACS Symposium Series 370. Washington, DC: American Chemical Society; 55–66.

Reineccius GA (1991). Carbohydrates for flavor encapsulation. *Food Technol*. 45:144–147.

Reineccius GA (1995). Liposomes for controlled release in the food industry. In: Risch SJ, Reineccius GA (eds). *Encapsulation and Controlled Release of Food Ingredients*. ACS Symposium Series 590. Washington, DC: American Chemical Society; 113–131.

Reineccius GA, Risch S (1986). Encapsulation of artificial flavours by b-cyclodextrin. *Perfum Flavor*. 11:1–11.

Reineccius TA, Reineccius GA, Peppard TL (2002). Encapsulation of flavors using cyclodextrins: comparison of flavor retention in alpha, beta and gamma types. *J Food Sci*. 67:3271–3279.

Reineccius TA (2013). *Source Book of Flavours*. Springer Science & Business Media.

Re-MI (1998). Microencapsulation by spray-drying. *Drying Technol*. 16:1195–1236.

Richard J, Benoît JP (2000). Microencapsulation. In: *Techniques de l'ingenieur*. J. 2210, Paris:Techniques de l'ingenieur; 1–20.

Riaz M (1996). Liposome preparation method. *Pak J Pharm Sci*. 9(1):65–77.

Risch SJ (1995). Encapsulation: overview of uses and techniques. In: Rish SJ, Reineccius GA (eds). *Encapsulation and Controlled Release of Food Ingredient*. Washington, DC: American Chemical Society; 2–7.

Rizvi SSH, Mulvaney SJ, Sokhey AS (1995). The combined application of supercritical fluid and extrusion technology. *Trends Food Sci Technol*. 6(7):232–240.

Roos KB (2003). Effect of texture and microstructure on flavour retention and release. *Int Dairy J*. 13:593–605.

Rosenberg M, Kopelman IJ, Talmon Y (1990). Factors affecting retention in spray-drying microencapsulation of volatile material. *J Agric Food Chem*. 38:1288–1294.

Rosenberg M, Talmon Y, Kopelman IJ (1984). The microstructure of spray-dried microcapsules. *Food Microstruct*. 7(1):15.

Rowe DJ, Winkel C (2009). *Stability of Aroma Chemicals in Chemistry and Technology of Flavors and Fragrances*. Hoboken: Wiley Publishers.

Sanchez C, Renard D (2002). Stability and structure of protein-polysaccharide coacervates in the presence of protein aggregates. *Int J Pharm*. 242:319–324.

Salles C, Chagnon MC, Feron G, Guichard E, Laboure H, Morzel M, Semon E, Tarrega A, Yven C (2011). In-mouth mechanisms leading to flavor release and perception. *Crit Rev Food Sci Nutr*. 51(1):67–90.

Schalmeus W (1995). Centrifugal extrusion encapsulation. In: Rish SJ, Reineccius GA (eds). *Encapsulation and Controlled Release of Food Ingredient*. Washington, DC: American Chemical Society; 96–103.

Schieren H, Rudolph S, Findelstein M, Coleman P, Weissmann G (1978). Comparison of large unilamellar vesicles prepared by a petroleum ether vaporization method with multilamellar vesicles: ESR, diffusion and entrapment analyses. *Biochim Biophys Acta*. 542(1):137–153.

Shahidi F, Han X-Q (1993). Encapsulation of food ingredients. *Crit Rev Food Sci Nutr* 33(6):501–547.

Shaheen SM, Ahmed FRS, Hossen MN, Ahmed M, Amran MS, Anwar Ul-Islam M (2006). Liposome as a carrier for advanced drug delivery. *Pak J Bio Sci*. 9(6):1181–1191.

Sharma DK, Tiwari BD (2001). Microencapsulation using spray-drying. *Indian Food Ind*. 20:48–51.

Shepherd GM (2012). *Neurogastronomy. How the Brain Creates Flavor, and Why it Matters*. New York: Columbia University Press.

Shepherd GM (2006). Smell images and the flavour system in the human brain. *Nature*. 444:316–321.

Small DM (2012). Flavour is in the brain. *Physiol Behav*. 107:540–552.

Small DM, Prescott J. (2005). Odor/taste integration and the perception of flavor. *Exp Brain Res*. Springer-Verlag. 10.1007/s00221-005-2376-9.

Soottitantawata A, Takayamaa K, Okamuraa K, Muranakaa D, Yoshiia H, Furutaa T, Ohkawarab M, Linko P (2005). Microencapsulation of l-menthol by spray drying and its release characteristics. *Innov Food Sci Emerg Technol*. 6:163–170.

Spence C (2015). Multisensory flavor perception. *Cell*. 161(1):24–35.

Song H, Geng HQ, Ruan J, Wang K, Bao CC, Wang J, Peng X, Zhang XQ, Cui DX (2011). Development of polysorbate 80/phospholipid mixed micellar formation for docetaxel and assessment of its in vivo distribution in animal models. *Nanoscale Res Lett*. 6:354.

Steinbock B, Vichailkul PP, Steinbock O (2001). Nonlinear analysis of dynamic binding in affinity capillary electrophoresis demonstrated for inclusion complexes of b-cyclodextrin. *J Chromatogr A*. 943:139–146.

Taylor AH (1983). Encapsulation systems and their applications in the flavor industry. *Food Flavor Ingredient Process Packag*. 4:48–52.

Teixeira MI, Andrade LR, Farina M, Rocha-Leao MHM (2004). Characterization of short chain fatty acid microcapsules produced by spray drying. *Mater Sci Eng C* 24:653–658.

Tolstuguzov VB, Rivier V (1997). Encapsulated Particles in Protein From a Polysaccharide-Containing Dispersion. European Patent Application, EP 0 797 925 A1. Nestle.

Turchiuli C, Cuvelier M-E, Giampaoli P, Dumoulin E. (2013). Aroma encapsulation in powder by spray drying, and fluid bed. Agglomeration and coating. In: Yanniotis S et al. (eds). *Advances in Food Process Engineering Research and Applications*. New York, NY: Springer Science + Business Media.

Turton R, Cheng XX (2007). Cooling processes and congealing. In: Swarbrick J (ed). *Encyclopedia of Pharmaceutical Technology*. New York, NY, USA: Informa Healthcare; 761–773.

Uhlemann H, Morl L (2000). *Wirbelschicht-Sprühgranulation*. Berlin: Springer.

Uhlemann J, Schleifenbaum B, Bertram HJ (2002). Flavor encapsulation technologies: An overview including recent developments. *Perfum Flavor*. 27:52–61.

Vemuri S, Rhodes CT (1995). Preparation and characterization of liposomes as therapeutic delivery systems: a review. *Pharm Acta Helv*. 70:95–111.

Vemuri S, Yu CD, Wangsatorntanakun V, Roosdorp N (1990). Large-scale production of liposome by a microfluidizer. *Drug Dev Ind Pharm*. 16(15):2243–2256.

Villota R, Hawkes JG (1994). Flavoring in extrusion: an overview. In: ACS symposium series (USA).

Wampler DJ (1992). Flavor encapsulation: a method for providing maximum stability for dry flavor systems. *Cereal Foods World*. 37:817–820.

Watanabe Y, Fang X, Minemoto Y, Adachi S, Matsuno R (2002). Suppressive effect of saturated L-ascorbate on the oxidation of linoleic acid encapsulated with maltodextrin or gum arabic by spray-drying. *J Agric Food Chem*. 50:3984–3987.

Weiss J, Decker EA, McClements DJ, Kristbergsson K, Helgason T, Awad T (2008). Solid lipid nanoparticles as delivery systems for bioactive food components. *Food Biophys*. 3(2):146–154.

Whorton C (1995). Factors influencing volatile release from encapsulation matrices. In: Rish SJ, Reineccius GA (eds). *Encapsulation and Controlled Release of Food Ingredients*. Washington, DC: American Chemical Society; 134–142.

Wu Y, Duan S, Zhao L, Gao Z, Song S, Xu W, Zhang C, Ma C, Wang S (2016). Aroma characterization based on aromatic series analysis in table grapes. *Sci Rep*. 6:Article number: 31116.

Xiao Z, Li W, Zhu G, Zhou R, Niu Y (2016). Study of production and the stability of styralyl acetate nanocapsules using complex coacervation. *Flavour Fragr J*. 31(4):283–289.

Yuliani S, Torley PJ, D'Arcy B, Nicholson T, Bhandari B (2006). Effect of extrusion parameters on flavour retention, functional and physical properties of mixtures of starch and D-limonene encapsulated in milk protein. *Int J Food Sci Technol*. 41(Supplement 2):83–94.

Yuliani S, Torley PJ, D'Arcy B, Nicholson T, Bhandari B (2005). Extrusion of mixtures of starch and d-limonene encapsulated with b-cyclodextrin: flavour retention and physical properties. *Food Res Int*. 39, 318–331.

Zeller BL, Saleeb FZ, Ludescher RD (1999). Trends in development of porous carbohydrate food ingredients for use in flavor encapsulation. *Trends Food Sci Technol*. 9:389–394.

Zhanga LG, Zhanga CC, Nia LN, Yang LN, Wang YJ (2011). Rectification extraction of Chinese herbs' volatile oils and comparison with conventional steam distillation. *Sep Purif Technol*. 77:261–268.

Zhang Y (2011). Relations between size and function of substance particles. *Nano Biomed Eng*. 3(1):1–16.

Zheng S, Alkan-Onyuksel H, Beissinger RL, Wasan DT (1999). Liposome microencapsulation without using any organic solvent. *J Dispers Sci Technol*. 20:1189–1203.

Zuidam NJ, Shimoni E (2010). *Encapsulation Technologies for Active Food Ingredients and Food Processing*. Springer-Verlag.

Nanoencapsulated Nutraceuticals: Pros and Cons

T. Anand, N. Ilaiyaraja, Mahantesh M. Patil,
Farhath Khanum, and Rakesh Kumar Sharma

Contents

11.1 Introduction 274
11.2 Definition of nutraceutical..275
    11.2.1 Categories of nutraceuticals 276
        11.2.1.1 Dietary supplements..276
        11.2.1.2 Functional foods 276
        11.2.1.3 Medicinal foods 277
        11.2.1.4 Pharmaceuticals 277
11.3 Nutraceutical nomenclature in Canada and the United States 277
    11.3.1 Nomenclature in Canada..277
    11.3.2 Nomenclature in the United States 277
11.4 International market potential and quality issues..............................277
    11.4.1 Market potential of nutraceuticals 277
    11.4.2 Global quality issues 278
11.5 Natural antioxidants..278
    11.5.1 Vitamins C and E..278
    11.5.2 Polyphenols 279
    11.5.3 Quercetin 279
11.6 Nanotechnology in drug delivery systems...................................280
    11.6.1 Advantages of nanotechnology...280
    11.6.2 Goals of nanotechnology 280
11.7 Nanoparticle preparation..280
11.8 Nanoencapsulation of bioactive agents/ingredients 284
    11.8.1 Encapsulation materials 285
        11.8.1.1 Chitosan 285
        11.8.1.2 PLGA...286
        11.8.1.3 TPP 287
        11.8.1.4 Phospholipids..287
11.9 Methods of producing chitosan nanoparticles 288
    11.9.1 Coacervation/precipitation 288
    11.9.2 Emulsion-droplet coalescence..288

	11.9.3	Reverse micelles	289
	11.9.4	Ionotropic gelation	290
11.10	Applications of chitosan nanoparticles		291
11.11	Activity of nanoparticles		291
	11.11.1	Controlled release of liposomes	292
	11.11.2	Antimicrobial activity of nanoemulsions	293
	11.11.3	Protein-based nanoencapsulates	294
	11.11.4	Polysaccharide-based nanoencapsulates	295
	11.11.5	Inorganic-based nanoencapsulates	295
11.12	Characterization of nanoparticles by analytical methods		295
	11.12.1	Electron microscopy: Transmission electron microscopy and scanning electron microscopy	296
	11.12.2	Dynamic light scattering	296
	11.12.3	*In-vitro* evaluation of nanoparticles	297
	11.12.4	Chemical method	299
11.13	Potential health benefits		300
	11.13.1	Advantages of nanoencapsulation	300
		11.13.1.1 Nanoencapsulation of antioxidants and flavoring agents	300
		11.13.1.2 Nanoencapsulation of anti-cancer agents	302
11.14	Disadvantages of nanoencapsulation		302
11.15	Conclusions		303
References			303

11.1 Introduction

Functional foods and dietary supplements are products derived from edible sources that provide positive health benefits beyond basic nutrition. A concern of the research community is that bioactive compounds may not be fully absorbed into the body as they get degraded before they reach the target organ moreover the efficacy of bioactive compounds may be reduced due to instability caused by unfavorable acidic conditions, enzymatic reactions, and low permeability in the gastrointestinal tract. Therefore, the delivery and release of bioactive compounds from ingestion to digestion have been identified as the most important criteria for improving absorption.

The controlled release of bioactive compounds involves the interaction of wall materials in the digestive system and their ability to sustain the shield at control rate dissolution. Core release rate control relies on the pattern of polymer breakdown in the gastrointestinal tract. Therefore, the selection of wall materials for the control of the delivery system corresponding to compound release is a serious challenge for creating an efficient encapsulation system. Physicochemical properties and behaviors of shield structures under digestive conditions play an important role in determining the degradation of wall materials and subsequent release of compound.

Selection of wall material is very important for the shield (e.g., using a bioadhesive wall material) is that the residue time of the bioactive shell can be prolonged in the small intestine, which helps increase the rate of absorption. Sustained release rate allows the encapsulated core to be expelled slowly over a long period of time in the gastrointestinal tract, and the release rate of core material becomes time independent. Therefore, sustained release ensures a steady amount of core material is released over time as well as improving the bioavailability of core material in responding to the targeted time. A sustained compound release formula can be achieved by sustaining the polymer time via polymer hydrogel or by incorporating the bioactive compound into the shield matrix. This mechanism is particularly designed for site-specific core delivery. It localizes the site of action so that the nutrient is more readily available.

The prefix *nano* is derived from the Greek word *nanos* meaning "dwarf." Today nano is used as a prefix to describe 10^{-9} (one billionth) of a measuring unit. Nanotechnology is the field of research and fabrication that is on a scale of 1–100 nm. For its first three decades the nanosciences were dedicated mainly to study and fabricate materials at the nano level. Nanotechnological research approaches and tools were then developed to address numerous issues in many disciplines, particularly biology.

Nanotechnology has brought new potential applications to many fields such as electronics and biomaterials production and the pharmaceutical industry. Nanotechnology has played an important role in new methods of drug delivery that have been developed recently. A general definition of a nanostructure is a structure 1–100 nm in size in at least in one dimension. However, one spatial dimension in the size up to 1000 nm (equal to 1 μm) is considered a nanomaterial because it still has physical and chemical properties different from the bulk. Forming these types of structures enables a new drug delivery approach, and new formulations can be made using different techniques.

11.2 Definition of nutraceutical

Nutraceutical is a broad term to describe any product derived from food sources professed to provide extra health benefits in addition to the basic nutritional value found in foods. These products may claim to prevent chronic diseases, improve health, delay the aging process, increase the life span, or support the structure or function of the body. They can be considered nonspecific biological therapies that promote general well-being, control symptoms, and prevent malignant processes. The term *nutraceutical* combines two words, nutrient (a nourishing food component) and pharmaceutical (a medical drug). The term was coined in 1989 by Stephen DeFelice, founder and chairman of the Foundation for Innovation in Medicine, an American organization located in Cranford, New Jersey.

The philosophy behind nutraceuticals is a focus on prevention. As the Greek physician Hippocrates, known as the father of medicine, said, "Let food be your medicine." The role of nutraceuticals in human nutrition is one of the most important areas of investigation, with wide-ranging implications for consumers, health-care providers, regulators, food producers, and distributors.

The Indians, Egyptians, Chinese, and Sumerians are among the many early civilizations that used foods as medicines (Wildman, 2001). The modern nutraceutical market began to develop in the 1980s in Japan. In contrast to the natural herbs and spices used as folk medicine for centuries throughout Asia, the nutraceutical industry has developed alongside the expansion and exploration of modern technology, and at the same rapid pace.

11.2.1 Categories of nutraceuticals

Nutraceuticals and related products can be classified on the basis of their natural sources, pharmacological conditions, and chemical constitution of the products. They are typically grouped into four categories: dietary supplements, functional foods, medicinal foods, and pharmaceuticals.

11.2.1.1 Dietary supplements

A *dietary supplement* is a product that contains nutrients derived from food products. Dietary supplements can be extracts or concentrates. They are found in many forms such as capsules, tablets, soft gels, liquids, and powders.

Although dietary supplements are regulated by the U.S. Food and Drug Administration (FDA) as foods, their regulation differs from drugs and other foods. Dietary supplements do not have to be approved by the FDA before marketing, but companies must register their manufacturing facilities with the FDA. Aside from a few well-defined exceptions, dietary supplements may only be marketed to support the structure or function of the body. They may not claim to treat a disease or condition, and they must include a label that states, "These claims have not been evaluated by the FDA. This product is not intended to diagnose, treat, cure, or prevent any disease."

11.2.1.2 Functional foods

Functional foods are designed to allow consumers to eat enriched foods close to their natural state, rather than take dietary supplements manufactured in liquid or capsule form. Functional foods have been either enriched or fortified, a process called nutrification. This practice restores the nutrient content in a food back to levels similar to those that existed before the food was processed. Sometimes complementary nutrients are added, such as vitamin D to milk, vitamin A to butter, or vitamin B complex to grain flour.

Health Canada defines a functional food as "ordinary food that has components or ingredients added to give a specific medical or physiological benefit, other than a purely nutritional effect." In Japan, all functional foods must meet three established requirements. Foods should be (1) present in their naturally occurring form, rather than a capsule, tablet, or powder; (2) consumed in the diet as often as daily; and (3) should regulate a biological process in hopes of preventing or controlling disease (Hardy 2000).

11.2.1.3 Medicinal foods

A *medical food* is formulated to be consumed or administered internally, under the supervision of a qualified physician. Its intended use is for specific dietary management of a disease or condition for which distinctive nutritional requirements are established by medical evaluation.

11.2.1.4 Pharmaceuticals

Pharmaceuticals are medically valuable components produced from modified agricultural crops or animals. Proponents of this concept maintain that using crops (and possibly even animals) as pharmaceutical factories is much more cost effective than conventional methods, with higher revenue for agricultural producers.

11.3 Nutraceutical nomenclature in Canada and the United States

11.3.1 Nomenclature in Canada

Under Canadian law, a nutraceutical can either be marketed as a food or as a drug; the terms "nutraceutical" and "functional food" have no legal distinction. Both refer to "a product isolated or purified from foods that is generally sold in medicinal forms not usually associated with food and it is demonstrated to have a physiological benefit or to provide protection against chronic disease" ("Nutraceuticals/Functional Foods and Health Claims on Foods: Policy Paper").

11.3.2 Nomenclature in the United States

The term nutraceutical has no meaning in U.S. law. Depending on its ingredients and the claims with which it is marketed, a product is regulated as a drug, dietary supplement, food ingredient, or food.

11.4 International market potential and quality issues

11.4.1 Market potential of nutraceuticals

In 2017 Zion Market Research published a market analysis report, "Dietary Supplements Market by Ingredients (Botanicals, Vitamins, Minerals, Amino

Acids, Enzymes) for Additional Supplements, Medicinal Supplements and Sports Nutrition Applications, Global Industry Perspective, Comprehensive Analysis and Forecast, 2016–2022" (Zion Market Research, 2017). According to this report, the global dietary supplements market was US$ 132.8 billion in 2016. It was forecasted to reach US$ 220.3 billion in 2022.

11.4.2 Global quality issues

There are significant product quality issues in the global market (Hasler, 2005). Nutraceuticals from the international market may claim to use organic or exotic ingredients, yet the lack of regulation may compromise the safety and effectiveness of products. Companies trying to achieve a wide profit margin may create unregulated products overseas with low-quality or ineffective ingredients.

11.5 Natural antioxidants

Antioxidants present in plants, including algae and mushrooms, are excellent natural additives to foodstuffs for their ion or hydrogen donating, metal chelating, and chain-breaking capabilities. Natural antioxidants are vitamins, polyphenols, and carotenoids. These groups of molecules may also exhibit additional properties (Carocho and Ferreira, 2013a). The main vitamins with antioxidant potential include vitamin C (ascorbic acid) and vitamin E (tocopherol), which are already used as food additives.

11.5.1 Vitamins C and E

Vitamin C is an essential vitamin for humans that can only be acquired through diet (Davey et al., 2000). This molecule is an effective scavenger of the superoxide radical anion, hydrogen peroxide, hydroxyl radical, singlet oxygen, and reactive nitrogen oxide, thus avoiding oxidative stress in the human body. Ascorbic acid is one of the most frequently used antioxidants in the food industry. It is used in the meat, beverage, fish, and bakery industries. By accepting oxygen in the food, and oxidizing itself to dehydro-ascorbic acid, the available oxygen is reduced, which helps preserve the food.

Apart from this mechanism, ascorbic acid also acts as an anti-browning agent by reconverting quinones back to the phenolic form and avoiding flavor deterioration in beverages (Davey et al., 2000; Carocho and Ferreira, 2013a). In particular, vitamin C exerts activity against lipid peroxidation and rancidification by donating its phenolic hydrogen to the peroxyl radicals forming tocopheroxyl radicals which, despite also being radicals, are unreactive and unable to continue the oxidative chain reaction.

Both vitamins C and E can work in synergism with the regeneration of vitamin E through vitamin C from the tocopheroxyl radical to an intermediate, reinstating yet again its antioxidant potential. This is why the two are usually utilized together to extend the shelf life of foodstuffs (Carocho and Ferreira, 2013a).

11.5.2 Polyphenols

Polyphenols, secondary metabolites of plants, are also excellent antioxidants. Among their various effects (antimicrobial, antimutagenic, anticancer, antitumor, anti-inflammatory), they also scavenge free radicals, chelate metals, quench oxygen atoms, and can act as ion or hydrogen donors. A total of 8,000 polyphenols are available in nature in the form of polyphenols. They are divided into eight groups: hydroxybenzoic acids, hydroxycinnamic acids, coumarins, lignans, chalcones, flavonoids, lignins, and xanthones (Spencer et al., 2008). Some polyphenols exhibit good antioxidant activity as pure compounds incorporated in foodstuffs, while others depend on synergism to carry out the protective effects. The focus is on the compounds that exert the antioxidant effect, but at the same time synergisms can be research opportunities that may be beneficial to the food industry (Carocho and Ferreira, 2013a,b). Polyphenols have been used as antioxidants in the fish and meat industries. Carcasses are dipped into polyphenolic extracts to delay oxidation and bacterial contamination (Fan et al., 2008; Kumudavally et al., 2008; Maqsood et al., 2013). Other approaches have been tested successfully. Natural extracts rich in polyphenols or pure compounds are incorporated into food to avoid rancidity, spoilage, and bacterial colony formation for a longer time when compared to the controls (Yao et al., 2004; Serra et al., 2008; Day et al., 2009; Bansal et al., 2013).

11.5.3 Quercetin

The antioxidant activity of the quercetin molecule is higher than other well-known antioxidant molecules such as ascorbyl, trolox, and rutin (Budhian et al., 2008) because of the number and position of the free hydroxyl groups in the quercetin (Cao et al., 1997). The flavonoid glycosides are extensively hydrolyzed in the small intestine or by bacterial activity in the colon to produce the quercetin aglycones. They are also metabolized into a glucuronidated or sulfated form of quercetin. McIvor et al. (2011) concluded that the antioxidant activity of quercetin is well recognized. It has an appropriate structure for free radical scavenging and ion chelation activity. These protective effects may be due to the antioxidant effects of catechin and quercetin. Kumar et al. (2011) investigated the possible anti-inflammatory effects of physiologically attainable quercetin concentrations. It increased antioxidant capacity *in-vivo* and displayed anti-inflammatory effects *in-vitro*.

11.6 Nanotechnology in drug delivery systems

11.6.1 Advantages of nanotechnology

The first advantage of nanoparticles is their small size. The small dimensions of nanoparticle delivery systems allow the particles to penetrate tissue barriers that large particles are unable to permeate (Jiang et al., 2007). The particles must be sufficiently diminutive to avoid detection and clearance by the system. *In-vivo* and *in-vitro* studies have shown that both the size and shape of nanostructures determine their ability to cross the threshold of cells and tissues (Chithrani et al., 2006; Geng et al., 2007).

11.6.2 Goals of nanotechnology

Nanoparticle delivery systems have the same general goal and benefit that drug delivery systems try to achieve: to provide a sustained and controlled release of the agent. Nanoparticle delivery systems aim to provide a more stable concentration of therapeutic agent that avoids maximums and minimums and provides treatment over a longer period of time. The rate of compound release from nanodelivery systems is controlled by the mass transfer characteristics of the particle carrier, which are affected by factors such as porosity and particle degradation, and can be manipulated to alter the compound concentration profile within the patient. This increases efficiency by reducing the number of times medication needs to be administered. This also significantly improves patient comfort and quality of life because many drugs require intrusive methods of application, such as repeated injection and intravenous administration. Another potential benefit of developing drug delivery systems, which can avoid macrophage detection and provide a controlled release, is that many drugs, such as chemotherapy drugs, have a variety of adverse effects on patients that reducing patient exposure can limit (Jiang et al., 2007).

11.7 Nanoparticle preparation

The use of nanosize vehicles for the protection and controlled release of nutrients and bioactive food ingredients is a growing area of interest in the food science and technology fields for several reasons. Nanoparticles can be incorporated into food products easily without sedimentation, without being noticed by the consumer, and/or with an enhanced bioavailability. The preparation of nanoparticles might be based on downsizing encapsulates prepared by classical technologies, as discussed above, or by using new techniques. In this section, we highlight some examples to demonstrate the potential of some of the techniques and concepts that have evolved in recent years (Table 11.1).

Table 11.1 Methods of Nanoencapsulation of Bioactive Compounds

Nanoencapsulation Technique	Material Used	Bioactive Compound	Property	References
Emulsification	Maltodextrin; modified starch (Hi-Cap 100) Tween 40, Tween 80, Span® 80, and sodium dodecyl sulphate, Tween 20	Curcumin	Enhances anti-inflammation activity	J.C. Wang et al. (2008); X. Wang et al. (2008)
Coacervation	Gelatin, maltodextrin, and tannins; Tween 60; glutaraldehyde Gelatin, acacia, and hydrolysable tannins; hydroxylethyl, cellulose; glutaraldehyde Gelatin, acacia, and tannins; Tween 60; glutaraldehyde	Capsaicin	Masks pungent odor, provides biocompatibility and biodegradation; Improves efficiency and delays release property; Masks pungent odor and improves stability	J.C. Wang et al. (2008); X. Wang et al. (2008); Xing et al. (2004); Jincheng et al. (2010)
Inclusion complexation	β-Lactoglobulin and low methoxyl pectin	Docosahexaenoic acid	Formation of transparent solution, improves colloidal stability, protects against degradation, and is useful for enrichment of acid drinks	Zimet and Livney (2009)
Nanoprecipitation	Monomethoxy poly (ethylene glycol)-poly (3-caprolactone) micelles Poly (lactide-co-glycolide); emulsifiers: polyethylene glycol-5000 Ethyl cellulose and methyl cellulose, poly(D,L-lactic acid) and poly(D,L-lactic-coglycolic acid); gelatin or Tween 20 Poly (ethylene oxide)-4-methoxycinnamoylphthaloylchitosan, poly(vinylalcohol-co-vinyl-4-methoxycinnamate), poly(vinylalcohol), and ethyl cellulose β-Cyclodextrin (β-CD). Poly(D,L-lactide-co-caprolactone) (PLC) copolymers, Poly(D,L-lactic-co-glycolic acid)-poly(ethylene glycol) (PLGA-PEG) Lipid-core nanocapsules	Curcumin Curcumin Curcumin β-Carotene Astaxanthin Myricetin Tamoxifen Pomegranate polyphenols Dihranol Scutellaria baicalensis Resveratrol and curcumin	Improves solubility; Improves bioavailability, bioactivity, encapsulation efficiency and enhancing the cellular uptake; Improves oral bioavailability and sustainability; Improves physical, chemical stability, and bioavailability; Improves solubility and bioavailability; Increases antioxidant potency; Anti-cancer; Potent antioxidants; Topical treatment of psoriasis anti-inflammatory; Quercetin (QUE) and resveratrol (RES); Lipoic acid	Gou et al. (2011); Anand et al. (2010); Suwannateep et al. (2011); Ribeiro et al. (2008); Tachaprutinum et al. (2009); Chakraborty et al. (2014); Perez et al. (2012); Shirode et al. (2015); Savian et al. (2015); Choi et al. (2014); Coradini et al. (2014)

(Continued)

Table 11.1 (Continued) Methods of Nanoencapsulation of Bioactive Compounds

Nanoencapsulation Technique	Material Used	Bioactive Compound	Property	References
	Lecithin	α-Tocopherol	Antioxidants	Cadena et al. (2013)
	Lipid-core nanocapsules		Promotes anti-obesity	Külkamp et al. (2011)
	Cyclodextrin		Antioxidant	Byun et al. (2011)
	Soybean lecithin		Antioxidant	
	Polycaprolactone			
Emulsification–solvent evaporation	Chitosan cross-linked with tripolyphosphate; Span 80 and Tween 80; acetic acid and ethanol	Curcumin	For controlled release	Sowasod et al. (2008)
	Hydroxyl propyl methyl cellulose and polyvinyl pyrrolidone; D-α-tocopheryl polyethylene glycol 1000 succinate, Tween 80, Tween 20, cremophor-RH 40, pluronic-F68, pluronic-F127	Curcumin	Enhances absorption and prolongs rapid clearance of curcumin	Dandekar et al. (2010)
	Poly-D,L-lactide and polyvinyl alcohol	Quercetin	Improves controlled release and encapsulation efficiency	Kumari et al. (2010)

(*Continued*)

Table 11.1 (Continued) Methods of Nanoencapsulation of Bioactive Compounds

Nanoencapsulation Technique	Material Used	Bioactive Compound	Property	References
Supercritical antisolvent precipitation	Hydroxylpropyl methyl cellulose phthalate	Lutein	Bioactivity, promote to food industry and to avoid thermal/light degradation	Heyang et al. (2009)
Spray drying	Carbohydrate matrix and maltodextrin; acetone	Catechin	Increases stability, protects from oxidation and incorporation into beverages	Ferreira et al. (2007)
	Modified n-octenyl succinate starch; ethyl acetate	β-Carotene	Improves dispersibility, coloring strength, and bioavailability	De Paz et al. (2012)
	Maltodextrin; emulsifiers: Hi-Cap, whey protein concentrate, and Tween 20	D-Limonene	Increases retention, stability during process	Jafari et al. (2007)
	Maltodextrin; emulsifiers: modified starch (Hi-Cap)/whey protein concentrate	Fish oil	Minimizes un-encapsulated oil at the surface and maximizes encapsulation efficiency	Jafari et al. (2008)
Freeze drying	β-Cyclodextrin, polycaprolactone; pluronic F68; ethyl acetate	Fish oil	Prevents oxidation and masks odor	Choi et al. (2010)
	Poly-ε-caprolactone	Fish oil	Increases oxidative stability and encapsulation efficiency	Bejrapha et al. (2010)
	Chitosan, zein; Tween 20	α-Tocopherol	Improves stability and protects from environmental factors	Luo et al. (2011)
	Polyethylene glycol; Tween 80	Vitamin E	Increases stability and retention percentage; extends shelf life	Zhao et al. (2011)
	Chitosan and sodium tripolyphosphate	Catechin	Protects catechin from degradation	Dube et al. (2010)

11.8 Nanoencapsulation of bioactive agents/ingredients

Nanoencapsulation involves the incorporation of bioactive agents, including food ingredients, vitamins, antioxidants, slimming agents, and enzymes, into small capsules with submicron-size diameters. The range of applications for this technology in the food industry has been increasing because of the advantages that nanoencapsulation confers on the encapsulated material. These include enhancing the stability of the encapsulated material by protecting it from humidity, heat, pH variations, and other extreme conditions, as well as masking unwanted odors or tastes. Various types of encapsulation technologies can be employed toward this end. They consist of nanospheres, nanoparticles, nanoemulsions, nano-cochleates, and liposomes.

Liposomes or lipid vesicles are closed continuous structures that enclose an internal aqueous compartment that is separated from the external medium by one or more concentric lipid bilayers. They can accommodate water-soluble moieties in their aqueous spaces and, if required simultaneously, lipid-soluble moieties in their lipid phases. This explains the application of lipid-based nano-encapsulation for the incorporation of nutrients, particularly those with unpleasant odors/tastes or with sensitive structures. Lipid vesicles have been applied in cheese making and in the preparation of food emulsions such as spreads, margarine, and mayonnaise (Gibbs et al., 1999). Another type of lipid-based encapsulation technology utilizes cochleates, which are used as drug and nutrient delivery systems (Zarif et al., 2000).

Lipid-based vesicles not only mask unwanted odors and/or flavors but also enhance the taste of the formulation, for example, by incorporation of sweeteners (e.g., maltitol) into their structures. Various techniques are employed to form these carrier systems, including spray drying, dehydration–rehydration, reverse-phase evaporation, and sonication. However, these preparation techniques suffer from one or more of the following shortcomings: (1) application of toxic organic solvents such as methanol, acetone, and chloroform; (2) difficult to scale up; (3) insufficient stability and short shelf life; and (4) lengthy and expensive preparation procedures.

Mozafari et al. (2002) have developed a procedure called the heating method for the manufacture of carrier systems, which do not suffer from the above-mentioned disadvantages. The carrier systems produced by this heating method were found to be stable, nontoxic, and efficient in encapsulation and release of nucleic acids, antioxidants (glutathione-reduced and vitamin E), and antineoplastic agents (Mozafari et al., 2004). The carrier systems prepared by the heating method can be formulated using endogenous materials present in the human body. In addition, by employing techniques such as the heating method, food/nutrient delivery formulations can be manufactured without utilizing harmful solvents. Another advantage of the formulation is the ability to impart controlled release of the ingredients at the desired location in the body.

11.8.1 Encapsulation materials

Nanoencapsulation materials, such as chitosan, PLGA, TPP, and phospholipids, are used to encapsulate the bioactive compounds. The properties and uses of these materials are discussed below.

11.8.1.1 Chitosan

Chitin is the major structural component of crustacean exoskeletons. Chitin is second only to cellulose as the most abundant polysaccharide, making chitosan a readily available and cheap biopolymer (Kumar, 2000). Chitin is structurally similar to cellulose, but it is chemically inert, which limits its applications. Chitosan is a polysaccharide that has been used in a variety of industries, including wastewater treatment, cosmetics, food and nutrition, photography, textiles, and the medical and pharmaceutical industries. Within the medical and pharmaceutical industries, chitosan has been applied to ophthalmology, artificial skin generation, chitosan-based wound dressings, and drug delivery (Mazzola, 2003).

Chitosan consists of repeating β-(1-4)-linked D-glucosamine and N-acetyl-D-glucosamine units generated by the alkaline N-deacetylation of chitin (Kumar, 2000; Kurita, 2006). Chitosan is available in molecular weights ranging from 3,000 to 20,000 Da and in percentage of deacetylation ranging from 66 to 95% (Shaji et al., 2010). The main difference between chitosan and cellulose is that chitosan is composed of 2-amino-2-deoxy-h-d-glucan combined with glycosidic linkages and the primary amine groups are the source of the properties that make chitosan useful in pharmaceutical applications. Chitosan is a weak base polysaccharide and a cation in acidic solutions, due to its free amine groups which become protonated, making it water soluble. Interest in the application of natural polymers as components of drug delivery systems has greatly increased over the past decade. Natural polymers have distinct advantages over their synthetic counterparts. Natural polymers are nontoxic, biodegradable, and allow for cell-specific targeting. Chitosan is biocompatible with living organisms and tissues because it does not cause any allergic or rejection reactions. It is biodegradable. It breaks down gradually, leaving harmless amino sugars that the body absorbs (Shaji et al., 2010).

The mucoadhesive (Agnihotri et al., 2004; Lehr et al., 1992; He et al., 1998) and low toxicity (He et al., 1998; Huang et al., 2004) properties of chitosan have been well documented and are a major reason why chitosan drug delivery systems have been so intensely investigated. Chitosan's mucoadhesive properties increase the residual time at the site of absorption, thus increasing the efficiency of drug delivery. Chitosan's free amine groups and cationic nature allow for ionic cross-linking with multivalent anions for the production of nanoparticles. Its positive charge also enables particles to act as permeation enhancers across epithelial cell membranes. Chitosan is of particular interest in the field of nanodrug delivery because of these many benefits and advantageous properties. Chitosan as a base component has been explored for use in buccal, intestinal, nasal, periodontal, and wound healing, and many other drug delivery systems (Prabaharan and Mano, 2005). Products have taken the form of beads (Chandy and Sharma, 1992), microcapsules, microspheres (Kurita, 2006), nanoparticles (Fernandez-Urrusuno et al., 1999; Ma et al., 2002), and tablets (Upadrashta et al., 1992).

11.8.1.2 PLGA

Poly(lactic-co-glycolic acid) (PLGA) is a co-polymer used in therapeutic devices owing to its biodegradability and biocompatibility. PLGA is synthesized by ring-opening co-polymerization of two different monomers, the cyclic dimers (1,4-dioxane-2,5-diones) of glycolic acid and lactic acid. Polymers can be synthesized as either random or block copolymers, thereby imparting additional polymer properties. Common catalysts used in the preparation of this polymer include tin(II) 2-ethylxanoate, tin(II) alkoxides, or aluminium isopropoxide. During polymerization, successive monomeric units (of glycolic or lactic acid) are linked together in PLGA by ester linkages, yielding a linear, aliphatic polyester as a product (Samadi et al., 2013).

Depending on the ratio of lactide to glycolide used for the polymerization, different forms of PLGA can be obtained. These are usually identified by the molar ratio of the monomers used (e.g., PLGA 75:25 identifies a copolymer whose composition is 75% lactic acid and 25% glycolic acid). The crystallinity of PLGAs varies from fully amorphous to fully crystalline depending on block structure and molar ratio. PLGAs typically show a glass transition temperature in the range of 40–60°C. PLGA can be dissolved by a wide range of solvents, depending on composition. Higher lactide polymers can be dissolved using chlorinated solvents whereas higher glycolide materials require the use of fluorinated solvents such as 1,1,1,3,3,3-hexafluoro isopropanol (HFIP).

PLGA degrades by hydrolysis of its ester linkages in the presence of water. The time required for degradation of PLGA is related to the ratio of monomers used in production: the higher the content of glycolide units, the lower the time required for degradation as compared to predominantly lactide materials. An exception to this rule is the copolymer with a 50:50 monomer ratio, which exhibits faster degradation (about 2 months). In addition, polymers that are end-capped with esters (as opposed to the free carboxylic acid) demonstrate longer degradation half-lives.

PLGA is a biodegradable polymer. It undergoes hydrolysis in the body to produce the original monomers, lactic acid and glycolic acid. These two monomers under normal physiological conditions are by-products of various metabolic pathways in the body. Since the body effectively deals with the two monomers, there is minimal systemic toxicity associated with using PLGA for drug delivery or biomaterial applications. The possibility to tailor the polymer degradation time by altering the ratio of the monomers used during synthesis has made PLGA a common choice in the production of a variety of biomedical devices such as grafts, sutures, implants, prosthetic devices, surgical sealant films, and micro- and nanoparticles. PLGA is the best chemical compound for drug delivery. There is great interest in using PLGA-based nanotechnology in medical applications such as sustained drug release, drug delivery, diagnostics, and treatment. Its unique properties include biocompatibility, bioavailability, and variable degradation kinetics, high drug loading capability, stability, and extended drug release over other carriers such as liposomes (Smola et al., 2008; Peter et al., 2004; McElvaney et al., 1991; Labiris and Dolovich, 2003). PLGA protects the encapsulated drug from enzymatic degradation and changes the pharmacokinetics of the drug. It provides a wide range of degradation rates from months to years depending upon its composition and molecular weight (McElvaney et al., 1991). FDA-approved products using PLGA as carriers include Nutropin Depot for growth deficiencies, Sandostatin LAR for acromegaly, and Trelstar Depot for prostate cancer (Cryan, 2005). Many other PLGA-based formulations are currently at the preclinical stage.

The biodegradation rate of PLGA polymers is dependent on the lactide/glycolide, molecular weight, degree of crystallinity, and the transition glass temperature (Tg) of the polymer (Uhrich et al., 1999). The release profile of PLGA nanoparticles can be divided into four different phases: initial burst, induction period, sustained release period, and final release period. Polymers containing a 50:50 ratio of lactic and glycolic acid have faster hydrolytic activities than those with other ratios of the monomers. PLGA nanoparticles can be used safely for oral, nasal, pulmonary, parenteral, transdermal, and intraocular routes of administration (Lu et al., 2009). Different techniques are used to prepare the PLGA nanoparticles. The emulsification solvent evaporation technique is most commonly used, due to its simplicity and high encapsulation efficiency. The single emulsion method is only suitable for hydrophobic drugs and leads to very poor encapsulation efficiency for protein or peptide drugs. The oil-in-oil (o/o) emulsification technique, which is known as the non-aqueous emulsion method, is a new and efficient method for encapsulation of hydrophilic drugs. The double emulsion method is also suitable for encapsulating hydrophilic drugs with high efficiency (Mundargi et al., 2008).

11.8.1.3 TPP

Tripolyphosphate is generally used along with chitosan in preparing nanoparticles based on ionic gelation method. Sodium tripolyphosphate (TPP) is a cross-linking agent. It is shown that favorable micromixing conditions are created on top of the membrane surface to form chitosan–TPP nanoparticles.

11.8.1.4 Phospholipids

Phospholidipids such as phosphotidyl choline are generally used to prepared nanosomes, and these lipids are charged ones.

11.9 Methods of producing chitosan nanoparticles

Several methods can be used to produce chitosan nanoparticles, including coacervation/precipitation, emulsion-droplet coalescence, ionotropic gelation, and reverse micellar methods. When selecting a method the requirements for particle size, thermal, structural and chemical stability, residual toxicity, polydispersity and the release kinetics of the final product must be considered. Method selection also depends on the intended therapeutic agent for delivery as well as the nanoparticle patient application procedure. The production methods described below are currently the most commonly used procedures.

11.9.1 Coacervation/precipitation

The coacervation/precipitation method is based on the physical and chemical properties of chitosan, which make it insoluble in alkaline pH medium. This insolubility in alkaline medium means that chitosan precipitates/coacervates out of solution when mixed with an alkaline solution. One method of applying these principles to generate chitosan nanoparticles is to blow chitosan solution into an alkali solution using a compressed air nozzle. Sodium hydroxide, sodium hydroxide-methanol, or ethanediamine alkali solutions are commonly used. The particles are separated and purified either through filtration, centrifugation, or both and are then washed successively with hot and cold water (Shaji et al., 2010). The particle size is controlled by altering the compressed air pressure and/or the spray-nozzle diameter. The release mechanics can be altered with the addition of a cross-linking agent, such as glutaraldehyde, to harden the particles. This method is most commonly used for microparticle production, but it can also be applied for nanoparticle production. The method that is most frequently used for chitosan nanoparticle production is known as complex coacervation. This method involves the rapid mixing of chitosan solution with drug-loaded alkali solution.

11.9.2 Emulsion-droplet coalescence

The emulsion-droplet coalescence method employs both precipitation and emulsion cross-linking principles. Unlike the emulsion cross-linking methods applied to produce chitosan microparticles, which cross-links stable droplets, this method induces precipitation by merging chitosan droplets with sodium hydroxide droplets. This method was first utilized by Tokumitsu et al. (1999). The first step is to produce a stable emulsion containing aqueous chitosan and therapeutic agent solution, which is generated in liquid paraffin oil. The same process is used to produce another stable emulsion containing chitosan aqueous solution in sodium hydroxide. The emulsions are then mixed using high-speed stirring, which causes random collisions and coalescence of droplets from each emulsion, precipitating chitosan droplets and producing nanoparticle-encapsulated gadopentetic acid. This method is used for gadolinium neutron capture therapy (Tokumitsu et al., 1999).

Increasing the gadopentetic acid concentration in the chitosan solution increased encapsulation efficiency but did not increase the particle size. The investigation produced nanoparticles with a mean particle size of 452 nm and 45% gadopentetic acid loading using 100% deacetylated chitosan. This is the best example of selecting the production process based on all factors, including drug properties. Gadopentetic acid is a bivalent anionic compound interacts electrostatically with the chitosan amine groups; however, this interaction would not happen if a particle production method that utilizes a cross-linking agent that blocks the free amine groups of chitosan was used. Using the coalescence method optimizes encapsulation efficiency.

11.9.3 Reverse micelles

Reverse micelles are mixtures of water, oil, and surfactant that are thermodynamically stable and have a dynamic behavior. Viewed up close at the microscopic scale, the structure of reverse micelles consists of aqueous and oil volumes separated by surfactant films. As the observer zooms out to a more macroscopic scale reverse micelles appear homogeneous and isotropic.

The advantage of reverse micellar formation is that ultrafine polymeric nanoparticles with narrow size distributions are produced. Traditional emulsion polymerization methods result in formation of larger nanoparticles (>200 nm) with broad size distributions. The reverse micelle aqueous core acts as a nanoreactor in the preparation of ultrafine nanoparticles. The minute dimensions and narrow size distribution of nanoparticles produced through the reverse micellar method are due to the small size of the reverse micellar droplets themselves, which are usually between 1 and 10 nm in dimension, and their high degree of monodispersity (Shaji et al., 2010).

Reverse micelles experience continuous coalescence and reseparation on a timescale that fluctuates from milliseconds to microseconds due to their constant Brownian motion. Thus, through a rapid dynamic equilibrium the system is able to preserve its size, polydispersity, and thermodynamic stability. When performing the reverse micelle method, surfactant is first dissolved in an organic solvent to create the reverse micelles. Then the chitosan and bioactive compound aqueous solution are added under constant vortexing to avoid any turbidity.

The system can be standardized based on the amount of aqueous phase. The aqueous phase is generally maintained at a level where the entire mixture is an optically transparent micro-emulsion. The aqueous phase also dictates the particle size. With an increased amount of water larger nanoparticles are obtained. Like other types of nanoparticle systems, the maximum compound loading varies between different therapeutic agents, but reverse micelles maximum drug loading can be found by gradually adding compound until the clear microemulsion is converted into a translucent solution.

The next step is to add a cross-linking agent under constant stirring and mix overnight to allow for complete cross-linking to occur. A transparent dry mass is collected by evaporating the organic solvent, after which the mass is dispersed in water and an appropriate salt is added to precipitate out the surfactant. The mixture is then centrifuged to collect the compound-loaded nanoparticles. Finally, after being resuspended in aqueous solution, the mixture is immediately dialyzed through a dialysis membrane for approximately 1 h. The liquid is then freeze dried, resulting in a dry powder.

11.9.4 Ionotropic gelation

Ionotropic gelation is a process that cross-links a polyelectrolyte with a counter ion, forming a hydrogel. The structures of hydrogels are maintained by hydrogen bonding, hydrophobic forces, ionic forces, or molecular entanglements (Prestwicha et al., 1997). This technique has been applied using a variety of materials, including gellan gums, alginates, carboxymethyl cellulose, and chitosan to create micro- and nanoparticles for encapsulation and controlled release of therapeutic compound. Several methods of ionotropic gelation can be applied, depending on the material used, the strength of the counter ion, and the desired particle size.

Syringe dropping and air atomization are used for bead formation (Patil et al., 2010) and flush mixing is used for nanoparticle formation (Fernandez-Urrusuno et al., 1999). The application of ionotropic gelation to form micro- and nanoparticles, which uses chitosan's positive charge to cross-link it with an anion, has attracted considerable attention because the process is very simple and takes place under mild conditions. "Mild conditions" refers to the lack of possible toxic reagents and other undesirable effects associated with chemical cross-linking compared to reversible physical cross-linking by electrostatic interaction.

The most well-documented chitosan nanoparticles produced by ionotropic gelation are chitosan-tripolyphosphate nanoparticles, which many research groups have explored as a potential drug delivery system (Fernandez-Urrusuno et al., 1999; Ma et al., 2002). Potential anions are tripolyphosphate (TPP), sodium alginate, κ-carrageenan, and hexadesyl sulphate. In the ionotropic gelation process chitosan is first dissolved in aqueous acidic solution, obtained using acetic acid, which generates the chitosan cations. The chitosan solution is added dropwise under constant stirring to polyanionic solution (i.e., TPP). The chitosan precipitates to form spherical particles through the ionic interaction between the oppositely charged species (Fernandez-Urrusuno et al., 1999). Prepared insulin-loaded chitosan-TPP nanoparticles are added to the TPP solution prior to mixing with the chitosan solution (Figure 11.1).

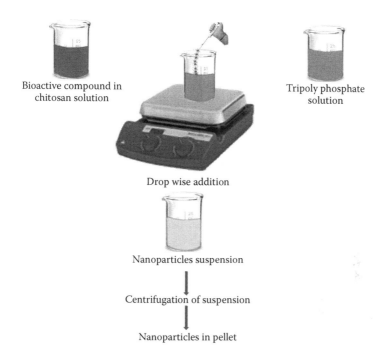

Figure 11.1 Preparation of nanoparticles by the chitosan-tripolyphosphate method.

11.10 Applications of chitosan nanoparticles

As mentioned earlier, chitosan has been used in a large number of drug delivery systems covering a variety of applications with different therapeutic goals and application regions of the human body. Chitosan nanoparticles have contributed to drug delivery research in the fields of cancer therapy, gene delivery, and ocular delivery. One example of their application to cancer therapy is the gadopentetic acid delivery system, mentioned in 11.9.2, which was used for delivery to tumor sites for gadolinium neutron capture therapy (Tokumitsu et al., 1999). Another cancer therapy system that has been explored is chitosan nanoparticles loaded with doxorubicin-dextran conjugate, which showed reduced side effects and increased therapeutic efficiency when applied to solid tumors (Mitra et al., 2001).

11.11 Activity of nanoparticles

Nanoparticles are used as antimicrobials, anticancer, flavor capsules, and other drug delivering agents. Examples are discussed below.

11.11.1 Controlled release of liposomes

Nanoemulsions are stable colloidal systems within nanometric size (≤ 100 nm) formed by dispersing one liquid in another immiscible liquid using suitable emulsifiers (Burguera and Burguera, 2012). Compared with microemulsions, nanoemulsions are optically transparent and demonstrate better shelf stability. The droplet size distribution remains after water dilution (Fathi et al., 2012). Depending on the desired structure and functionality, nanoemulsions can be prepared using high-energy methods (high-pressure homogenization, microfluidization, and ultrasonication) and low-energy methods (solvent diffusion). High-energy methods produce intense disruptive forces minimizing droplet size to form emulsions, while low-energy methods promote spontaneous emulsification by mixing all the emulsion ingredients (Burguera and Burguera, 2012; Donsì et al., 2011; Ghosh et al., 2014).

Liposomes take several forms: (1) oil in water (o/w), where the oil droplets are dispersed in the aqueous phase and the interphase is stabilized by emulsifiers; (2) the multiple emulsions oil-in-water-in-oil (o/w/o) and water-in-oil-in-water (w/o/w), where, for example, nanometer-size water droplets contained within large oil droplets are dispersed within an aqueous phase (w/o/w); and (3) multilayer emulsions, which consist of oil droplets surrounded by nanometric size layers of different polyelectrolytes (Weiss et al., 2006). The o/w nanoemulsions can encapsulate and deliver poorly water-soluble compounds, improving physical stability of the active compound and increasing its active distribution in food matrices.

The nanometric droplet size has the advantage of increasing interactions of the active compound with the cell membrane of bacteria, affecting the stability of the lipid membrane and resulting in leakage of bacteria intracellular constituents. This nonspecific action mechanism decreases the development of resistant microbial strains. Ideally, an optimal delivery system for antimicrobial compounds that could have applications in the food industry would enhance the mass transfer rates to the sites of action, in order to maximize the antimicrobial activity and to use concentrations that are low enough to minimally alter the quality of the product, but at the same time sufficient to inhibit microbial growth within the limits of food regulations (Maswal and Dar, 2014). Some food antimicrobial agents have been encapsulated in nanoemulsions.

Essential oils are generally recognized as safe (GRAS) food additives according to the FDA. They display activity against human pathogenic and food spoilage microorganisms, but water solubility constraints, evaporation, and sensory properties have limited their incorporation into food products (Kim et al., 2013). Encapsulation of essential oils at the nanoscale level is an available and efficient approach that increases their physical stability and reduces the mass transfer resistance of the active molecules to the sites of action.

Terjung et al. (2012) developed nanoemulsions containing carvacrol and eugenol with triacylglyceride or Tween 80® by high-pressure homogenization and ultrasonication.

11.11.2 Antimicrobial activity of nanoemulsions

The antimicrobial activity of carvacrol emulsions was tested against microorganisms such as *Escherichia coli* C 600 and *Listeria innocua*. Emulsions with a mean oil droplet size of 3000 nm at a concentration of 800 ppm completely inhibited *L. innocua*, while only a delay in growth was observed for 80-nm emulsions. This was attributed to the fact that antimicrobial nanoemulsions were less active than macroemulsions due to an increased sequestering of antimicrobials in emulsion interfaces and a decreased solubilization in excess Tween 80 micelles (Terjung et al., 2012). Carvacrol, limonene, and cinnamaldehyde were encapsulated in sunflower oil–based nanoemulsions obtained by high-pressure homogenization and stabilized by different emulsifiers.

Emulsifiers that solubilized the essential oil in the aqueous phase (lecithin and pea proteins) slightly promoted bacteriostatic action (Donsì et al., 2012). Basil oil (*Ocimum basilicum*) containing 88% of estragole was encapsulated in a nanoemulsion formulated with Tween 80 and water by the ultrasonic emulsification method. The nanoemulsion showed antibacterial activity against *E. coli* even after being diluted. For example, 10-fold and 100-fold dilutions completely inactivated *E. coli* after 45 min, while the 1000-fold dilution achieved a reduction of 40% after 60 min. Fluorescence microscopy and Fourier transform infrared spectroscopy (FTIR) results showed that nanoemulsion promotes bacterial cell membrane alterations (Ghosh et al., 2014).

Lemongrass oil (LO) was encapsulated in a carnauba-shellac wax (CSW)-based nanoemulsion by high pressure homogenization (Kim et al., 2013; Song et al., 2014) and alginate nanoemulsions by ultrasonication and microfluidization (Salvia-Trujillo et al., 2014). The LO-loaded CSW-based nanoemulsions inhibited growth by 8.18 log CFU/g the total population of *E. coli* O157:H7 and *L. monocytogenes* after 2 h. After five months of storage, the unloaded and LO-loaded CSW nanoemulsions applied as an edible coating to apples inhibited the growth a decrease of 0.8 and 1.4 log CFU of aerobic bacteria, respectively. The coatings inhibited the development of yeast and molds. Moreover, the coating inhibited the growth of *Salmonella typhimurium* and *E. coli* O157:H7 strains on apples and plums, respectively. The application of nano-emulsions also preserved various physicochemical qualities of fruits (Kim et al., 2013; Song et al., 2014).

Lipid-based nanoencapsulation systems are among the most rapidly developing fields of nanotechnology application in food systems. Lipid-based nanoencapsulation systems have several advantages, including the ability to entrap material with different solubilities and the use of natural ingredients on an industrial scale (Bummer, 2004; Mozafari et al., 2006; Taylor et al., 2005).

Lipid-based nanocarriers can also be used for targeted delivery of their contents to specific areas within the gastrointestinal tract or food matrix.

When referring to nanoscale lipid vesicles, the term *nanoliposome* was introduced (Mozafari et al., 2006) to describe lipid vesicles whose diameter ranges up to tens of nanometers. These so-called nanoliposomes have similar structural, physical, and thermodynamic properties as liposomes. The manufacture of nanoliposomes (as of liposomes) requires high energy for the dispersion of lipid/phospholipid molecules in the aqueous medium. One of the most promising lipid-based nanodelivery systems for food applications is nanosized self-assembled liquids (NSSLs) (Garti et al., 2005).

NSSL vehicles tackle shortcomings of microemulsion systems. Mixtures of food-grade oils (two or more food-grade, nonionic hydrophilic emulsifiers), co-solvent (polyol), and co-emulsifiers that self-assemble to form mixed reverse micelles ("the concentrate") can be inserted into o/w nanodroplets. This system is transformed into discontinuous structures by dilution with an aqueous phase, progressively and continuously, without phase separation. These reversed micelles can solubilize compounds that are poorly soluble in water or in the oil phase. NSSLs can be used to solubilize hydrophobic substances several times their normal solubility. The use of solid lipid nanoparticles (SLNs) and nanostructured lipid carriers (NLCs, which are structured SLNs) have been developed in the last two decades for mainly pharmaceutical purposes (Müller et al., 2000; Radtke et al., 2005), but may also find their way into foods. NLCs may be prepared by melting lipid(s), dissolving a lipophilic agent into the molten lipid, followed by hot high-pressure homogenization of the molten lipid phase in the presence of aqueous surfactant solution at 5–10°C above the melting temperature of the lipid.

Alternatively, one may use a cold homogenization technique, in which the lipid with the agent is ground into 50- to 100-mm microparticles and then homogenized in an aqueous surfactant solution at a temperature well below the melting temperature of the lipid (<10–20°C below the melting temperature of the lipid). The cold homogenization procedure might be used to prepare SLN with temperature-sensitive or even hydrophilic agents. One may spray-dry such a suspension of SLN in the presence of a water-soluble carrier material to obtain a dried powder with a particle size between 20 and 100 mm (Shefer and Shefer, 2003), with an option to entrap an extra water-soluble active in the water-soluble coating.

11.11.3 Protein-based nanoencapsulates

Semo et al. (2007) used self-assembled casein micelles as nanocapsular vehicles. These authors realized that casein micelles are in effect nanocapsules created by nature to deliver nutrients such as calcium, phosphate, and protein to the neonate. Thus, they suggested using casein micelles as a self-assembled system for nanoencapsulation and stabilization of hydrophobic nutraceutical substances for enrichment of food products. Vitamin D_2 was used as a model for hydrophobic nutraceutical compounds. The reassembled micelles had average diameters of 146 and 152 nm without and with vitamin D_2, respectively, similar to normal casein micelles, which are typically 150 nm on average. The vitamin concentration in the micelle was about 5.5 times more than in the serum. The use of protein–polysaccharide interaction to form encapsulation systems based on complex coacervation was downsized to the nanoscale by Huang and Jiang (2004). They used coacervates formed by gelatin type A-carrageenan complexes as an inexpensive encapsulation method for the green tea catechin, i.e., epigallocatechin gallate (EGCG) at the micro- and nanoscale levels.

Nanoparticles can also be formed by combining molecular inclusion of lipophilic active agent such as vitamin D or docosahexaenoic acid (DHA) with β-lactoglobulin with complex coacervation by electrostatic interactions between this β-lactoglobulin and polysaccharides such as pectin at pH 4.5 (Zimit and Livney, 2009).

11.11.4 Polysaccharide-based nanoencapsulates

Vitamin E nanoparticles were prepared by using modified starch because it is stable in beverage and did not alter the beverage appearance (Chen and Wagner, 2004). Particles were produced by dissolving starch sodium octenyl succinate in distilled water with vitamin E acetate added slowly and homogenized with a high shear mixer until the emulsion droplet size was below 1.5 mm. The crude emulsion was then further homogenized until the emulsion droplets reached the target particle size. They were then spray-dried to yield a powder containing about 15% vitamin E acetate.

11.11.5 Inorganic-based nanoencapsulates

One of the few examples of inorganic-based nanoencapsulated is silica nanoparticles obtained by sol-gel synthesis of silica gels in W/O microemulsions to encapsulate enzymes (Cellesi and Tirelli, 2006). The process published, however, was not food grade, although this might be possible in theory. Another example is two-dimensional, layered double hydroxide nanohybrids, composed of vanillic acid, zinc, and aluminium oxides (molar ratios of 1.5:2:1–1.5:4:1) (Hong et al., 2008).

11.12 Characterization of nanoparticles by analytical methods

There is evidence that physicochemical properties, such as size and surface chemistry, can dramatically affect nanoparticle behavior in biological systems (Kobayashi et al., 2003; Oberdorster et al., 2005; Ogawara et al., 1999) and may, in part, determine the bio-distribution, safety, and efficacy of a particle.

Preclinical physicochemical characterization of a nanoparticle includes (but is not limited to) measurement of size and shape, surface chemistry, and aggregation/agglomeration state. These properties are dependent on environmental conditions and must therefore be assessed not only in their "as-dosed" form but also in a biological context. Many techniques are available for this characterization of nanoparticles in the literature (Patri et al., 2006; Powers et al., 2006). Nanoparticle size characterization is often complicated by the poly dispersity of samples, but characterizing nanoparticles using different methods, such as both TEM and DLS, can provide size information on nanoparticles that might be relevant to their therapeutic performance.

Several techniques are used to understand these characterization parameters in nanoparticles, including transmission electron microscopy (TEM), scanning electron microscopy (SEM), dynamic light scattering (DLS), *in-vitro* evaluation, and chemical method.

11.12.1 Electron microscopy: Transmission electron microscopy and scanning electron microscopy

Nanoparticles are frequently irresolvable by optical microscopy, so electron microscopy (EM) is required to measure nanoparticle size and shape. Scanning electron microscopy (SEM) can be used to obtain information on the size, size distribution, and shape of nanoparticles. Transmission electron microscopy (TEM) often uses more powerful electron beams than SEM. It is therefore higher resolution and provides greater detail at the atomic scale such as information about the crystal structure and granularity of a nanoparticle.

Several biological compounds (e.g., liposomes and proteins) are invisible to EM without heavy-metal staining procedures because these compounds do not deflect an electron beam sufficiently. For these "soft" particles, EM does not always provide a relevant measure of particle size. Even for electron-dense nanoparticles (e.g., metal colloids) with attached multifunctional components (e.g., polyethylene glycol (PEG)) targeting antibodies or drugs. EM often cannot image the surface groups without the use of cryogenic methods and therefore does not measure the physiologically relevant size readily. Furthermore, TEM requires high-vacuum and thin sample sections for electron-beam penetration. Sample preparation and drying might alter the physicochemical state of the nanoparticle or introduce artifacts.

11.12.2 Dynamic light scattering

One technique that can provide a measure of particle size in solution is dynamic light scattering (DLS), which is also known as photon correlation spectroscopy (PCS). In a DLS measurement, a beam of laser light is scattered off the nanoparticle solution and small, time-dependent fluctuations in the intensity of the scattered light are monitored with a photon detector.

The fluctuations are caused by the Brownian diffusion of the nanoparticles and the Stokes–Einstein equation relates the timescale of particle diffusion (i.e., the timescale of the scattered light fluctuations) to the equivalent-sphere hydrodynamic diameter of the particle, taking into account the viscosity of the sample solution and the temperature at which the measurement is performed (Berne and Pecora, 2000). An equivalent sphere is a rigid sphere that diffuses at the same rate as the analyte particle.

As DLS measures hydrodynamic diameter, it provides a fundamentally different measure of particle size than EM, which measures electron diffraction. DLS is very sensitive to biological molecules, such as polymers, proteins, and antibodies, because they cause significant frictional drag, which can dramatically influence the rate of the particle's motion under Brownian diffusion. However, because DLS measures the equivalent-sphere hydrodynamic diameter, it provides no information about particle shape. DLS also cannot be used to measure particle size if the sample absorbs at the wavelength of the laser used in the size-measuring instrument (otherwise a different laser must be used). Also, the intensity of scattered light from small spherical particles is proportional to particle diameter to the 10^6, so that larger particles scatter light much more efficiently than smaller particles. Therefore, even small traces of agglomerates or dust in a sample can skew a DLS measurement. Finally, unless conducted in-line with a separation/fractionation method, DLS cannot separate similar sized populations, such as monomers and dimers of the same species. Using multiple techniques, such as complementary size characterization by TEM and DLS, can be helpful in resolving ambiguities associated with either measurement technique alone.

For electron-dense nanoparticles, EM images of tissues from animal studies can provide a wealth of information on nanoparticle disposition into cells, tissues, and organs; however, it is difficult to identify particles definitively in EM images or to determine whether a multifunctional nanoparticle maintains its integrity through the various stages of metabolism. Unlike particle size and aggregation state, the elemental composition of a nanoparticle usually does not change in the biological environment (Powers et al., 2006).

11.12.3 *In-vitro* evaluation of nanoparticles

Understanding the biological mechanisms of action of nanoparticle therapeutics requires rigorous *in-vitro* biological characterization. The fewer variables and amplified reactions of *in-vitro* assays generally make their results more interpretable than their complementary *in-vivo* experiments. Unfortunately, nanoparticle-based therapeutics are frequently poorly suited to conventional in vitro pharmacological assays. For instance, many nanoparticles aggregate or adsorb proteins during in vitro assays. Other nanoparticles, such as gold colloids, scatter light and often invalidate colorimetric assays that rely on absorbance measurements. Similarly, some nanoparticles (e.g., quantum dots) have very large molar-extinction coefficients (ε) and their emission wavelengths can yield ambiguous results with colorimetric assays and light-based instrumentation. Other nanoparticles, such as dendrimers, may have catalytic properties that often interfere with standardized enzymatic tests, such as those that evaluate endotoxin contamination.

A variety of cell-based systems are available to evaluate biocompatability of nanomaterials with blood *in-vitro*. One measure of nanoparticle biocompatability is the effect of nanomaterials on the viability of adherent cell lines (Mickuviene et al., 2004). Cell viability is assessed by methods that measure loss of membrane integrity, metabolic activity, or monolayer adherence, or that monitor progression through the cell cycle. Membrane integrity assays, such as the trypan blue exclusion assay and lactate dehydrogenase (LDH)-leakage assay (Decker and Lohmann-Matthes, 1988; Korzeniewski and Callewaert, 1983), might be particularly illuminating for certain cationic nanoparticles because these nanoparticles disrupt cell membranes in some circumstances (Hong et al., 2004). Tetrazolium dye reduction, adenosine triphosphate (ATP), and H^3-thymidine incorporation assays measure cell viability through metabolic activity. Nanoparticles that are antioxidants, such as certain functionalized fullerenes, might interfere with these assays by enhancing tetrazolium-dye reduction and may result in an overestimation of cell viability. Progression through the cell cycle can be monitored by DNA staining and flow cytometry (Tuschl and Schwab, 2004). Cell-cycle effects have been demonstrated for a variety of nanoparticle samples, including carbon nanotubes (Cui et al., 2005).

Hemolysis (breakage of red blood cells) is a toxicity that can lead to life-threatening conditions, such as hemolytic anemia, jaundice, and renal failure. Several studies have reported on nanoparticle haemolysis using variations of a standard method in which the percentage of particle-induced hemolysis is evaluated by spectrophotometric detection of hemoglobin derivatives after incubating analyte particles with blood and separation of undamaged cells by centrifugation (Kim et al., 2005; Lim et al., 2004; Verma et al., 2005). However, few clear trends have been observed. This may be partly explained by variability in the experimental procedures of these studies: the incubation time, wavelength at which hemoglobin is quantified, source (human Vs rabbit), and type (whole blood Vs purified erythrocytes) of blood, all of which vary from one study to another. In addition to these variables, differences in centrifugation speeds and times and blood storage times and conditions can further complicate meaningful comparison of the results of disparate studies. One particular concern is that none of these hemolysis studies included controls that were nanoparticles themselves; such a control could be used potentially to identify interferences with the assay that are specific to nanoparticles.

Most of the nanoparticle drug delivery platforms are selected based on their biocompatibility and are not toxic or immunogenic intrinsically. For example, colloidal gold and liposomes are inert biologically. However, the body has a natural immune defense system against all small particulate materials, and the success or failure of a nanoparticle-based drug might depend on its ability to avoid or exploit this system. There are two main areas of consideration for a nanoparticle-based drug's processing and clearance: (1) rapid clearance of the nanoparticle will require large dosages in order to be effective (Moghimi et al., 2001) and (2) excessive particle uptake can potentially interfere with the cells and/or organs of the immune system, resulting in immunological side effects, such as immunosuppression or immunostimulation, and can potentially lead to organ damage (Sundstrom et al., 2004).

Nanoparticles are processed and cleared by the body through a variety of pathways, depending in part on their size and surface chemistries. One important pathway is through phagocytic uptake by cells specialized for engulfing and clearance of foreign bodies. The liver has been identified in several studies as the primary organ responsible for the capture of certain nanoparticles, often through phagocytosis by residual macrophages (Kupffer cells) (Ogawara et al., 2001; Ballou et al., 2004; Gharbi et al., 2005). Assays to evaluate cellular uptake by phagocytosis involve detection and quantification of particles inside cellular compartments. The standard assay for phagocytosis relies on light microscopy for particle detection; however, nanoparticles are frequently too small to be resolved by optical microscopy, so EM is required.

As mentioned previously, EM is sensitive primarily to electron-dense materials, such as metallic nanoparticles, and often cannot reliably image polymers, dendrimers, liposomes, or other "soft" materials similar in size to cellular organelles without heavy-metal staining procedures (e.g., with phosphotungstic acid) that might have unintended consequences on biological systems. In certain circumstances, nanoparticles can be fitted with fluorescent tags to enable the visualization of internalized particles with confocal microscopy (Prabha et al., 2002). However, the chemical attachment of a fluorophore might not be stable under physiological conditions and potentially creates a new molecular entity with different properties than those of the original nanoparticle. This limits direct confocal detection of phagocytosis to nanoformulations with intrinsic fluorescence (e.g., quantum dots, some solid lipid nanoparticles, and fullerene derivatives).

11.12.4 Chemical method

Phagocytosis can be detected in nanoparticles that are not fluorescent intrinsically and are invisible to EM through chemiluminesence that accompanies the release of reactive forms of oxygen when phagocytic cells are challenged (Campbell et al., 2001). Luminol (3-aminophthalhydrazide) releases light on reaction with oxidizing agents and light microscopy is then used to detect the luminol chemiluminescence (Gref et al., 2000). This assay also might be limited in its applicability to nanoparticles, many of which have optical properties that might quench or interfere with luminol signals inside cells. This assay also has limited sensitivity compared to the results of luminol chemiluminescence detection and direct detection by TEM of polyethylene glycol

(PEG)-coated and uncoated gold colloid in macrophages. The TEM images clearly show increased uptake of the uncoated particles compared with the PEG-coated particles; however, there is no discernable difference in luminol chemiluminescence between samples.

11.13 Potential health benefits

Nutraceuticals have attracted considerable interest due to their potential nutritional, safety, and therapeutic effects. They play a role in a plethora of biological processes, including antioxidant defenses, cell proliferation, gene expression, and safeguarding of mitochondrial integrity. Nutraceuticals may be used to improve health, prevent chronic diseases, postpone the aging process (and in turn increases life expectancy), or just support functions and integrity of the body. They are considered to be healthy sources of constituents to prevent life-threatening diseases such as diabetes and renal and gastrointestinal disorders, as well as various infections.

A wide range of nutraceuticals have played crucial roles in immune status and susceptibility to certain disease states. They also exhibit disease-modifying indications related to oxidative stress including allergies, Alzheimer's disease, cardiovascular diseases, cancer, eye conditions, Parkinson's disease, and obesity.

11.13.1 Advantages of nanoencapsulation

Nanoencapsulation offers tremendous potential in providing a means to deliver difficult-to-attain nutrients in fortified food products. Nutraceuticals that provide nutrients with a clear therapeutic or disease prevention effect are better delivered and more bioavailable through nanoencapsulation. Consumers appreciate these potential benefits. Their concerns are mainly about food modification in general and thus is not particular to nanotechnology, but the emphasis on modification/fortification of foods may divert resources from providing organic, fresh, and unprocessed foods.

11.13.1.1 Nanoencapsulation of antioxidants and flavoring agents

The instability and low bioavailability of polyphenols limit their applications in the food industry. Epigallocatechin gallate (EGCG) and soybean seed ferritin deprived of iron (apoSSF) were used as a combined double-shell material to encapsulate rutin flavonoid molecules (Yang et al., 2016). Green tea polyphenols have been reported to have many biological properties. Despite the many potential benefits of green tea extracts, their sensitivity to high temperature, pH, and oxygen is a major disadvantage hindering their effective utilization in the food industry. The stability of green tea extracts was improved by spray-drying using different carrier materials including maltodextrin (MD), gum arabic (GA), and chitosan (CTS), and their combinations at different ratios. Storage stability of encapsulated catechin extracts under different temperature conditions was improved remarkably improved compared to nonencapsulated extract powder (Zokti et al., 2016). Blackberry (*Rubus fruticosus*) juice contains compounds with antioxidant activity, which can be protected by different biopolymers used in the microencapsulation. MD, GA, and whey protein concentrate (WPC) were used to encapsulate the blackberry juice and retain its antioxidant properties of encapsulated blackberries using a spray-drying technique (Díaz et al., 2015).

Alginate, chitosan, calcium chloride, and trans-cinnamaldehyde were used to synthesize the nanoparticles. Trans-cinnamaldehyde-incorporated chitosan-alginate nanoparticles were synthesized using the ionic gelation and polyelectrolyte complexation technique. Optimized nanoparticles showed increased stability (6 weeks) and translucency in solution. The final radical scavenging effect of loaded particles in apple juice was 62%. Trans-cinnamaldehyde was just as available to react in free form as it was in inclusion complexes. These particles were used as an flavoring agent. Their antimicrobial and antioxidant compounds have the potential to improve the effectiveness and efficiency of delivery in food systems (Loquercio et al., 2015). Eugenol and trans-cinnamaldehyde are natural compounds known to be highly effective antimicrobials; however, both are hydrophobic molecules, which limits their use within the food industry. Nanoparticles were synthesized using poly (DL-lactide-co-glycolide) (PLGA) to entrapped eugenol and trans-cinnamaldehyde as an antimicrobial delivery application. The emulsion evaporation method was used to form the nanoparticles in the presence of poly (vinyl alcohol) (PVA) as a surfactant. The inclusion of antimicrobial compounds into the PLGA nanoparticles was accomplished in the organic phase. Synthesis was followed by ultrafiltration (performed to eliminate the excess of PVA and antimicrobial compound) and freeze-drying. The nanoparticles were characterized for their antimicrobial efficiency. All loaded nanoparticle formulations proved to be efficient in inhibiting the growth of *Salmonella* spp. (Gram-negative bacterium) and *Listeria* spp. (Gram-positive bacterium) with concentrations ranging from 20 to 10 mg/mL. Nanoencapsulation of lipophilic antimicrobial compounds has great potential for improving the effectiveness and efficiency of delivery in food systems (Gomes et al., 2011).

Vanilla oil (VO) was nanoencapsulated using the complex coacervation approach, to control release of VO and enhance its thermostability for spice application in the food industry. The nanocapsules were spherical in shape with good dispersibility when moderate viscosity CS was used. Moreover, VO could remain about 60% in the microcapsules after release for 30 days, which demonstrated the flavor microcapsules had good potential to serve as a high-quality food spice with long residual action and high thermostability (Yang et al., 2014).

11.13.1.2 Nanoencapsulation of anti-cancer agents

Nanoencapsulation of calcitriol (1,25-dihydroxyvitamin D_3) was employed to inhibit cancer growth. Different polymer and oil ratios were used to load the nanoparticles. Calcidiol/calcitriol-loaded nanoparticles had good encapsulation efficiencies (around 90%) associated with sustained release over 7 days and enhanced stability. Calcitriol, the active metabolite of vitamin D_3, is a potential anticancer agent but with high risk of hypercalcemia, which limits the achievement of effective serum concentrations. The growth inhibitory efficiency of these nanoparticles was carried out *in-vitro* on human breast adenocarinoma cells (MCF-7). The nano-encapsulation of vitamin D_3 active metabolites might offer a new and potentially effective strategy for vitamin D_3–based chemotherapy overcoming its actual limitations (Almouazen et al., 2013).

Curcumin liposomes were prepared to evaluate the *in-vitro* skin permeation and *in-vivo* antineoplastic effect of curcumin using liposomes as the transdermal drug-delivery system. Soybean phospholipids (SPC), egg yolk phospholipids (EPC), and hydrogenated soybean phospholipids (HSPC) were selected for the preparation of different kinds of phospholipids composed of curcumin-loaded liposomes: C-SPC-L (curcumin-loaded SPC liposomes), C-EPC-L (curcumin-loaded EPC liposomes), and C-HSPC-L (curcumin-loaded HSPC liposomes). An in vitro skin penetration study indicated that C-SPC-L most significantly promoted drug permeation and deposition, followed by C-EPC-L, C-HSPC-L, and curcumin solution. Moreover, C-SPC-L displayed the greatest ability of all loaded liposomes to inhibit the growth of B16BL6 melanoma cells. A significant effect on antimelanoma activity was observed with C-SPC-L compared to treatment with curcumin solution a. The above study results suggest that C-SPC-L is a promising transdermal carrier for curcumin in cancer treatment (Chen et al., 2012). Curcumin is a poorly water-soluble compound. For bioactivity lipophilic phytochemicals need encapsulation. A low-cost, low-energy, and organic solvent-free encapsulation technology utilizing the pH-dependent solubility properties of curcumin and self-assembly properties of sodium caseinate (NaCas) can be employed. Curcumin needs to be deprotonated and dissolved, while NaCas gets dissociated at pH 12 and 21°C. The subsequent neutralization enables the encapsulation of curcumin in self-assembled casein nanoparticles. The curcumin encapsulated in casein nanoparticles showed significantly improved anti-proliferation activity against human colorectal and pancreatic cancer cells (Pan et al., 2014).

11.14 Disadvantages of nanoencapsulation

Legislation is the biggest barrier to commercialization of nano-based foods. It is difficult to provide guidelines for regulation and standards. The lack of comprehensive knowledge of toxicity of nanomaterials is a broader issue, but it becomes specific in the case of encapsulation for foods precisely because the technology aims at increasing bioavailability. European research in this area is quite strong, and companies involved in nano- and microencapsulation are visible. But without a clearer regulatory landscape, growth in this area and the exploitation of research findings in nanoencapsulation will be considerably limited. With regard to exposure for consumers, the fate of the vectors that carry the encapsulated nutraceuticals is unclear. The short- and longer-term effects need to be further explored. Knowledge about the toxicity of nano-vectors as well as their overall bioavailability throughout the body is still relatively sparse. During manufacturing of nano-structured delivery systems, workers may potentially be exposed to nanomaterials. Upon disposal or after excretion from the human body, exposure of surface water and soil is possible. Hazard and risk assessment of nanomaterials should be completed on a case-to-case basis.

11.15 Conclusions

Developments in nanotechnology continue to emerge, and its applicability to the food industry continues to grow. Most aspects of incremental nanotechnology are likely to enhance product quality and choice and will be perceived as progressive changes in standard and accepted technology. There are a few issues, particularly regarding the accidental or deliberate use of nanoparticles in food, or food-contact materials, which may cause consumer concern.

It is particularly important to ensure that consumers have the option to choose to use the products of nanotechnology and that they have information available to assess the benefits and risks of such products. The success of these advancements will depend on consumer acceptance and the exploration of regulatory issues. Food producers and manufacturers can make great strides in food safety using nanotechnology and consumers will reap benefits as well. The level of interest in companies conducting research in nanotechnology and its application to food products is certain to increase.

The future belongs to new products and new processes with the goal of customizing and personalizing the products. Improving the safety and quality of food is the first step. Designing and producing food by shaping molecules and atoms is the future of the food industry worldwide.

References

Agnihotri SA, Mallikarjuna NN, Aminabhavi TM (2004). Recent advances on chitosan-based micro- and nanoparticles in drug delivery. *J Control Release*. 100:5–28.

Almouazen E, Bourgeois S, Jordheim LP, Fessi H, Briançon S (2013). Nano-encapsulation of Vitamin D_3 active metabolites for application in chemotherapy: Formulation study and in vitro evaluation. *Pharm Res*. 30:1137–1146.

Anand P, Nair HB, Sung B, Kunnumakkara AB, Yadav VR, Tekmal RR, Aggarwal BB (2010). Design of curcumin-loaded PLGA nanoparticles formulation with enhanced cellular uptake, and increased bioactivity in vitro and superior bioavailability in vivo. *Biochem Pharmacol.* 79:330–338.

Ballou B, Lagerholm BC, Ernst LA, Bruchez MP, Waggoner AS (2004). Noninvasive imaging of quantum dots in mice. *Bioconj Chem.* 15:79–86.

Bansal S, Choudhary S, Sharma M, Kumar SS, Lohan S, Bhardwaj V, Jyoti S (2013). Tea: A native source of antimicrobial agents. *Food Res Int.* 53:568–584.

Bejrapha P, Min SG, Surassmo S, Choi MJ (2010). Physicothermal properties of freeze-dried fish oil nanocapsules frozen under different conditions. *Drying Technol.* 28:481–489.

Berne BJ, Pecora R (2000). *Dynamic Light Scattering: With Applications to Chemistry, Biology, and Physics*. New York: Dover.

Budhian A, Siegel SJ, Winey KI (2008). Controlling the in vitro release profiles for a system of haloperidol-loaded PLGA nanoparticles. *Int J Pharm.* 346:151–159.

Bummer PM (2004). Physical chemical considerations of lipid-based oral drug delivery-solid lipid nanoparticles. *Crit Rev Ther Drug Carrier Syst.* 21:1–20.

Burguera JL, Burguera M (2012). Analytical applications of emulsions and microemulsions. *Talanta.* 96:11–20.

Byun Y, Whiteside S, Cooksey K, Darby D, Dawson PL (2011). α-Tocopherol-loaded polycaprolactone (PCL) nanoparticles as a heat-activated oxygen scavenger. *J Agric Food Chem.* 59:1428–1431.

Cadena PG, Pereira MA, Cordeiro RB, Cavalcanti IM, Neto BB, Maria do Carmo CB, Santos-Magalhães NS (2013). Nanoencapsulation of quercetin and resveratrol into elastic liposomes. *Biochem Biophys Acta Biomem.* 1828:309–316.

Campbell PA, Canono BP, Drevets DA (2001). Measurement of bacterial ingestion and killing by macrophages. *Curr Protoc Immunol.* Chapter 14:6. doi:10.1002/0471142735.im1406s12

Cao G, Sofic E, Prior RL (1997). Antioxidant and prooxidant behavior of flavonoids: Structure-activity relationships. *Free Radic Biol Med.* 22:749–760.

Carocho M, Ferreira IC (2013)a. A review on antioxidants, prooxidants and related controversy: Natural and synthetic compounds, screening and analysis methodologies and future perspectives. *Food Chem Toxicol.* 51:15–25.

Carocho M, Ferreira IC (2013)b. The role of phenolic compounds in the fight against cancer—A review. *Anti-Cancer Agents Med Chem.* 13:1236–1258.

Cellesi F, Tirelli N (2006). Sol–gel synthesis at neutral pH in W/O microemulsion: A method for enzyme nanoencapsulation in silica gel nanoparticles. *Colloids Surf. A Physicochem Eng Asp.* 288:52–61.

Chakraborty S, Basu S, Basak S (2014). Effect of β-cyclodextrin on the molecular properties of myricetin upon nano-encapsulation: Insight from optical spectroscopy and quantum chemical studies. *Carbohydr Polym.* 99:116–125.

Chandy T, Sharma CP (1992). Chitosan beads and granules for oral sustained delivery of nifedipine: In vitro studies. *Biomaterials.* 13:949–952.

Chen Y, Wu Q, Zhang Z, Yuan L, Liu X, Zhou L (2012). Preparation of curcumin-loaded liposomes and evaluation of their skin permeation and pharmacodynamics. *Molecules.* 17:5972–5987.

Chen CC, Wagner G (2004). Vitamin E nanoparticle for beverage applications. *Chem Eng Res Des.* 82:1432–1437.

Chithrani BD, Ghazani AA, Chan WC (2006). Determining the size and shape dependence of gold nanoparticle uptake into mammalian cells. *Nanoletters.* 6:662–668.

Choi W, No RH, Kwon HS, Lee HY (2014). Enhancement of skin anti-inflammatory activities of *Scutellaria baicalensis* extract using a nanoencapsulation process. *J Cosmet Laser Ther.* 16:271–278.

Coradini K, Lima FO, Oliveira CM, Chaves PS, Athayde ML, Carvalho LM, Beck RCR (2014). Co-encapsulation of resveratrol and curcumin in lipid-core nanocapsules improves their in vitro antioxidant effects. *Eur J Pharm Biopharm.* 88:178–185.

Cryan S-A (2005). Carrier-based strategies for targeting protein and peptide drugs to the lungs. *AAPS J.* 7:E20–E41.

Cui D, Tian F, Ozkan CS, Wang M, Gao H (2005). Effect of single wall carbon nanotubes on human HEK293 cells. *Toxicol Lett.* 155:73–85.

Dandekar PP, Jain R, Patil S, Dhumal R, Tiwari D, Sharma S, Vanage G, Patravale V (2010). Curcumin-loaded hydrogel nanoparticles: Application in anti-malarial therapy and toxicological evaluation. *J Pharm Sci.* 99:4992–5010.

Davey MW, Montagu MV, Inzé D, Sanmartin M, Kanellis A, Smirnoff N, Fletcher J (2000). Plant L-ascorbic acid: Chemistry, function, metabolism, bioavailability and effects of processing. *J Sci Food Agric.* 80:825–860.

Day L, Seymour RB, Pitts KF, Konczak I, Lundin L (2009). Incorporation of functional ingredients into foods. *Trends Food Sci Technol.* 20:388–395.

De Paz E, Martín Á, Estrella A, Rodríguez-Rojo S, Matias AA, Duarte CM, Cocero MJ (2012). Formulation of β-carotene by precipitation from pressurized ethyl acetate-on-water emulsions for application as natural colorant. *Food Hydrocoll.* 26:17–27.

Decker T, Lohmann-Matthes ML (1988). A quick and simple method for the quantitation of lactate dehydrogenase release in measurements of cellular cytotoxicity and tumor necrosis factor (TNF) activity. *J Immunol Methods.* 115:61–69.

Díaz, DI, Beristain CI, Azuara E, Luna G, Jimenez M (2015). Effect of wall material on the antioxidant activity and physicochemical properties of *Rubus fruticosus* juice microcapsules. *J Microencapsul.* 32:247–254.

Donsì F, Annunziata M, Vincensi M, Ferrari G (2012). Design of nanoemulsion-based delivery systems of natural antimicrobials: Effect of the emulsifier. *J Biotechnol.* 159:342–350.

Donsì F, Sessa M, Mediouni H, Mgaidi A, Ferrari G (2011). Encapsulation of bioactive compounds in nanoemulsion-based delivery systems. *Proc Food Sci.* 1:1666–1671.

Dube A, Ng K, Nicolazzo JA, Larson I (2010). Effective use of reducing agents and nanoparticle encapsulation in stabilizing catechins in alkaline solution. *Food Chem.* 122:662–667.

Fan W, Chi Y, Zhang S (2008). The use of a tea polyphenol dip to extend the shelf life of silver carp (*Hypophthalmicthys molitrix*) during storage in ice. *Food Chem.* 108:148–153.

Fathi M, Mozafari MR, Mohebbi M (2012). Nanoencapsulation of food ingredients using lipid based delivery systems. *Trends Food Sci Technol.* 23:13–27.

Fernandez-Urrusuno R, Calvo P, Remuñán-López C, Vila-Jato JL, Alonso MJ (1999). Enhancement of nasal absorption of insulin using chitosan nanoparticles. *Pharm Res.* 16:1576–1581.

Ferreira I, Rocha S, Coelho M (2007). Encapsulation of antioxidants by spray-drying. *Chem Eng Trans.* 11:713–717.

Garti N, Spernath A, Aserin A, Lutz R (2005). Nano-sized self-assemblies of nonionic surfactants as solubilization reservoirs and microreactors for food systems. *Soft Matter.* 1:206–218.

Geng Y, Dalhaimer P, Cai S, Tsai R, Tewari M, Minko T, Discher DE (2007). Shape effects of filaments versus spherical particles in flow and drug delivery. *Nat Nanotech.* 2:249–255.

Gharbi N, Pressac M, Hadchouel M, Szwarc H, Wilson S R, Moussa F (2005). Fullerene is a powerful antioxidant in vivo with no acute or subacute toxicity. *Nano Lett.* 5:2578–2585.

Ghosh V, Mukherjee A, Chandrasekaran N (2014). Eugenol-loaded antimicrobial nanoemulsion preserves fruit juice against, microbial spoilage. *Colloids Surf B Biointerfaces.* 114:392–397.

Gibbs F, Kermasha S, Alli I, Catherine N, Mulligan B (1999). Encapsulation in the food industry: A review. *Int J Food Sci Nutr.* 50:213–224.

Gomes C, Moreira RG, Castell-Perez E (2011). Poly (DL lactide-co-glycolide) (PLGA) nanoparticles with entrapped trans-cinnamaldehyde and eugenol for antimicrobial delivery applications. *J Food Sci.* 76:N16–N24.

Gou M, Men K, Shi H, Xiang M, Zhang J, Song J, Qian Z (2011). Curcumin-loaded biodegradable polymeric micelles for colon cancer therapy in vitro and in vivo. *Nanoscale.* 3:1558–1567.

Gref R, Lück M, Quellec P, Marchand M, Dellacherie E, Harnisch S, Müller RH (2000). "Stealth"corona-core nanoparticles surface modified by polyethylene glycol (PEG): Influences of the corona (PEG chain length and surface density) and of the core composition on phagocytic uptake and plasma protein adsorption. *Colloids Surf B Biointerfaces.* 18:301–313.

Hardy G (2000). Nutraceuticals and functional foods: Introduction and meaning. *Nutrition.* 16:688–689.

Hasler MC (2005). *Regulation of Functional Foods and Nutraceuticals: A Global Perspective.* Ames, IA: IFT Press, Chicago, and Blackwell Publishing.

He P, Davis SS, Illum L (1998). In vitro evaluation of the mucoadhesive properties of chitosan microspheres. *Int J Pharm.* 166:75–88.

Heyang J, Fei X, Cuilan J, Yaping Z, Lin H (2009). Nanoencapsulation of lutein with hydroxypropylmethyl cellulose phthalate by supercritical antisolvent. *Chin J Chem Eng.* 17:672–677.

Hong MM, Oh JM, Choy JH (2008). Encapsulation of flavor molecules, 4-hydroxy-3-methoxy benzoic acid, into layered inorganic nanoparticles for controlled release of flavor. *J Nanosci Nanotechnol.* 8:5018–5021.

Hong S, Bielinska AU, Mecke A, Keszler B, Beals JL, Shi X, Banaszak Holl MM (2004). Interaction of poly (amidoamine) dendrimers with supported lipid bilayers and cells: Hole formation and the relation to transport. *Bioconj Chem.* 15:774–782.

Huang M, Khor E, Lim LY (2004). Uptake and cytotoxicity of chitosan molecules and nanoparticles: Effects of molecular weight and degree of deacetylation. *Pharm Res.* 21:344–353.

Huang Q, Jiang Y (2004). Enhancing the stability of phenolic antioxidants by nanoencapsulation. Abstracts of Papers, 228th ACS National Meeting, Philadelphia, PA, August 22–26.

Jafari SM, He Y, Bhandari B (2007). Encapsulation of nanopartricles of D-limonene by spray drying: Role of emulsifiers and emulsifying agent. *Drying Technol.* 25:1079–1089.

Jafari SM, Assadpoor E, Bhandari B, He Y (2008). Nanoparticle encapsulation of fish oil by spray drying. *Food Res Int.* 41:172–183.

Jiang W, Kim BYS, Rutka JT, Chan WCW (2007). Advances and challenges of nanobased drug delivery systems. *Expert Opin Drug Deliv.* 4:621–633.

Jincheng W, Xiaoyu Z, Siahao C (2010). Preparation and properties of nanoencapsulated capsaicin by complex coacervation method. *Chem. Eng. Comm.* 197:919–933.

Kim D, El-Shall H, Dennis D, Morey T (2005). Interaction of PLGA nanoparticles with human blood constituents. *Colloids Surf B Biointerfaces.* 40:83–91.

Kim IH, Lee H, Kim JE, Song KB, Lee YS, Chung DS, Min SC (2013). Plum coatings of lemongrass oil-incorporating carnauba wax-based nanoemulsion. *J Food Sci.* 78:1551–1559.

Kobayashi H, Kawamoto S, Jo SK, Bryant HL, Brechbiel MW, Star RA (2003). Macromolecular MRI contrast agents with small dendrimers: Pharmacokinetic differences between sizes and cores. *Bioconj Chem.* 14:388–394.

Korzeniewski C, Callewaert DM (1983). An enzyme-release assay for natural cytotoxicity. *J Immunol Methods.* 64:313–320.

Külkamp IC, Rabelo BD, Berlitz SJ, Isoppo M, Bianchin MD, Schaffazick SR, Pohlmann AR, Guterres SS (2011). Nanoencapsulation improves the in vitro antioxidant activity of lipoic acid. *J Biomed Nanotechnol.* 7:598–607.

Kumar A, Glam M, El-Badri N, Mohapatra S, Haller E, Park S, Patrick L, Nattkemper L, Vo D (2011). Initial observations of cell-mediated drug delivery to the deep lung. *Cell Transplant.* 20:609–618.

Kumar M, (2000). A review of chitin and chitosan applications. *React Funct Polym.* 46:1–27.

Kumari A, Yadav SK, Pakade YB, Singh B, Yadav SC (2010). Development of biodegradable nanoparticles for delivery of quercetin. *Colloids Surf B Biointerfaces.* 80:184–192.

Kumudavally KV, Phanindrakumar HS, Tabassum A, Radhakrishna K, Bawa AS (2008). Green tea—A potential preservative for extending the shelf life of fresh mutton at ambient temperature (25 ± 2°C). *Food Chem.* 107:426–433.

Kurita K, (2006). Chitin and chitosan: Functional biopolymers from marine crustaceans, *Marine Biotechnol.* 8:203–226.

Labiris NR, Dolovich MB (2003). Pulmonary drug delivery. Part I: Physiological factors affecting therapeutic effectiveness of aerosolized medications. *Br J Clin Pharmacol.* 56:588–599.

Lehr CM, Bouwstra JA, Schacht EH, Junginger HE (1992). In vitro evaluation of mucoadhesive properties of chitosan and some other natural polymers. *Int J Pharm.* 78:43–48.

Lim SJ, Lee MK, Kim CK (2004). Altered chemical and biological activities of all-trans retinoic acid incorporated in solid lipid nanoparticle powders. *J Cont Rel.* 100:53–61.

Loquercio A, Castell-Perez E, Gomes C, Moreira RG (2015). Preparation of chitosan-alginate nanoparticles for trans-cinnamaldehyde entrapment. *J Food Sci.* 80:N2305–N2315.

Lü J-M, Wang X, Marin-Muller C, Wang H, Lin PH, Yao Q, Chen C (2009). Current advances in research and clinical applications of PLGA-based nanotechnology. *Expert Rev Mol Diagn.* 9:325–341.

Luo Y, Zhang B, Whent M, Yu L, Wang Q (2011). Preparation and characterization of zein/chitosan complex for encapsulation of α-tocopherol, and its *in vitro* controlled release study. *Colloids Surf B Biointerfaces.* 85:145–152.

Ma Z, Yeoh H, Lim L (2002). Formulation pH modulates the interaction of insulin with chitosan nanoparticles. *J Pharm Sci.* 91:1396–1404.

Maqsood S, Benjakul S, Shahidi F (2013). Emerging role of phenolic compounds as natural additives in fish and fish products. *Crit Rev Food Sci Nutr.* 53:162–179.

Maswal M, Dar A (2014). Formulation challenges in encapsulation and delivery of citral for improved food quality. *Food Hydrocoll.* 37:182–195.

Mazzola L (2003). Commercializing nanotechnology. *Nat Biotechnol.* 21:1137–1143.

McElvaney NG, Hubbard RC, Birrer P, Crystal RG, Chernick MS, Frank MM, Caplan DB (1991). Aerosol α1-antitrypsin treatment for cystic fibrosis. *Lancet.* 337(8738):392–394. doi: 10.1016/0140-6736(91)91167-S.

McIvor RA, Tunks M, Todd DC (2011). COPD. *Clin Evid.* 6:pii: 1502.

Mickuviene I, Kirveliene V, Juodka B (2004). Experimental survey of non-clonogenic viability assays for adherent cells in vitro. *Toxicol In Vitro.* 18:639–648.

Mitra S, Gaur U, Ghosh PC, Maitra AN (2001). Tumour targeted delivery of encapsulated dextran-doxorubicin conjugate using chitosan nanoparticles as carrier. *J Cont Rel.* 74:317–323.

Moghimi SM, Hunter AC, Murray JC (2001). Long-circulating and target-specific nanoparticles: Theory to practice. *Pharmacol Rev.* 53:283–318.

Mozafari MR, Flanagan J, Matia-Merino L, Awati A, Omri A, Suntres ZE, Singh H (2006). Recent trends in the lipid-based nanoencapsulation of antioxidants and their role in foods. *J Sci Food Agric.* 86:2038–2045.

Mozafari MR, Reed CJ, Rostron C, Kocum C, Piskin E (2002). Construction of stable anionic liposome-plasmid particles using the heating method: A preliminary investigation. *Cell Mol Biol Lett.* 7:923–927.

Mozafari MR, Reed CJ, Rostron C, Martin DS, (2004). Transfection of human airway epithelial cells using a lipid-based vector prepared by the heating method. *J Aerosol Med.* 17:100.

Müller RH, Mäder K, Gohla S (2000). Solid lipid nanoparticles (SLN) for controlled drug delivery: A review of state of the art. *Eur J Pharm Biopharm.* 50:161–177.

Mundargi RC, Ramesh Babu V, Rangaswamy V, Patel P, Aminabhavi TM (2008). Nano/micro technologies for delivering macromolecular therapeutics using poly(D,L-lactide-co-glycolide) and its derivatives. *J Control Release*. 125:193–209.

"Nutraceuticals/Functional Foods and Health Claims on Foods: Policy Paper." Health Canada. June 24, 2013. Retrieved January 30, 2014.

Oberdorster G, Oberdorster E, Oberdorster J, (2005). Nanotoxicology: An emerging discipline evolving from studies of ultrafine particles. *Environ Health Perspect*. 113:823–839.

Ogawara K, Furumoto K, Takakura Y, Hashida M, Higaki K, Kimura T (2001). Surface hydrophobicity of particles is not necessarily the most important determinant in their in vivo disposition after intravenous administration in rats. *J Control Release*. 77:191–198.

Ogawara K, Yoshida M, Higaki K (1999). Hepatic uptake of polystyrene microspheres in rats: Effect of particle size on intrahepatic distribution. *J Control Release*. 59:15–22.

Pan K, Luo Y, Gan Y, Baek SJ, Zhong Q (2014). pH-driven encapsulation of curcumin in self assembled casein nanoparticles for enhanced dispersibility and bioactivity. *Soft Matter*. 10:6820.

Patil JS, Kamalapur MV, Marapur SC, Kadam DV (2010). Ionotropic gelation and polyelectrolyte complexation: The novel techniques to design hydrogel particulate sustained, modulated drug delivery system: A review, *J Nanomat Biostruct*. 5:241–248.

Patri AK, Dobrovolskaia MA, Stern ST, McNeil SE (2006). Preclinical characterization of engineered nanoparticles intended for cancer therapeutics. In Amiji MM (ed). *Nanotechnology for Cancer Therapy*. Boca Raton, FL: CRC Press; 105–139.

Pérez E, Benito M, Teijón C, Olmo R, Teijón JM, Blanco MD (2012). Tamoxifen-loaded nanoparticles based on a novel mixture of biodegradable polyesters: Characterization and in vitro evaluation as sustained release systems. *J Microencapsul*. 29:309–322.

Peter HM, Brüske-Hohlfeld I, Salata OV (2004). Nanoparticles—Known and unknown health risks. *J Nanobiotechnol*. 2:11. doi:10.1186/1477-3155-2-12

Powers KW, Brown SC, Krishna VB, Wasdo SC, Moudgil BM, Roberts SM (2006). Research strategies for safety evaluation of nanomaterials. Part VI. Characterization of nanoscale particles for toxicological evaluation. *Toxicol Sci*. 90:296–303.

Prabaharan M, Mano JF (2005). Chitosan-based particles as controlled drug delivery systems. *Drug Deliv*. 12:41–57.

Prabha S, Zhou WZ, Panyam J, Labhasetwar V (2002). Size-dependency of nanoparticle-mediated gene transfection: Studies with fractionated nanoparticles. *Int J Pharm*. 244:105–115.

Prestwicha GD, Marecaka DM, Marecekb JF, Vercruyssea KP (1997). Controlled chemical modification of hyaluronic acid: Synthesis, applications, and biodegradation of hydrazide derivatives, *J Control Release*. 53:93–103.

Radtke M, Souto EB, Müller RH (2005). Nanostructured lipid carriers: A novel generation of solid lipid drug carriers. *Pharm Tech Eur*. 17, 45–50.

Ribeiro HS, Chua BS, Ichikawab S, Nakajima M (2008). Preparation of nanodispersions containing β-carotene by solvent displacement method. *Food Hydrocoll*. 22:12–17.

Salvia-Trujillo L, Rojas MA, Soliva-Fortuny R, Martin-Belloso O (2014). Impact of microfluidization or ultrasound processing on the antimicrobial activity against *Escherichia coli* of lemongrass oil-loaded nanoemulsions. *Food Control*. 37:292–297.

Samadi N, Abbadessa A, Di Stefano A, van Nostrum CF, Vermonden T, Rahimian S, Teunissen EA, van Steenbergen MJ, Amidi M, Hennink WE (2013). The effect of lauryl capping group on protein release and degradation of poly(D,L-lactic-co-glycolic acid) particles. *J Control Release*. 172:436–443.

Savian AL, Rodrigues D, Weber J, Ribeiro RF, Motta MH, Schaffazick SR, Adams AI, de Andrade DF, Beck RC, da Silva CB (2015). Dithranol-loaded lipid-core nanocapsules improve the photostability and reduce the in vitro irritation potential of this drug. *Mater Sci Eng C Mater Biol Appl*. 46:69–76.

Semo E, Kesselman E, Danino D, Livney YD (2007). Casein micelle as a natural nanocapsular vehicle for nutraceuticals. *Food Hydrocoll.* 21:936–942.

Serra AT, Matias AA, Nunes AVM, Leit˜ao MC, Brito D, Bronze R, Silva S, Pires A, Crespo MT, Romao MVS, Duarte CM (2008). In vitro evaluation of olive- and grape-based natural extracts as potential preservatives for food. *Innov Food Sci Emerg Technol.* 9:311–319.

Shaji J, Jain V, Lodha S (2010). Chitosan: A novel pharmaceutical excipient. *Int J Pharm Appl Sci.* 1:11–28.

Shefer A, Shefer SD (2003). Multi component controlled release system for oral care, food products, nutracetical, and beverages. U.S. Patent No. US20030152629.

Shirode AB, Bharali DJ, Nallanthighal S, Coon JK, Mousa SA, Reliene R (2015). Nanoencapsulation of pomegranate bioactive compounds for breast cancer chemoprevention. *Int J Nanomed.* 9:475–484.

Smola M, Vandamme T, Sokolowski A (2008). Nanocarriers as pulmonary drug delivery systems to treat and to diagnose respiratory and non respiratory diseases. *Int J Nanomed.* 3:1–19.

Song L, Zhi Z, Pickup JC (2014). Nanolayer encapsulation of insulin-chitosan complexes improves efficiency of oral insulin delivery. *Intl J Nanomed.* 9:2127–2136.

Sowasod N, Charinpanitkul ST, Tanthapanichakoon W (2008). Nanoencapsulation of curcumin in biodegradable chitosan via multiple emulsion/solvent evaporation. *Int J Pharm.* 347:93–101.

Spencer JP, Abd El Mohsen MM, Minihane AM, Mathers JC (2008). Biomarkers of the intake of dietary polyphenols: Strengths, limitations and application in nutrition research. *Br J Nutr.* 99:12–22.

Sundstrom JB, Mao H, Santoianni R, Villinger F, Little DM, Huynh TT, Mayne AE, Hao E, Ansari AA (2004). Magnetic resonance imaging of activated proliferating rhesus macaque T cells labeled with superparamagnetic monocrystalline iron oxide nanoparticles. *J Acquir Immune Defic Syndr.* 35:9–21.

Suwannateep N, Banlunara W, Wanichwecharungruang SP, Chiablaem K, Lirdprapamongkol K, Svasti J (2011). Mucoadhesive curcumin nanospheres: Biological activity, adhesion to stomach mucosa and release of curcumin into the circulation. *J Control Release.* 151:176–182.

Tachaprutinun A, Udomsup T, Luadthong C, Wanichwecharungruang S (2009). Preventing the thermal degradation of astaxanthin through nanoencapsulation. *Int J Pharm.* 374:119–124.

Taylor TM, Davidson PM, Bruce BD, Weiss J, (2005). Liposomal nanocapsules in food science and agriculture. *Crit Rev Food Sci Nutr.* 45:587–605.

Terjung N, Löffler M, Gibis M, Hinrichs J, Weiss J (2012). Influence of droplet size on the efficacy of oil-in-water emulsions loaded with phenolic antimicrobials. *Food Funct.* 3:290–301.

Tokumitsu H, Ichikawa H, Fukumori Y (1999). Chitosan-gadopentetic acid complex nanoparticles for gadolinium neutron capture therapy of cancer: Preparation by novel 87 emulsion droplet coalescence technique and characterization. *Pharm Res.* 16:1830–1835.

Tuschl H, Schwab CE (2004). Flow cytometric methods used as screening tests for basal toxicity of chemicals. *Toxicol In Vitro.* 18:483–491.

Uhrich KE, Cannizzaro SM, Langer RS, Shakesheff KM (1999). Polymeric systems for control. *Drug Rel Chem Rev.* 99:3181–3198.

Upadrashta SM, Katikaneni PR, Nuessle NO (1992). Chitosan as a tablet binder. *Drug Dev Ind Pharm.* 18:1701–1708.

Verma AK, Sachin K, Saxena A, Bohidar HB (2005). Release kinetics from bio-polymeric nanoparticles encapsulating protein synthesis inhibitor-cycloheximide, for possible therapeutic applications. *Curr Pharm Biotechnol* 6:121–130.

Wang JC, Chen SH, Xu ZC (2008). Synthesis and properties research on the nanocapsulated capsaicin by simple coacervation method. *J Dispers Sci Technol.* 29:687–695.

Wang X, Jiang Y, Wang YW, Huang MT, Hoa CT, Huang Q (2008). Enhancing anti-inflammation activity of curcumin through O/W nanoemulsions. *Food Chem.* 108:419–424.

Weiss J, Takhistov P, McClements DJ (2006). Functional materials in food nanotechnology. *J Food Sci.* 71:R107–R116.

Wildman, RE (Ed.). (2001). *Handbook of Nutraceuticals and Functional Foods*, 1st ed. CRC Series in Modern Nutrition. Boca Raton, FL: CRC Press.

Xing F, Cheng G, Yi K, Ma L (2004). Nanoencapsulation of capsaicin by complex coacervation of gelatin, acacia, and tannins. *J Appl Pol Sci* 96:2225–2229.

Yang R, Sun G, Zhang M, Zhou Z, Li Q, Strappe P, Blanchard C (2016). Epigallocatechin gallate (EGCG) decorating soybean seed ferritin as a rutin nanocarrier with prolonged release property in the gastrointestinal tract. *Plant Foods Hum Nutr.* 71:277–285.

Yang Z, Peng Z, Li J, Li S, Kong L, Li P, Wang Q (2014). Development and evaluation of novel flavour microcapsules containing vanilla oil using complex coacervation approach. *Food Chem.* 145:272–277.

Yao LH, Jiang YM, Shi J, Tomás-Barberán FA, Datta N, Singanusong R, Chen, SS (2004). Flavonoids in food and their health benefits. *Plant Food Hum Nutr.* 59:113–122.

Yu L, Banerjee IA, Gao X, Nuraje N, Matsui H (2005). Fabrication and application of enzyme-incorporated peptide nanotubes. *Bioconjug Chem.* 16:1484–1487.

Zarif L, Graybill JR, Perlin D, Mannino RJ (2000). Cochleates: New lipid-based drug delivery system. *J Lipid Res.* 10:523–538.

Zhao L, Xiong H, Peng H, Wang Q, Han D, Bai C, Liu Y, Shi S, Deng B (2011). PEG-coated lyophilized pro-liposomes: Preparation, characterizations and in vitro release evaluation of vitamin E. *Eur Food Res Tech.* 232:647–654.

Zimet P, Livney YD (2009). Beta-lactoglobulin and its nanocomplexes with pectin as vehicles for ω-3 polyunsaturated fatty acids. *Food Hydrocoll.* 23:1120–1126.

Zion Market Research. 2017. Dietary Supplements Market by Ingredients (Botanicals, Vitamins, Minerals, Amino Acids, Enzymes) for Additional Supplements, Medicinal Supplements and Sports Nutrition Applications, Global Industry Perspective, Comprehensive Analysis and Forecast, 2016–2022. April 25, 2017.

Zokti JA, Sham B, Mohammed AS, Abas F (2016). Green tea leaves extract: Microencapsulation, physicochemical and storage stability study. *Molecules.* 21(8):pii: E940.

Index

Page numbers followed by f and t indicate figures and tables, respectively.

A

α-ASTREE electronic tongue, 127–128, 127f; see also Sensory evaluation strategy for taste masking
Acceptable daily intake (ADI), 188
Acetic acid, 56, 57
  bacteria, 239
Acetone, 284
Acidophilus-bifidus milk (A/B milk), 62; see also Fermented milk drink products
Acidophilus milk, 61–62; see also Fermented milk drink products
Acidophilus-yeast milk, 62; see also Fermented milk drink products
Acromegaly, 287
Actilight, 54
Actimask, 139–140
Active pharmaceutical ingredient (API), 43, 125
Activia, 58
Additivity, 209
Air atomization, 290
Airborne odor molecules, 240
α-lactoglobulin, 295
Albion, 44
Algae
  of nutritional interest, 86
  source of polyunsaturated fatty acids, 88
Algal polysaccharides, 86
Alginate, 301
Alleles, naturally occurring, 153–154
Allergies, 300
Allium sativum, 14–15
Alzheimer's disease, 300
Amino acid substitutions, 154
AMP-activated protein kinase (pAMPK), 186
Ampicillin, 133

Amylose, 140
Analgesic effects
  citral, 181–182
  menthol, 191
Animal-based functional foods; see also Functional foods
  beef, 21
  eggs enriched with PUFA, 21
  fish, 19–21
Anthelmintic effects, geraniol, 187
Anthracyclines, 84
Antiadipogenic effects, citral, 182–183
Anti-browning agent, 278
Anti-cancer agents
  nanoencapsulation of, 302
Anticarcinogens, 14
Anti-inflammatory activity
  citral, 181–182
  geraniol, 186–187
  menthol, 192
Antimicrobial activity
  citral, 183
  geraniol, 184–185
  menthol, 192–193
Antineoplastic agents, 284
Antioxidant activity, geraniol, 185
Antioxidants, 83, 284
Antioxidants and flavoring agents
  nanoencapsulation of, 300–301
Antipruritic activity, menthol, 190–191
Antithrombogenic property, 86
Antitumor activity
  geraniol, 185–186
  menthol, 192
Appearance, 208; see also Nutraceuticals, flavor problems with
Aqueous alcoholic extraction process, 97–98; see also Herbal products extraction methods

311

Aroma
  components, 239–240
  defects in probiotics/prebiotics, 65–66;
    see also Flavors in probiotics/
    prebiotics
  defined, 217
  detection, 242
Arthrospira platensis, 83
Artificial flavoring substances, 237t
Artificial flavors, 40
Artificial sweeteners, 129, 209
Ascorbic acid, 278
Ascorbyl, 279
Assimilation, 212
Attitudinal determinants, cognitive and, 157
Avena sativa, 7
Ayran; see also Fermented dairy products
  about, 58
  starter culture characteristics for, 58–59
Azeotropic mixture, 237

B

Bacterial colony formation, 279
Bacterial nutraceuticals, 83
Barley (*Hordeum vulgare*), 7
Barley bran, 7
Basil oil (*Ocimum basilicum*), 293
Basolateral anionic transporter,
  bicarbonate-dependent, 192
Bath sonication, 253
B16BL6 melanoma cells, 302
Beef, 21; see also Animal-based functional
  foods
Bentonite, 136
Benzodiazepines, 193
Beverage-based functional foods, 13;
  see also Functional foods
β-glucan, 7
Bifidobacterium, 11, 56, 57
Bifidus milk, 62; see also Fermented milk
  drink products
Bioactive agents/ingredients,
  nanoencapsulation of, 281t–283t
Bioactive molecules, 80
Bioactive peptides, 88
Biomimetic sensors, 128; see also Sensory
  evaluation strategy for taste
  masking
Biotechnological production process, 239
Bitter blockers, 131–132, 131t; see also
  Taste-masking techniques

Bitter components on sensory perception
  of food
  bitter perception
    bitter taste perception, 152–153
    bitter taste perception mechanisms,
      151–152
    bitter taste receptors as potential
      therapeutic targets, 153
    family and twin studies, 154–155
    misconception about zones of
      tongue, 150–151
  consumer acceptance, challenges in/
    successful marketing of products
    about, 157–158
    role of sensory science in decision
      making, 158–159
    sensory evaluation and quality of
      food, 159
  consumer acceptance of functional
    foods
    cognitive and attitudinal
      determinants, 157
    socio-demographic determinants,
      156–157, 157t
  genetics of sweet/bitter perception
    cross-species comparisons, 154
    family and twin studies, 154–155
    naturally occurring alleles, 153–154
  overview, 146–147
  perception of taste, 149–150, 150f
  taste, importance of, 149
  taste, types of
    bitter, 147–148
    salty, 147
    sour, 147
    sweet, 147
    umami, 148–149
  taste improvement technology
    about, 159–160
    food technology to create new
      flavors, 162
    masking by encapsulation, 161–162
    suppression of mixture, 160–161
    taste inhibition, 160
    umami in food science, role of, 162
  taste perception/behavior
    bitter taste and food rejection, 156
    population, challenges in, 155
Bitter compounds, 134, 137
Bitterness
  poisonous, 154
  unusual, 156

Bitter perception
  bitter taste perception, 152–153
  bitter taste perception mechanisms, 151–152
  bitter taste receptors as potential therapeutic targets, 153
  family and twin studies, 154–155
  misconception about zones of tongue, 150–151
Bitter taste, 125, 202
  about, 147–148
  and food rejection, 156
  receptors as therapeutic targets, 153
Blackberry (*Rubus fruticosus*) juice, 301
Botanical extracts for nutraceutical industry, *see* Natural ingredients for nutraceutical industry
Boza, 63
Bran, 9
Brazil, 4
Breakfast cereal foods, 10–11; *see also* Grains/cereals
Breastfeeding, 36
Broccoli and cruciferous vegetables, 15
Brownian diffusion of the nanoparticles, 297
Brown rice, 9; *see also* Grains/cereals
Buckwheat, 9–10; *see also* Grains/cereals
Burst release, 261
Business performance, 158
Buttermilk, 63; *see also* Probiotic beverages, dairy-based

C

Cacik/tzatziki, 57, 63; *see also* Probiotic beverages, dairy-based
Calcitriol (1,25-dihydroxyvitamin D_3), 302
Calcium chloride, 301
Calcium deficiency, 37; *see also* Nutritional deficiency on infants/children
Calcium supplements, taste of, masking, 45
Camouflage technology, 140
Canada, nutraceuticals nomenclature in, 277
Cancer, 300
Carbohydrates, 84, 86–88, 223; *see also* Marine nutraceuticals matrices, 249
Carbonated ayran, 58
Carbon nanotubule field-effect transistor (CNT-FET) bioelectronic sensors, 128; *see also* Sensory evaluation strategy for taste masking
Cardiovascular diseases, 300
Carnauba-shellac wax (CSW)-based nanoemulsion, 293
Carnivorous mammals, consumption of foods, 149
Carotenoids, 79, 89, 174
Carrageenan, 87
Carvacrol, 293
Casein micelles, 294–295
Catechin, 301
Cereal-based functional foods, 8f
Cheese making, 284
Chemical methods, 299–300; *see also* Nanoencapsulation of flavors
  interfacial polymerization, 258, 259t
  molecular inclusion, 256–258
Chemotherapy drugs, 280
Chewing, 240
Children
  flavor on supplement intake and consumption in, 38–39
  nutritional deficiencies on, *see* Nutritional deficiency on infants/children
Chilli-eating cultures, 211
Chitin, 285
Chitooligosaccharide (COS), 87
Chitosan (CTS), 285–286, 301
Chitosan nanoparticles
  applications of, 291
  methods of producing, 284–287
    coacervation/precipitation, 288
    emulsion-droplet coalescence, 288–289
    ionotropic gelation, 290
    reverse micelles, 289–290
Chloroform, 284
Chocolate, 13; *see also* Confectionery-based functional foods
Cinnamaldehyde, 293
Citral (3,7-dimethyl-2,6-octadienal); *see also* Terpene flavors, nutraceutical applications of
  about, 180
  biological properties of
    analgesic/anti-inflammatory activities, 181–182
    antiadipogenic effects, 182–183
    antimicrobial effects, 183
    CNS effects, 180–181
  stability enhancement of, 229
Citrus fruits, 17, 17t

Clevenger distillation, 238
Clinical trials on natural ingredients/
   botanical extracts, 106
2-C-methyl-D-erythritol-4-phosphate (MEP)
   route, 168
CNS effects, citral, 180–181
Coacervation/precipitation, 251–252, 288;
   see also Chitosan nanoparticles;
   Physico-chemical methods
   phase separation, 224–225
Coating material, 263
Cocoa (*Theobroma cacao*), 12; see also
   Confectionery-based functional
   foods
Co-crystallization, 251; see also
   Nanoencapsulation of flavors
Co-enzyme Q_{10}, 19
Coffee aroma, 229
Cognitive and attitudinal determinants, 157
Cohobation, distillation with, 100–101
Cold homogenization technique, 294
Cold receptors, 204
Collagen in pork/beef, 88
Colorimetric assays, 298
Columanganese, 99
Combination release profiles, 261
Complexation, 135; see also Taste-masking
   techniques
Concentrated infusion, 94
Condensation, 238
Conditioned taste aversion, 205
Confectionery-based functional foods;
   see also Functional foods
   chocolate, 13
   cocoa (*Theobroma cacao*), 12
   food bars, 12–13
Confocal microscopy, 299
Conjugated linolic acid (CLA), 21
Consumer acceptance
   challenges in/successful marketing
      of products
      about, 157–158
      sensory evaluation and quality
         of food, 159
      sensory science in decision making,
         role of, 158–159
   of functional foods
      cognitive and attitudinal
         determinants, 157
      socio-demographic determinants,
         156–157, 157t
Consumer orientation, 158
Continuous hot extraction; see also Herbal
   products extraction methods
   advantages, 97

description, 95–96
disadvantages, 97
process, 96
Controlled flavor release; see also Release
   mechanisms
   about, 260–262, 261t
   by degradation, 262
   by diffusion, 262
   by melting, 263
   nanocapsules, characterization of
      dynamic light scattering, 263
      electron microscopy, 263
      electron spin resonance (ESR), 263
      nuclear magnetic resonance (NMR),
         264
      small angle neutron scattering
         (SANS), 264
   stability of encapsulated flavors, 263
   by swelling, 262–263
Convective diffusion, 262
Cordyceps mushroom, 18–19
Core flavor perception, 240
Cornmint essential oil, 188
Cough syrup, 41
Countercurrent extraction process, 98;
   see also Herbal products
   extraction methods
Cover and run-down method, 92
Cranberry, 17
Critical micelle concentrations (CMC), 255
Cross-linking agent, 288, 290
Cruciferous vegetables, 15
Crushed garlic, 15
Crystallization process, 139
Cucurbita pepo, 19
Curcumin liposomes, 302
Customization, 77
Cyclamate, 43
Cyclic oligosaccharides, 135, 256
Cyclin-dependent kinases (CDK), 186
Cyclodextrins (CD), 135, 256
Cyclosporin sirolimus, 84
Cystic fibrosis conductivity regulator
   (CFTR), 191
Cytochrome P450, 183

D

Dairy-based functional foods, 11; see also
   Functional foods
Dairy-based probiotic beverages; see also
   Flavors in probiotics/prebiotics
   buttermilk, 63
   cacik, 63
   lassi, 63

Damascenones, 174
Decoction, 95; see also Herbal products extraction methods
Degradation, flavor release by, 262
Dehydration-rehydration, 284
Dehydro-ascorbic acid, 278
Demasking of iron supplements, advanced, 44
Dementholized cornmint essential oil, 188
Denaturation, 56
Dendrimers, 298
Density functional theory (DFT), 132
3-deoxyanthocyanidins in sorghum, 10
Deoxyxylulose-5-phosphate (DXP), 168
Detergent removal method, 255; see also Solvent dispersion methods
Devereux, 66
Diacetyl, 57
Dietary iodine, 80
Dietary supplements, 276
Diffusion, 262
Dimethylallyl diphosphate (DMAPP), 168
2,5-dimethyl pyrazine, 229
Direct immersion solid-phase microextraction (DI-SPME), 103
Direct steam distillation, 100; see also Distillation
Distillation
  about process, 98–99
  in production of natural flavoring substances, 237–238
  types of
    direct steam distillation, 100
    distillation with cohobation, 100–101
    hydrodiffusion, 99
    steam and water distillation, 100
    water/hydrodistillation technique, 99–100
Diterpenes, 168
Divinylbenzene (DVB), 104
DNA staining, 298
Docosahexaenoic acid (DHA), 20, 45, 295
Double maceration, 91
Dried figs, 18
Drink, sensory/affective responses to, 202–206, 202t
D-tryptophan, 132
Dynamic extraction mode, 102
Dynamic light scattering (DLS), 263, 296–297

E

Eggs enriched with PUFA, 21; see also Animal-based functional foods
Egg yolk phospholipids (EPC), 302

Eicosapentaenoic acid (EPA), 20, 45
Electronic-tongue (e-tongue), 230
Electron microscopy, 263, 296
Electron spin resonance (ESR), 263
Electrospray, 227–228
Emulsion-droplet coalescence, 288–289; see also Chitosan nanoparticles
Emulsion evaporation method, 228f, 301
Emulsion extrusion method, 226f
Emulsion technology, 227
Encapsulation
  defined, 236
  of flavors via extrusion, 248
  masking by, 161–162
  method of, 244f, 259t
  uses of, 243f
Encapsulation, flavor
  applications of, 242–243, 243f
  challenges, 242
Encapsulation materials, 285–287; see also Nanoencapsulation of bioactive agents/ingredients
  chitosan, 285–286
  phospholidipids, 287
  poly(lactic-co-glycolic acid) (PLGA), 286–287
  tripolyphosphate (TPP), 287
Endotoxin contamination, 298
Enzymatic synthesis, 54
Enzymes, 89–90, 284; see also Marine nutraceuticals
Epigallocatechin gallate (EGCG), 295, 300
Epigallocatechin-3-gallate (EGCG), 13
Escherichia coli C 600, 293
Essential fatty acids
  deficiency risk /taste/odor of, 47t–48t
Essential oil, 219
Established nutraceuticals, 124
Estragole, 293
Ethanediamine alkali solutions, 288
Ethanol injection, 254; see also Solvent dispersion methods
Ether injection (solvent vaporization), 254; see also Solvent dispersion methods
Eugenol, 229, 293, 301
European market
  nutraceutical, for plant extracts, 77–78
Exopolysaccharides, 87
Expectancy
  flavor-related problems, solutions to, 212–213
Exposure
  flavor-related problems, solutions to, 211–212

Extracellular signal-regulated kinase (ERK), 186
Extraction process, 237
Extractor method (British/Indian pharmacopeias), 93
Extraoral bitter taste receptors, 15
Extrusion; *see also* Nanoencapsulation of flavors
  about, 248–249
  double-capillary extrusion devices
    centrifugal extrusion device, 249–250
    coaxial double capillary device, 249
    recycling centrifugal extrusion, 250
  microemulsion technique, 225–226
  simple extrusion, 249

F

Family and twin studies
  bitter perception, 154–155
  sweet/bitter perception, genetics of, 154–155
Farnesyl pyrophosphate (FPP), 168
Fatty acids, 19, 88; *see also* Marine nutraceuticals
Fatty fish, 20
Fecal enzymes, 11
Feeding, 35
Female functional food consumers, in USA/Europe, 157t
Fermentation products, 57
Fermented dairy products; *see also* Flavors in probiotics/prebiotics
  ayran, 58–59
  dairy-based probiotic beverages, miscellaneous, 63
  fermented milk drink, 61–63
  kefir, 59–61
  kumiss, 61
  yogurt, 56–58
Fermented foods derived from rice, 9
Fermented milk drink products; *see also* Flavors in probiotics/prebiotics
  acidophilus milk, 61–62
  acidophilus-yeast milk, 62
  bifidus milk and acidophilus–bifidus milk (A/B milk), 62
  mil-mil, 62
  sweet acidophilus milk, 62
  yakult, 62–63
Fermented milks, 55
  products, flavor/aroma defects in, 67t
Fig, 18

Film coating, 132, 133t; *see also* Taste-masking techniques
Finger millet, 10; *see also* Grains/cereals
First-order release, 260
Fish, 19–21; *see also* Animal-based functional foods
  liver oil, 19, 20f
  source of fatty acids, 88
Fishy odor, masking of, 45
Fixed restrictors, 102
Flavonoids, 79, 89
Flavors; *see also* Nanoencapsulation of flavors
  classification of, 130t
  compounds, 239–240
  compounds based on functional group present, 219
  defined, 236, 237f, 240
  encapsulation
    applications of, 242–243, 243f
    challenges, 242
  enhancers, 219
  extrinsic (outside mouth) and intrinsic (inside mouth) properties of, 202t
  importance of, 125
  network, 241
  perception of, 240–242
  products, evaluation of, 229–230
  stability, 222
  types of, 237t
  used in food industry, 218t
Flavorants, 130, 130t; *see also* Taste-masking techniques
Flavor-boosters, 162
Flavoring agents
  and antioxidants, nanoencapsulation of, 300–301
  defined, 39
  pediatric, regulatory aspects of, 41–44
  types/applications, 39–41, 40t, 41t; *see also* Nutritional deficiency on infants/children
FlavoRite technology, 139
Flavor nanotechnology
  flavor products, evaluation of, 229–230
  nanoencapsulation in flavor compounds, applications of, 228–229
  nanoencapsulation of flavor components
    coacervation phase separation, 224–225
    electrospray, 227–228
    emulsion technology, 227
    extrusion microemulsion technique, 225–226
    freeze drying, 224

spray drying, 223
supercritical fluid technology, 226–227
nanotechnology in flavors, 222, 222f, 223t
overview, 218–221, 218t, 219t, 221t
Flavor–nutrient learning, 205
Flavor perception, biopsychology of
flavor-related problems, solutions to
expectancy, 212–213
exposure, 211–212
masking, 209–211
nutraceuticals, flavor problems with
appearance, 208
smell, 207–208
somatosensation, 208
taste, 206–207, 206t
overview, 201–202
sensory/affective responses to food and drink, 202–206, 202t
Flavor release; see also Controlled flavor release
by degradation, 262
by diffusion, 262
by melting, 263
by swelling, 262–263
Flavor-release profile, 261t
Flavors in probiotics/prebiotics
about, 55–56
fermented dairy products
ayran, 58–59
dairy-based probiotic beverages, miscellaneous, 63
fermented milk drink, 61–63
kefir, 59–61
kumiss, 61
yogurt, 56–58
fermented nondairy probiotic beverages, 63–65, 64f
flavor/aroma defects in probiotics/prebiotics, 65–66
overview, 51–55, 55f
testing of flavor/sensory defects, 66–68, 67t
Flavors of terpenoid origin, sensory qualities/nutraceutical applications
overview, 167–168
terpene flavors, nutraceutical applications of
about, 174, 175t–179t
citral (3,7-dimethyl-2,6-octadienal), 180–183
geraniol, 184–187
menthol, 187–193
terpenes
defined, 167

sensory qualities of terpenes, 169–174, 170f, 171t–173t
types of, 168
Flaxseed (*Linum usitassimum*), 7–8
Flow cytometry, 298
Fluid bed spray coating, 248; see also Nanoencapsulation of flavors
Fluidized bed method, 247
Fluorescence microscopy, 293
Food
bars, 12–13; see also Confectionery-based functional foods
habits, 146
quality, 159
rejection, bitter taste and, 156
science, 146
sensory/affective responses to, 202–206, 202t
technology, 159
creating new flavors, 162
Food and Drug Administration (FDA), 77, 276
Formulcoat, 140
Fortified foods, 46
Fourier transform infrared spectroscopy (FTIR), 293
Free-nerve ending receptors, 203
Freeze-dried kefir, 60
Freeze-drying technique, 224, 250, 301; see also Nanoencapsulation of flavors
Freeze-protectant for liposomes (lyophilization), 255–256; see also Solvent dispersion methods
Freeze-thawed liposomes, 254; see also Mechanical dispersion methods
French pressure cell technique, 254; see also Mechanical dispersion methods
Fresh food flavor, 244
Fresh infusions, 94
Fruit- and vegetable-based functional foods; see also Functional foods
broccoli and cruciferous vegetables, 15
citrus fruits, 17, 17t
cordyceps mushroom, 18–19
cranberry, 17
fig, 18
garlic (*Allium sativum*), 14–15
pumpkin (*Cucurbita pepo*), 19
red wine and grapes, 17–18
strawberries, 18
tomato, 15

Fruit juice–based probiotics, 64
Fucoidan, 87
Fucoxanthin, 87
Functional foods
    animal-based
        beef, 21
        eggs enriched with PUFA, 21
        fish, 19–21
    beverage-based
        tea, 13
    in Canada, 276–277
    chemical structures of, 126f
    concept of, 2–3
    confectionery-based
        chocolate, 13
        cocoa (*Theobroma cacao*), 12
        food bars, 12–13
    consumer acceptance of
        cognitive and attitudinal
            determinants, 157
        socio-demographic determinants,
            156–157, 157t
    dairy-based, 11
    defined, 2, 276–277
        by U.S. Institute of Food Technology,
            124
    fruit- and vegetable-based functional
        foods
        broccoli and cruciferous vegetables,
            15
        citrus fruits, 17, 17t
        cordyceps mushroom, 18–19
        cranberry, 17
        fig, 18
        garlic (*Allium sativum*), 14–15
        pumpkin (*Cucurbita pepo*), 19
        red wine and grapes, 17–18
        strawberries, 18
        tomato, 15
    in Japan, 277
    legume-based functional foods
        soybeans, 13–14
    margarine, 12
    nutraceuticals
        concept of, 4–5, 5t
        market scenario for, 5–6
        safety aspects of, 6
    and off-tastes, 126t
    prebiotics, 12
    primary market for, 4
    sources of, 3t
        barley (*Hordeum vulgare*), 7
        breakfast cereal foods, 10–11
        brown rice, 9
        buckwheat, 9–10
        finger millet, 10
        flax (*Linum usitassimum*), 7–8
        maize, 10
        oats (*Avena sativa*), 7
        psyllium, 9
        sorghum (*Sorghum bicolor*), 10
        wheat, 6
    sources of, and health effects, 3t
    world market for, 3–4
Functional protein, 134
Fungi (Phycomycetes), 88
Furcellaran, 87

G

Gadolinium neutron capture therapy, 288
Gadopentetic acid, 288, 289
    delivery system, 291
Galactanes, 54
Garlic (*Allium sativum*), 14–15
Gastrointestinal tract (GIT), 52
Gastroprotective activity, geraniol, 187
Gelatin type A-carrageenan complexes,
    295
Generally recognized as safe (GRAS), 52,
    139
    food additives, 292
Geraniol; *see also* Terpene flavors,
    nutraceutical applications of
    about, 184
    antifungal activity of, 185
    antiproliferative effects of, 185, 186
    biological properties of
        anthelmintic effects, 187
        anti-inflammatory activity, 186–187
        antimicrobial activity, 184–185
        antioxidant activity, 185
        antitumor activity, 185–186
        gastroprotective activity, 187
        percutaneous absorption,
            enhancement of, 187
Geranylgeranyl pyrophosphate (GGPP),
    168
Geranyl pyrophosphate (GPP), 168
Global nutraceutical industry, 77
Global quality issues, 278
Glucosamine, 80
Glucose, 129
Glutamates in crystal form, 148
Glutamic acid, 148
Glutaraldehyde, 288
Good health, 33
G-protein-coupled receptors (GPCR), 152

G proteins, 242
Grading methods, 158
Grainfields Whole Grain Probiotic, 65
Grains/cereals; see also Functional foods
  barley (*Hordeum vulgare*), 7
  breakfast cereal, 10–11
  brown rice, 9
  buckwheat, 9–10
  finger millet, 10
  flaxseed (*Linum usitassimum*), 7–8
  maize, 10
  oats (*Avena sativa*), 7
  psyllium, 9
  sorghum (*Sorghum bicolor*), 10
  wheat, 6
Grapes, red wine and, 17–18
Green tea, 13
  catechin, 295
  polyphenols, 300–301
Gum arabic (GA), 301

H

Half-order release, 260
Head-space solid-phase microextraction (HS-SPME), 103
Health Canada, 276–277
HEARTBAR, 12
Heat-activated receptors, menthol and, 190
Hemiterpenes, 168
Hemolysis, 298
Hemolytic anemia, 298
Herbal nutraceuticals, 78–79
Herbal products extraction methods; see also Natural ingredients for nutraceutical industry
  aqueous alcoholic extraction process, 97–98
  continuous hot extraction, 95–97
  countercurrent extraction process, 98
  decoction, 95
  distillation, 98–101
  infusion, 93–95
  maceration, 90–91
  microextraction techniques, 103–106
  percolation, 91–93
  supercritical fluid extraction (SCF), 101–103
Heterogeneous degradation, 262
1,1,1,3,3,3-hexafluoro isopropanol (HFIP), 286
Hollow fiber liquid-phase microextraction (HF-LPME), 106
Homeostatic mechanisms, 81

Homogeneous degradation, 262
Homogenization, 59
Hordeum vulgare, 7
Hot-melt extrusion (HME), 136–137; see also Taste-masking techniques
Hot receptors, 204
Human breast adenocarinoma cells (MCF-7), 302
Human umbilical vein endothelial cell (HUVEC), 89
Hydrodiffusion, 99; see also Distillation
Hydrodistillation, 100
Hydrogel aqueous coating material, 140
Hydrogenated soybean phospholipids (HSPC), 302
Hydrophilic drugs, 287
3-hydroxy-3-methylglutarylcoenzyme A(HMG-CoA)-reductase, 186
Hyperthyroidism, iodine-induced, 80

I

Imbibition, 92
Infants, nutritional deficiencies on, see Nutritional deficiency on infants/children
Infusion, 93–95; see also Herbal products extraction methods
Inoculation, 65
Inorganic-based nanoencapsulates, 295
Insent taste-sensing system, 127; see also Sensory evaluation strategy for taste masking
Interfacial polymerization, 258, 259t; see also Chemical methods
International Food Information Council (IFIC), 156, 157
International market potential, 277–278
Inulin, 54
In-vitro evaluation, 297–299
Iodine, 80–81; see also Minerals
Ion-exchange resins (IER), 134–135, 135t; see also Taste-masking techniques
Ionic gelation, 301
Ionones, 174
Ionotropic gelation, 290; see also Chitosan nanoparticles
Iron, 81; see also Minerals
Iron deficiency, 44, 81
  anemia, 36–37; see also Nutritional deficiency on infants/children
Iron-deficient humans, 81
Iron supplements, advanced demasking of, 44

Isoflavones, 13
Isopentenyl diphosphate (IPP), 168
Isoprenoids, 79, 173

J

Jameed, 57
Jaundice, 298

K

Kefir; *see also* Fermented dairy products
  about, 59–60, 59f
  defects in, 66
  starter culture characteristics for, 60–61
Keratinocytes, 190
KLEPTOSE®Linecaps, 140
K-opioid receptor antagonists, 191
Krill, 88
Kumiss, 61; *see also* Fermented dairy products

L

L. monocytogenes, 293
Lactate dehydrogenase (LDH)-leakage assay, 298
Lactic acid bacteria (LAB), 51, 64, 83
Lactobacillus, 11
Lactobacillus delbrueckii, 65
Large-scale extractor method, 93
Lassi, 63; *see also* Probiotic beverages, dairy-based
Lecithin, 293
Legislation in nano-based foods, 302–303
Legume-based functional foods, 13–14; *see also* Functional foods
Legumes, 79
Lemongrass oil (LO), 293
Light microscopy, 299
Limonene, 229, 293
Linum usitassimum, 7–8
Lipid-based encapsulation technology, 284
Lipid-based nanoencapsulation, 293–294
Lipid nanoparticles, 137; *see also* Taste-masking techniques
Lipid peroxidation, 278
Lipophilic active agent, 295
Liposome entrapment; *see also* Physico-chemical methods
  about, 252–253
  liposome and nanoliposome manufacture, 253
  liposome preparation/core material loading, 253–256

Liposomes
  controlled release of, 292–293
  or lipid vesicles, 284
Liquid–liquid extraction (LLE), 105
Liquid-phase microextraction (LPME), 105–106; *see also* Microextraction techniques
Listeria innocua, 293
Listeria spp., 301
Listerine, 41
Local anesthetic activity, menthol, 192
L-tryptophan, 132
Luminol (3-aminophthalhydrazide), 299–300
  chemiluminescence, 299
Lycopene, 15
Lyophilization, 224, 255–256; *see also* Solvent dispersion methods

M

Maceration, 90–91; *see also* Herbal products extraction methods
Magnesium supplements, taste of, masking, 45
Maize, 10; *see also* Grains/cereals
Malted barley, 65
Maltitol, 284
Maltodextrin (MD), 301
Manganese, 82; *see also* Minerals
Manipulation of expectancies, 207
Mare's milk, 61
Margarine, 12; *see also* Functional foods
Marine carbohydrates, 84
Marine nutraceuticals; *see also* Nutraceuticals, natural sources-based
  carbohydrates, 84, 86–88
  enzymes/vitamins/minerals, 89–90
  fatty acids, 88
  obtained from marine organisms, 85t
  peptides, 88
  phenolic compounds/prebiotics, 89
  proteins, 88
Marine oils, 88
Marine polysaccharides, 84, 87
Masking
  by encapsulation, 161–162
  of fishy odor/taste of omega 3 fatty acids, 45
  flavor-related problems, solutions to, 209–211
  taste of calcium/magnesium supplements, 45
Mass transfer, 280, 292–293

Mechanical dispersion methods; see also
Physico-chemical methods
  freeze-thawed liposomes, 254
  french pressure cell technique, 254
  solvent dispersion methods
    detergent removal method, 255
    ethanol injection, 254
    ether injection (solvent vaporization), 254
    freeze-protectant for liposomes (lyophilization), 255–256
    reverse phase evaporation method, 254–255
    solid lipid nanoparticles (SLN), 256
  sonication, 253
Medicinal foods, 277
Melting, flavor release by, 263
Membrane integrity assays, 298
Menstruum, 90–91, 92
Menthol; see also Terpene flavors, nutraceutical applications of
  about, 187–189
  biological properties of
    analgesic effects, 191
    anti-inflammatory activity, 192
    antimicrobial activity, 192–193
    antipruritic activity, 190–191
    antitumor activity, 192
    local anesthetic activity, 192
    olfactory system, menthol and, 189
    respiratory system, activity on, 191–192
    skin penetration enhancement, 193
    thermoreceptors, modulation of, 189–190
  derived from peppermint, 79
  stereochemistry of, 191
Metallic nanoparticles, 299
Metamucil, 9
Methanol, 284
Mevalonic acid (MVA) pathway, 168
Microbes, 82–84; see also Nutraceuticals, natural sources-based
Microbial metabolites, 84
Microcaps technology, 138–139
Microcapsules, 133
Microencapsulation, 133–134, 161
Microextraction techniques; see also Herbal products extraction methods
  liquid-phase microextraction (LPME), 105–106
  solid-phase microextraction (SPME) technique, 103–104, 104f
  stir bar sportive extraction (SBSE) technique, 104–105

Microfluidization technique, 256
Micromask, 140–141
Microorganisms
  prevention of diseases, 83
Microspheres, 133
  porous, 38f, 137–138
Microstructures of spray-dried particles, 245
Microwave extraction, 238
Mil-mil, 62; see also Fermented milk drink products
Minerals; see also Nutraceuticals, natural sources-based
  deficiency risk /taste/odor of, 47t–48t
  iodine, 80–81
  iron, 81
  manganese, 82
  marine nutraceuticals and, 89–90
  zinc, 81–82
Mitogen-activated protein kinase (MAPK), 187
Molecular diffusion, 262
Molecular inclusion, 256–258; see also Chemical methods
Molecular stereochemistry, 169
Monoterpenes, 168
Monoterpenols, 169
Montmorillonite (MMT), 136
M-opioid receptor agonists, 191
Moraceae family, 18
Mouth process, 240
Mouthwashes, 41
Mucoadhesive, 285
Multilamellar vesicles (MLVs), 253
Multi-particulate rupture, 138; see also Taste-masking techniques

N

Nanoemulsion, 227
Nanoencapsulates
  inorganic-based, 295
  polysaccharide-based, 295
  protein-based, 294–295
Nanoencapsulation
  advantages of, 300–302
  of anti-cancer agents, 302
  of antioxidants and flavoring agents, 300–301
  of bioactive agents/ingredients, 281t–283t, 284–287
    encapsulation materials, 285–287
  classification of polymers used for, 223t
  disadvantages of, 302–303
  of flavor components; see also Flavor nanotechnology

coacervation phase separation, 224–225
electrospray, 227–228
emulsion technology, 227
extrusion microemulsion technique, 225–226
freeze drying, 224
spray drying, 223
supercritical fluid technology, 226–227
in flavor compounds, applications of, 228–229
lipid-based, 293–294
Nanoencapsulation of flavors
biotechnological production process, 239
chemical methods
interfacial polymerization, 258, 259t
molecular inclusion, 256–258
flavor, defined, 236, 237f
flavor compounds, 239–240
flavor encapsulation
applications of, 242–243, 243f
challenges, 242
methods of
about, 244, 244f
co-crystallization, 251
extrusion, 248–250
fluid bed spray coating, 248
freeze-drying technique, 250
physical/mechanical methods, 245–246, 246t
spray cooling/spray chilling, 246–248
natural flavoring substances, production of
about, 236–237
distillation process, 237–238
extraction process, 237
flavor, types of, 237t
microwave extraction, 238
pressurized liquid extraction (PLE), 239
solvent extraction, 238
supercritical fluid extraction, 238
perception of flavor, 240–242
physico-chemical methods
coacervation, 251–252
liposome entrapment, 252–256
release mechanisms
about, 258, 260
controlled flavor release, 260–264, 261t, 262
release rates, 260

Nanohybrid technology, 136; see also Taste-masking techniques
Nanoliposomes, 255, 294
Nanoparticles
activity of, 291–295
antimicrobial activity of nanoemulsions, 293–294
controlled release of liposomes, 292–293
inorganic-based nanoencapsulates, 295
polysaccharide-based nanoencapsulates, 295
protein-based nanoencapsulates, 294–295
characterization by analytical methods, 295–300
chemical method, 299–300
dynamic light scattering (DLS), 296–297
electron microscopy, 296
in-vitro evaluation, 297–299
preparation, 280
Nanosalt, 229
Nanosized selfassembled liquids (NSSLs), 294
Nano spray dryer, 246
Nanostructured lipid carriers (NLCs), 294
Nanotechnology
in flavors, 222, 222f, 223t
on food industry, 220, 221t
Nanotechnology in drug delivery systems, 280
advantages of, 280
goals of, 280
Natural antioxidants, 278–279
polyphenols, 279
quercetin, 279
vitamins C and E, 278–279
Natural flavor enhancers, 40
Natural flavoring substances, 237t
Natural flavoring substances, production of; see also Nanoencapsulation of flavors
about, 236–237
distillation process, 237–238
extraction process, 237
flavor, types of, 237t
microwave extraction, 238
pressurized liquid extraction (PLE), 239
solvent extraction, 238
supercritical fluid extraction, 238
Natural flavors, 40
Natural health, importance of, 79

Natural ingredients for nutraceutical
industry
  clinical trials on, 106
  common plant/herbal extracts/
    therapeutic applications, 106,
    107t–112t
  global nutraceutical industry, 77
  herbal products extraction methods
    aqueous alcoholic extraction
      process, 97–98
    continuous hot extraction, 95–97
    countercurrent extraction process, 98
    decoction, 95
    distillation, 98–101
    infusion, 93–95
    maceration, 90–91
    microextraction techniques,
      103–106
    percolation, 91–93
    supercritical fluid extraction (SCF),
      101–103
  nutraceutical markets for plant extracts
    European market, 77–78
    U.S. market, 78
  nutraceuticals, natural sources-based,
    classification of
    marine nutraceuticals, 84–90, 85t
    microbes, 82–84
    minerals, 80–82
    plants, 78–80
  overview, 76–77
  prospects, 106, 112
Natural monoterpenes flavors
  chemical structures of, 170f
Natural polymers, 132
Natural sesquiterpene flavors, chemical
  structures of, 170f
Natural sweeteners, 129
Nature-identical flavoring substances, 237t
Neural activity, 241
NIZO (Netherlands Institute for Dairy
  Research), 160, 161
Nodulisporium sylviforme, 84
Nondairy probiotic beverages, fermented,
  63–65, 64f; *see also* Flavors in
  probiotics/prebiotics
Nonflavonoid polyphenolics, 79
Non-nutritive sweeteners, 129
Novice alcohol drinkers, 211
Nuclear magnetic resonance (NMR), 264
Nutraceutical, defined, 275–277
Nutraceutical markets for plant extracts;
  see also Natural ingredients for
  nutraceutical industry
  European market, 77–78
  U.S. market, 78
Nutraceuticals; *see also* Functional foods
  in biological processes, 5
  categories, 276–277
    dietary supplements, 276
    functional foods, 276–277
    medicinal foods, 277
    pharmaceuticals, 277
  concept of, 4–5, 5t
  defined, 124, 275–276
  in disease prevention, 79
  global quality issues, 278
  and health effects, 4, 5t
  international market potential,
    277–278
  market scenario for, 5–6
  nomenclature in Canada, 277
  nomenclature in United States, 277
  potential health benefits, 300–302
    advantages of nanoencapsulation,
      300–301
    nanoencapsulation of anti-cancer
      agents, 302
  safety aspects of, 6
  types of, 124
Nutraceuticals, flavor problems with;
  see also Flavor perception,
    biopsychology of
  appearance, 208
  smell, 207–208
  somatosensation, 208
  taste, 206–207, 206t
Nutraceuticals, natural sources-based;
  see also Natural ingredients
    for nutraceutical industry
  marine nutraceuticals, 84–90, 85t
  microbes, 82–84
  minerals, 80–82
  plants, 78–80
Nutrients, 35, 146
Nutrification, 276
Nutritional deficiency, 33, 35
Nutritional deficiency on infants/children;
  see also Pediatric nutritional
    supplements, flavoring of/
    pediatric compliance
  calcium deficiency, 37
  iron deficiency anemia, 36–37
  vitamin A deficiency, 37–38
  vitamin D deficiency, 36
Nutritional status, 34
Nutritive sweeteners, 129
Nutropin Depot, 287

O

Oat-rich foods, 7
Oats (*Avena sativa*), 7; see also Grains/cereals
 cholesterol-lowering properties, 7
Obesity, 300
Observational learning, 205
Odorants, 240–241
Odoriferous monoterpenes, 171t–173t
Off-taste masking agents, 137; see also Taste-masking techniques
Oil-in-oil (o/o) emulsification technique, 287
Olfactory information, 241
Olfactory system, menthol and, 189
Oligosaccharides, 12, 53
Omega-3 fatty acid, 20, 45
Omega fatty acids, 45
Omnivores, consumption of foods, 149
Opadry, 139
Oral cavity, menthol for, 189
Oral supplements, 46
Organic solvent, 227
Organosulfur compounds, 15, 155
Orthonasal smell, 240
Over-the-counter (OTC) formulations, 140
OXPzero, 139
Oxygenated terpenes, 169

P

Packed columanganese, 99, 101
Palatability of oral pediatric medicines, 42
Paraffin oil, 288
Parkinson's disease, 300
Particle degradation, 280
Particle density, 245
Pasteurized products, 56
Patented techniques; see also Taste-masking techniques
 Actimask, 139–140
 Camouflage technology, 140
 FlavoRite technology, 139
 Formulcoat, 140
 KLEPTOSE®Linecaps, 140
 Microcaps technology, 138–139
 Micromask, 140–141
 Opadry, 139
 OXPzero, 139
Patient exposure, 280
Pea proteins, 293
Pectin, 59, 295
Pediatric flavoring agents, regulatory aspects of, 41–44

Pediatric medicine, 42
Pediatric nutritional supplements, 46
Pediatric nutritional supplements, flavoring of/pediatric compliance
 child nourishment problems/related diseases/nutritional supplementation solutions, 35
 discussion, 46, 47t–48t
 flavoring agents, types/applications, 39–41, 40t, 41t
 flavor on supplement intake/consumption in children, 38–39
 nutritional deficiencies on infants/children
  calcium deficiency, 37
  iron deficiency anemia, 36–37
  vitamin A deficiency, 37–38
  vitamin D deficiency, 36
 overview, 33–35
 regulatory aspects of pediatric flavoring agents, 41–44
 success
  advanced demasking of iron supplements, 44
  masking of fishy odor and taste of omega 3 fatty acids, 45
  masking taste of calcium and magnesium supplements, 45
Peptides, 88; see also Marine nutraceuticals
Perception
 of flavor, 240–242
 of taste, 149–150, 150f
Perceptual masking, 207
Percolation; see also Herbal products extraction methods
 about, 91–92
 modifications made to, 92
 percolators, types of, 92–93
Percolators, types of; see also Percolation
 extractor method (British/Indian pharmacopeias), 93
 small-scale percolators for laboratory use, 92
 Soxhlet apparatus, 93
Percutaneous absorption, enhancement of, geraniol, 187
Personalization, 77
Phagocytosis, 299–300
Pharmaceuticals, 277
Phenolic compounds, 89; see also Marine nutraceuticals
Phenols, 155
Phenylpropanoids, 175t–179t
Phenylthiocarbamide (PTC), 152, 154, 207
Phospholipidips, 287

Phosphotungstic acid, 299
Photon correlation spectroscopy (PCS), 296–297
Physical/mechanical methods, 245–246, 246t
Physico-chemical methods; *see also* Nanoencapsulation of flavors
   coacervation, 251–252
   liposome entrapment
      about, 252–253
      liposome and nanoliposome manufacture, 253
      liposome preparation/core material loading, 253–256
Phytochemicals, 79
Phytosterols, 174
Plain ayran, 58
Plants/herbs, 78–80; *see also* Nutraceuticals, natural sources-based
   nutraceutically important, 107t–112t
Polydimethylsiloxanes (PDMS), 103, 104
Polyelectrolyte complexation technique, 301
Polyelectrolytes, 292
Polyethylene glycol (PEG), 296
Poly(lactic-co-glycolic acid) (PLGA), 286–287
Polymer
   hybrids, 136
   matrix, 134
   types of, 133
Polyphenolics, 83
Polyphenolic substances, 79
Polyphenols, 13, 89, 279; *see also* Natural antioxidants
Poly (DL-lactide-co-glycolide) (PLGA), 301
Poly (vinyl alcohol) (PVA), 301
Polysaccharide-based nanoencapsulates, 295
Polysaccharides, 84
Polyterpenes, 168
Polyunsaturated fatty acid (PUFA), 45
Polyvinylacetaldiethylaminoacetate, 136
Poor diet in children, 35
Porosity, 280
Porous microspheres, 38f, 137–138
Potential nutraceuticals, 124
Prebiotics, 12, 89; *see also* Functional foods; Marine nutraceuticals
   defined, 53
Preservation of food flavors, 229
Pressurized liquid extraction (PLE), 239
Proanthocyanidin, 79
Probe sonication, 253

Probiotic beverages, dairy-based; *see also* Flavors in probiotics/prebiotics
   buttermilk, 63
   cacik, 63
   lassi, 63
Probiotic beverages, fermented nondairy, 63–65, 64f; *see also* Flavors in probiotics/prebiotics
Probiotic cabbage juice, 65
Probiotics
   defined, 51
   microorganisms, 52–53
Procarcinogens (aflatoxin B1), citral, 183
Procyanidins, 13
Prodrugs, 132, 132t; *see also* Taste-masking techniques
Product developers, 157
Product quality, 158
Progressive impregnation by saliva, 240
Propylthiouracil (PROP), 154, 207
Prostate cancer, 287
Proteins, 88; *see also* Marine nutraceuticals
   expression markers, 153
Protein-based nanoencapsulates, 294–295
Protein kinase C (PKC), 186
Protein–polysaccharide interaction, 295
Psychological techniques, use of, 207
Psyllium, 9; *see also* Grains/cereals
PUFA, 21
Pumpkin (*Cucurbita pepo*), 19
Purified enzymes, 239

Q

Quality assurance, 158
Quality of food, sensory evaluation and, 159
Quality score, 159
Quantum dots, 297, 299
Queen of cereals, 10
Quercetin, 279; *see also* Natural antioxidants

R

Raftiline HP®, 54
Raftilose, 54
Raita, 58
Rancidification, 278
Rancidity, 279
Rapid expansion of supercritical solutions (RESS), 226
Raspberry ketone, 219
Red blood count (RBC) generation, 36
Red wine and grapes, 17–18

Release mechanisms; *see also* Nanoencapsulation of flavors
  about, 258, 260
  controlled flavor release, 260–264, 261t, 262
  release rates, 260
Release rates, 260
Renal failure, 298
Respiratory system, activity on, menthol, 191–192
Restrictors, 102
Retronasal smell, 240
Reverse micelles, 289–290; *see also* Chitosan nanoparticles
Reverse phase evaporation method, 254–255, 284; *see also* Solvent dispersion methods
Rosemary essential oils, 245
Rutin, 279
  flavonoid, 300

S

Saccharine, 43
Saccharomyces cerevisiae, 82
Salmonella spp., 301
Salmonella typhimurium, 293
Salt, 130–131; *see also* Taste-masking techniques
Salty taste, 147, 203
Sandostatin LAR, 287
Scanning electron microscopy (SEM), 263, 296
Secondary electron imaging mode (SEI), 263
Self assembly of food proteins, 229
Sensory defects, testing of, 66–68, 67t; *see also* Flavors in probiotics/prebiotics
Sensory evaluation and quality of food, 159; *see also* Bitter components on sensory perception of food
Sensory evaluation strategy for taste masking; *see also* Taste-masking techniques
  about, 126–127
  α-ASTREE electronic tongue, 127–128, 127f
  biomimetic sensors, 128
  carbon nanotubule field-effect transistor (CNT-FET) bioelectronic sensors, 128
  insent taste-sensing system, 127
Sensory experience of food, 202
Sensory qualities of terpenes, 169–174, 170f, 171t–173t; *see also* Terpenes

Sensory science in decision making, role of, 158–159
Sensory stimulation, 211
Sesquiterpenes, 168, 171t–173t
Sesterterpenes, 168
S-galactofucan, 86
Shelf life of foodstuffs, 279, 284
Single-drop microextraction (SDME), 105
Siphonaxanthin, 89
Site-specific core delivery, 275
Skin penetration enhancement, menthol, 193
Slimming agents, 284
Small angle neutron scattering (SANS), 264
Small-scale percolators for laboratory use, 92
Small unilamellar vesicles (SUV), 253
Smell, 203, 207–208; *see also* Nutraceuticals, flavor problems with
Smoke flavoring, 237t
Socio-demographic determinants, 156–157, 157t
Socio-demographic factors on food choice, 157, 157t
Sodium, 131
Sodium caseinate (NaCas), 302
Sodium hydroxide, 288
Sodium hydroxide-methanol, 288
Sol-gel synthesis, 295
Solid lipid microparticle (SLMs), 247
Solid lipid nanoparticles (SLNs), 137, 256, 294; *see also* Solvent dispersion methods
Solid-phase microextraction (SPME) technique, 103–104, 104f; *see also* Microextraction techniques
Solvent dispersion methods; *see also* Mechanical dispersion methods
  detergent removal method, 255
  ethanol injection, 254
  ether injection (solvent vaporization), 254
  freeze-protectant for liposomes (lyophilization), 255–256
  reverse phase evaporation method, 254–255
  solid lipid nanoparticles (SLN), 256
Solvent extraction, 238
Solvent-free cold extrusion, 137; *see also* Taste-masking techniques
Solvent-free microwave extraction (SFME), 238
Solvent vaporization, 254; *see also* Solvent dispersion methods

Somatosensation; *see also* Nutraceuticals, flavor problems with
   pungency, 208
   texture and astringency, 208
Somatosensory system, 203
Sonication, 284
Sorghum (*Sorghum bicolor*), 10; *see also* Grains/cereals
Sour tastants, 202
Sour taste, 147
Soxhlet apparatus, 93
Soybean phospholipids (SPC), 302
Soybeans, 13–14, 79
Soybean seed ferritin deprived of iron (apoSSF), 300
Spatoglossum schroederi, 86
Spirulina (*Arthrospira platensis*), 8
Spoilage, 279
Spray chilling, 246–247
Spray congealing, *see* Spray cooling
Spray cooling; *see also* Nanoencapsulation of flavors
   advantages, 247
   disadvantages, 247–248
Spray drying, 161, 223, 284, 295, 300–301
   advantages/disadvantages of, 246t
Stability of encapsulated flavors, 263; *see also* Controlled flavor release
Stabilizers, 57, 59
Starter culture characteristics; *see also* Flavors in probiotics/prebiotics
   for ayran manufacturing/flavor, 58
   for kefir manufacture/flavor, 60
   for yogurt manufacture/flavor, 57
Static diffusion, 262
Static extraction mode, 102
Steam and water distillation, 100; *see also* Distillation
Steam distillation, 238
Steroids, topical, 193
Stir bar sportive extraction (SBSE) technique, 104–105; *see also* Microextraction techniques
Stirred yogurt, 57
Stokes–Einstein equation, 297
Strawberries, 18
Sucrose, 43, 129
Sulforaphane, 15
Sunflower oil–based nanoemulsions, 293
Supercritical antisolvent (SAS) method, 226
Supercritical carbon dioxide, 226
Supercritical fluid (SCF), 135–136, 226–227; *see also* Taste-masking techniques

Supercritical fluid (SCF) extraction; *see also* Herbal products extraction methods
   applications, 103
   components to, 101–103
   in production of natural flavoring substances, 238
Supplementation of micronutrients, 34
Suppression, 209–210
   of mixture, 160–161
Sustained release, 261
Sweet acidophilus milk, 62; *see also* Fermented milk drink products
Sweet/bitter perception, genetics of; *see also* Bitter components on sensory perception of food
   cross-species comparisons, 154
   family and twin studies, 154–155
   naturally occurring alleles, 153–154
Sweeteners; *see also* Taste-masking techniques
   non-nutritive, 129, 130t
   nutritive, 129, 130t
   and relative sweetness, 130t
   in yogurt, 57
Sweet tastes, 147, 202
Sweet-tasting proteins, 152; *see also* Proteins
Swelling, flavor release by, 262–263
Synbiotics, 54
Synthetic flavoring agents, 40
Synthetic flavors, 219, 222
Synthetic nonsteroidal anti-inflammatory drug, 182
Synthetic polymers, 132
Synthetic sweeteners, 152
Syringe dropping, 290
Syringe pump, 101

T

Tacrolimus, 84
Tannin, 79
Tarator, 57
Taste; *see also* Bitter components on sensory perception of food; Nutraceuticals, flavor problems with
   buds, 151, 151f
   flavor problems with nutraceuticals, 206–207, 206t
   importance of, 125, 149
   inhibition, 160
   masking of medicines, 42
   perception of, 149–150, 150f

preferences, 38
receptors, 146, 148
types of
   bitter, 147–148
   salty, 147
   sour, 147
   sweet, 147
   umami, 148–149
Taste Free, 44
Taste improvement technology
   about, 159–160
   food technology, creating new flavors, 162
   masking by encapsulation, 161–162
   suppression of mixture, 160–161
   taste inhibition, 160
   umami in food science, role of, 162
Taste-masking approaches, 159
Taste-masking techniques
   conventional techniques in
      bitter blockers, 131–132, 131t
      complexation, 135
      film coating, 132, 133t
      flavorants, 130, 130t
      ion-exchange resins (IERs), 134–135, 135t
      microencapsulation, 133–134
      prodrugs, 132, 132t
      salt, 130–131
      supercritical fluids (SCF), 135–136
      sweeteners, 129, 130t
   hot-melt extrusion (HME), 136–137
   lipid nanoparticles, 137
   multi-particulate rupture, 138
   nanohybrid technology, 136
   off-taste masking agents, 137
   overview, 124–126, 126t
   patented techniques
      Actimask, 139–140
      Camouflage technology, 140
      FlavoRite technology, 139
      Formulcoat, 140
      KLEPTOSE®Linecaps, 140
      Microcaps technology, 138–139
      Micromask, 140–141
      Opadry, 139
      OXPzero, 139
   porous microspheres, 38f, 137–138
   sensory evaluation strategy for taste masking
      about, 126–127
      α-ASTREE electronic tongue, 127–128, 127f
      biomimetic sensors, 128

carbon nanotubule field-effect transistor (CNT-FET) bioelectronic sensors, 128
insent taste-sensing system, 127
solvent-free cold extrusion, 137
Taste perception and behavior
   bitter taste and food rejection, 156
   population, challenges in, 155
Taxol (paclitaxel), 84
Taxomyces andreanae, 84
Tea, 13
Terpene flavors, nutraceutical applications of
   about, 174, 175t–179t
   citral (3,7-dimethyl-2,6-octadienal), 180–183
   geraniol, 184–187
   menthol, 187–193
Terpenes; *see also* Flavors of terpenoid origin, sensory qualities/nutraceutical applications
   bioactivity of, 168
   characteristics, 167
   defined, 167
   natural flavoring, 175t–179t
   with polar groups, 193
   sensory qualities of terpenes, 169–174, 170f, 171t–173t
   types of, 168
Terpenoids, 169
Tetraterpenes, 168
Tetrazolium-dye reduction, 298
Texture, 39
   in food, 203
   masking, 210
Theobroma cacao, 12
Thermal process flavoring, 237t
Thermal stability, 263
Thermogravimetric analysis (TGA), 263
Thermolabile materials, 224
Thermoreceptors, modulation of, menthol, 189–190
Thrombin, 86
Tomato, 15
Tongue
   and soft palate, 146
   taste buds, 151
Trans-cinnamaldehyde, 301
Transient receptor potential cation channel subfamily M member 8 (TRPM8), 181–182
Transient receptor potential (TRP) ion channels, 181

Transmission electron microscopy (TEM), 263, 296, 299–300
Tray columanganese, 99
Trelstar Depot, 287
Triacylglyceride, 293
Trigeminal senses, 39
Triggered release, 261
Triple maceration, 91
Tripolyphosphate (TPP), 287
Triterpenes, 155, 168
Trolox, 279
TRPV1 (transient receptor potential cation channel subfamily V member 1), 181, 182
Trypan blue exclusion assay, 298
Tween 80®, 293

U

Ultrafiltration, 301
Ultrasonic emulsification method, 293
Umami
   in food science, role of, 162
   identified by Japanese, 125
   primary type of taste, 148–149; *see also* Bitter components on sensory perception of food
Uncooked beef, 21
United States, nutraceuticals nomenclature in, 277
U.S. market
   nutraceutical, for plant extracts, 78

V

Vacuum maceration, 91
Vanilla oil (VO), 301
Vanillic acid, 295
Variable restrictors, 102
Vegetable-based functional foods, *see* Fruit- and vegetable-based functional foods
Vibration mesh spray technology, 246
Vita Biosa, 65
Vitamins, 89–90, 284; *see also* Marine nutraceuticals
   deficiency risk /taste/odor of, 47t–48t

Vitamin A deficiency, 37–38; *see also* Nutritional deficiency on infants/children
Vitamin A deficiency disorders (VADD), 37
Vitamin B-complex, 41
Vitamin D, 295
   deficiency, 36, 37; *see also* Nutritional deficiency on infants/children
Vitamin D_2, 295
Vitamin D_3, 302
Vitamin E, 278–279, 284
   nanoparticles, 295
Vitamins C, 278–279; *see also* Natural antioxidants

W

Wall materials in the digestive system, 274
   rate of absorption, 275
   selection, 275
Water distillation, 100; *see also* Distillation
Water/hydrodistillation technique, 99–100; *see also* Distillation
Wheat, 6; *see also* Grains/cereals
Whey protein concentrate (WPC), 301
World market for functional foods, 3–4; *see also* Functional foods

Y

Yakult, 62–63; *see also* Fermented milk drink products
Yogurt, 52; *see also* Fermented dairy products
   about, 56–57
   cheese, 57
   drinks, 58
   starter culture characteristics for, 57
   types/flavors, 57–58

Z

Zero order release rates, 260
Zinc, 81–82; *see also* Minerals
   deficiency, 81
   salt, 131
Zion Market Research, 277–278